Thermophysical Basis of Energy Technologies (TBET-2016)

MATEC Web of Conferences Volume 92 (2017)

Tomsk, Russia
26 - 28 October 2016

Editors:

G.V. Kuznetsov
P.A. Strizhak
A.O. Zhdanova

ISBN: 978-1-5108-3509-2

Printed from e-media with permission by:

Curran Associates, Inc.
57 Morehouse Lane
Red Hook, NY 12571

Some format issues inherent in the e-media version may also appear in this print version.

Printed by Curran Associates, Inc. (2017)

For additional information, please contact EDP Sciences – Web of Conferences at the address below.

EDP Sciences – Web of Conferences
17, Avenue du Hoggar
Parc d'Activité de Courtabœuf
BP 112
F-91944 Les Ulis Cedex A
France

Phone: +33 (0) 1 69 18 75 75
Fax: +33 (0) 1 69 28 84 91

contact-edps@webofconferences.org

Additional copies of this publication are available from:

Curran Associates, Inc.
57 Morehouse Lane
Red Hook, NY 12571 USA
Phone: 845-758-0400
Fax: 845-758-2633
Email: curran@proceedings.com
Web: www.proceedings.com

TABLE OF CONTENTS

MATEC Web of Conferences
Open access conference proceedings in Materials science, Engineering and Chemistry
A flexible, fast and efficient publication service, for an open access recording of your event and a wide dissemination of your proceedings
www.matec-conferences.org
edp open
Open Access Publishing from EDP Sciences

Statement of Peer review

In submitting conference proceedings to MATEC Web of Conferences, I certify to the Publisher that I adhere to the **Policy on Publishing Integrity** of the journal in order to safeguard good scientific practice in publishing.

1. All articles have been subjected to peer review administered by the proceedings editors.
2. Reviews have been conducted by expert referees, who have been requested to provide unbiased and constructive comments aimed, whenever possible, at improving the work.
3. Proceedings editors have taken all reasonable steps to ensure the quality of the materials they publish and their decision to accept or reject a paper for publication has been based only on the merits of the work and the relevance to the journal.

Title, date and place of the conference

Thermophysical basis of energy technologies - 2016
TBET - 2016
October 26-28, 2016
Tomsk, Russia

Proceedings editor(s):

Geniy V. Kuznetsov
Pavel A. Strizhak
Alena O. Zhdanova

Date and editor's signature

30.11.2016

Geniy V. Kuznetsov
Pavel A. Strizhak
Alena O. Zhdanova

MATEC Web of Conferences 92, 01001 (2017) DOI: 10.1051/matecconf/20179201001
Thermophysical Basis of Energy Technologies - 2016

Mathematical model of stacked one-sided arrangement of the burners

J.A. Oraz [1,*], *I*.E Korzilova[1], *A*.N. Ermolaev[2], *S*.A. Khaustov[1], and *A*.Y. Dolgih[1]

[1]National Research Tomsk Polytechnic University, 634050 Tomsk, Russia
[2]Federal State budget institution of higher education "Tyumen industrial University", 625000 Tyumen, Russia

Abstract. Paper is aimed at computer simulation of the turbulent methane-air combustion in upgraded U-shaped boiler unit. To reduce the temperature in the flame and hence NOx release every burner output was reduced, but the number of the burners was increased. The subject of studying: complex of characteristics with space-time fields in the upgraded steam boiler E-370 with natural circulation. The flare structure, temperature and concentrations were determined computationally.

1 Introduction

The power industry of the country is one of the key industries. Energy balance directly effects on the economic welfare of the state. Currently, the Russian power industry is experiencing a moral and technological crisis. One of the main problems is outdated equipment. According to experts [1–4] up to 80% of boilers and turbines are exhausted their resources. To replace existing power plants is necessary to design new equipment with modern technologies, taking into account the operating experience of the analogical power systems [4, 5].

The object of research is the boiler unit is a U-shaped layout with natural circulation for the combustion of natural gas with superheated steam capacity of 370 t / h. The subject of research are the basic geometrical characteristics of the steam boiler with natural circulation. Full-scale multiple-factor experimental research of described three-dimensional vortical turbulent combustion is quite expensive. Therefore, nowadays computer simulation with the use of wide approved application packages is used to research furnace aerodynamics [6, 7]. For example, burning turbulent flow was simulated in [5–7].

Modern computer technology makes simulation of the boiler unit operation possible even during the development. Computer experiment allows to study the processes of heat and mass transfer, combustion and aerodynamics. The results reveal the shortcomings of the developed design, predict the operation of the boiler unit, check the possibility of non-project fuel combustion. Using mathematical modeling the economic costs associated with the construction are minimized.

[*] Corresponding author: zhanibek.oraz@gmail.com

To date, one of the most powerful software products for mathematical modeling is software package ANSYS FLUENT [4, 6, 7]. This software module allows solving the problem of hydro and gas dynamics, taking into account the turbulence, heat transfer and chemical reactions.

2 The method of research and the initial data

Construction of the main elements of the boiler was performed in accordance with the experience of similar boilers designing. The sequence of calculation, construction of heating surfaces of the boiler layout selection, the combustion chamber and burners made in accordance with regulatory guidelines.

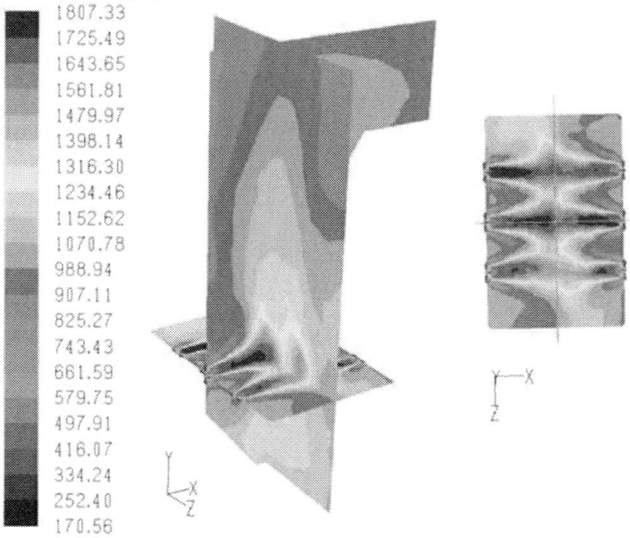

Fig. 1. The temperature distribution of the combustion medium (°C) before modernization.

To identify gaps in the developed design and predict the various boiler unit loads mathematical modeling of thermal processes in the combustion chamber of the boiler E-370 was made using FLUENT software package. Methods of modeling and mathematical models used are given in [4, 6, 7]. It is established that there is a large flare temperature, causing the high intensity of the generating of the nitrogen oxides and other harmful substances.

Table 1. Initial parameters.

Initial parameter		**unit**	**Elements of the horizontal gas duct**			
			SH2	SP	SH3	SH1
pipe diameter	d	mm	32×4	133×13	32×4	32×4
pipe arrangement	–	–	row arrangement			
pipe pitch	$\frac{S_1}{S_2}$	mm/mm	$\frac{530}{48}$	$\frac{0.74}{–}$	$\frac{100}{140}$	$\frac{80}{120}$
relative pipe pitch	$\frac{\sigma_1}{\sigma_2}$	–	$\frac{16.56}{1.5}$	$\frac{5.56}{–}$	$\frac{3.125}{1.5625}$	$\frac{2.5}{1.5625}$
number of entries	n	–	2	1	1	1
heating surface	H	m^2	455.429	39.375	1606.164	890.21

Turbulent combustion of natural gas in the reverse flame of fire-tube boiler was studied by means of the ANSYS Fluent 12.1.4 engineering simulation software. The following processes were simulated through the finite-element method: chemical reactions, turbulence, and heat exchange. The gas-phase was simulated according to the the Navier-Stokes differential equations in Euler's approximation. Aerodynamic calculations were performed using the differential equations of flow continuity and energy conservation. The radiant heat exchange calculation was conducted by the differential equations of spherical harmonics in first approximation (P1-model).

2.1 Modernization project

As a result of this work it has been upgraded E-370 boiler capacity of 260 MW with an estimated fuel consumption of 6.4 m^3 / s and a gross efficiency of 94.6 %. According to the adopted in the design of the combustion chamber of the boiler is not cold funnel, instead of it is designed for sloping (Fig. 2). On the front wall of the furnace are nine combination gas-oil burners in three levels.

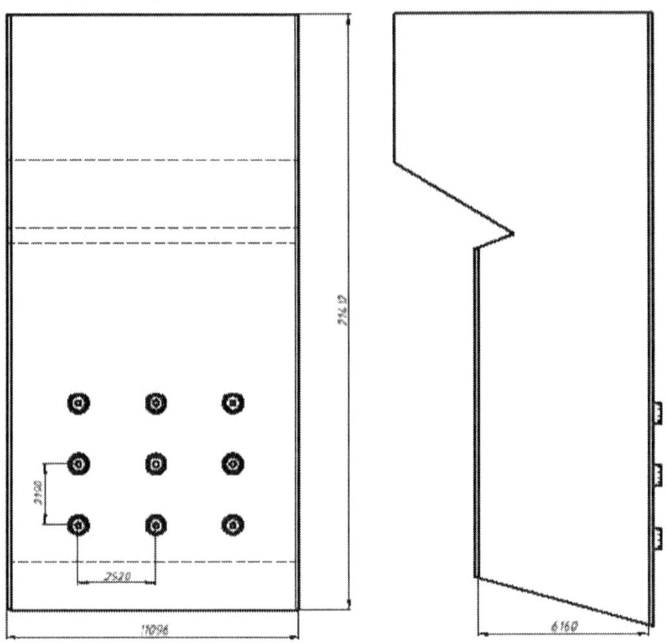

Fig. 2. Sketch of the upgraded furnace.

In the upgraded boiler burner output is reduced, but the number of the burners is increased (fig. 2). This technical solution allows to increase the efficiency of the boiler unit and to achieve the required temperature of exhaust gases.

Upgraded radiative-convective superheater has a four-stage steam superheating and executed by a mixed scheme. Two-step regulation of the superheated steam is applied using the temperature of its own condensate injection. The convective part consists of single-stage water economizer and recuperative air heater.

2.2 Initial data for mathematical modelling

Using computer-aided design three-dimensional model of the computational domain for modeling was built and converted to the tetragonal computational mesh (Fig. 3a) with its further transformation into a polyhedral mesh (Fig. 3b). As a result of the transformation the number of cells is declined, but the number of computational nodes is increased

(Table 2). These allow to increase the convergence and accuracy of the numerical simulation, as well as time of the calculation.

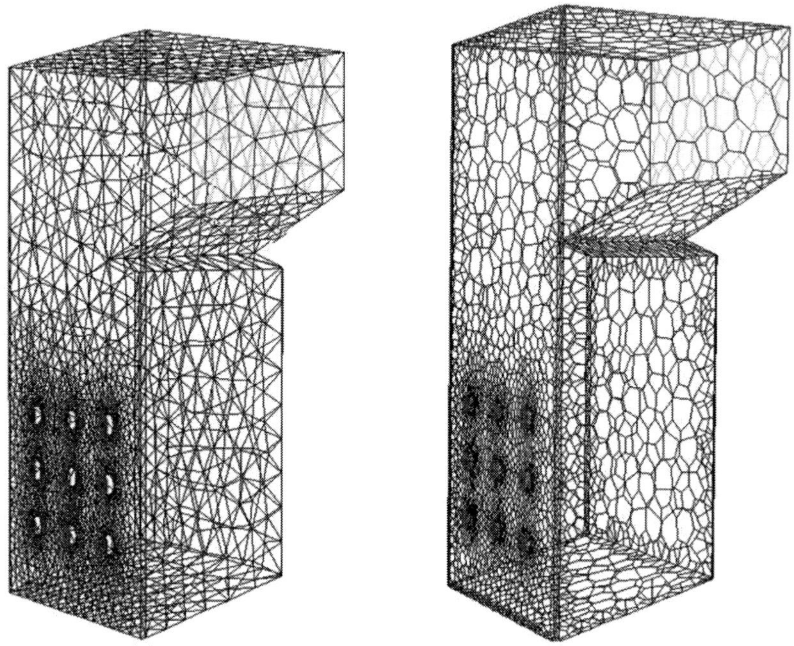

Fig. 3. Computational mesh for mathematical modeling : a) tetrahedral; b) polyhedral.

Table 2. Computational mesh features.

Mesh type	Nodes	Cells
Tetrahedral	237033	22288
polyhedral	213942	159388

3 Calculation results

One of the most important thermal characteristics of the boiler unit is the temperature of the heating medium (Fig. 1). Quantitative assessment of this indicator will set the heat absorption of each of the heating surface, consequently, to assess the reliability of the boiler.

Chart shows the changes of the flare temperature (fig. 5). The main steam superheating occurs in the platen heating surface. Decrease coolant temperature (jumps on the chart) due to the presence of its own condensate injection. Lowering the temperature of steam ensures the reliability of the superheater. Just before injection SHII and SH3 makes possible, if necessary, the coolant temperature control before being fed to the consumer.

To avoid possible overheating of the walls of furnace tubes of metal was analyzed long range burners. According to the results of analysis of the burner have been designed so that combustion of the core housed in the center of the combustion chamber (Fig. 5), except flame impingement on the screen panel, thereby preventing the burnout of pipes and ensuring combustion stability.

 DOI: 10.1051/matecconf/20179201001

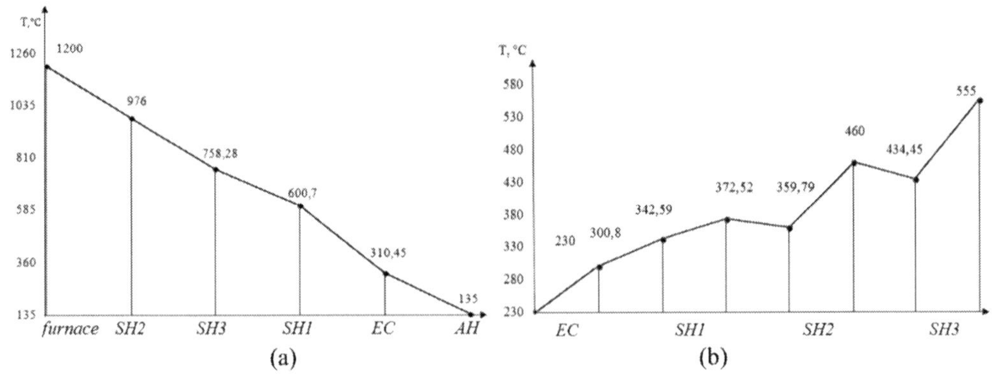

Fig. 4. Flue gas (a) and steam (b)temperature: SH –superheater, EC – economizer, AH – air heater.

Ensuring the environmental safety of the area of installation requires calculation of emissions. It is well known fact that natural gas combustion has large heat observed (Fig. 5) in the combustion nucleus [8]. In connection with that the problems associated with an increasing in emissions of nitrogen oxides are expected.

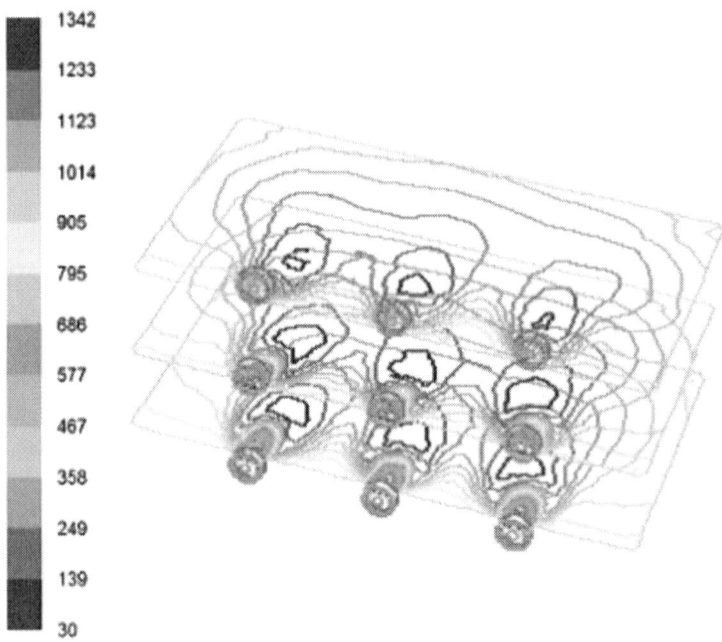

Fig. 5. temperature isolines in cross sections of the combustion chamber (°C).

4 Conclusion

In this paper, the numerical simulation of turbulent combustion of natural gas in large power ramjet burner was used to study the processes occurring in the combustion chamber. According to the results of numerical modelling it can be concluded that the selected arrangement of the multilevel burners reduces emissions of nitrogen oxides (fig. 6). The furnace outlet NO_x amount does not exceed the permissible value of 100 mg/m^3. There is a pronounced vortex aerodynamics in the furnace, which contribute to a sharp increase in the residence time of the fuel and oxidizer near the burner, and promotes intensive ignition and fuel burnup.

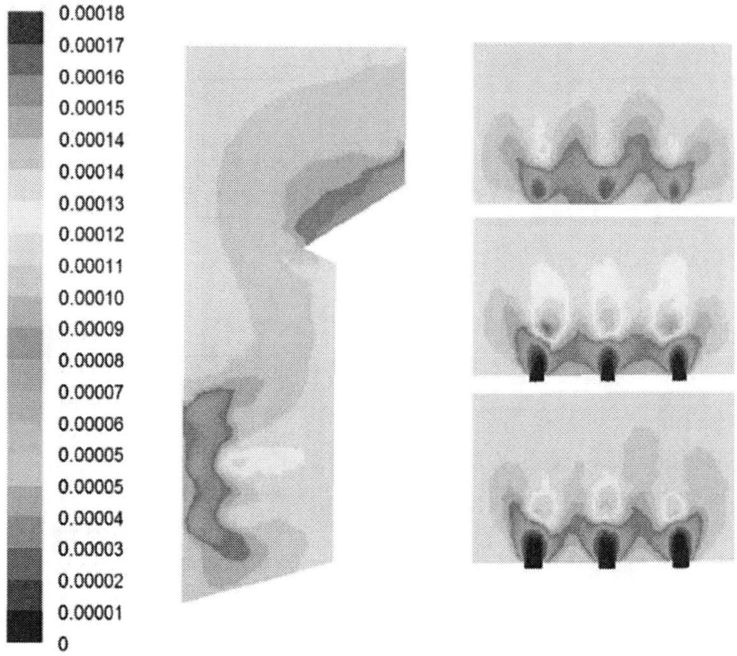

Fig. 6. Contour plot of the total NO_x mass fraction.

The upgraded boiler is a worthy replacement for the outdated boilers. The results obtained during the computational experiment, suggest that engineered boiler unit is designed in accordance with modern standards and regulations. These results confirm the effective of the modeling techniques for designing the boiler equipment.

The reported study was partially supported by the Ministry of education and science of The Russian Federation, state order No. 13.948.2014/K.

References

1. Z. Arsenijević, Z. Grbavčić, B. Grbić, N. Radić, R. Garić-Grulović, S. Miletić, G. Savčić, B. Dordević, J. Serb. Chem. Soc. **75**, 523 (2010)
2. K.A. Al-attab, J.C. Ho, Z.A. Zainal, Exp. Therm. Fluid Sci. **62**, 1 (2015)
3. X.-R. Chen, X.-S. Li, Z.-Y. Chen, Y. Zhang, K.-F. Yan, Q.-N. Lv, Energies **8**, 1242 (2015)
4. S. Dolgov, E. Savchenko, S. Khaustov, R. Tabakaev, A. Zavorin, EPJ Web of Conf. **110**, 01074 (2016)
5. S. A. Khaustov, A. V. Kazakov, G. A. Cherkashina, L. A. Sobinova, EPJ Web of Conf. **110**, 01028 (2016)
6. A. V. Kazakov, S. A. Khaustov, R. B. Tabakaev and Y. A. Belousova, IOP Conf. Series: Mater. Sci. Eng. **124**, 012110 (2016)
7. S. A. Khaustov, A. S. Zavorin, K. V. Buvakov, L. D. Kudryashova, A. V. Tshelkunova, EPJ Web Conf. **82**, 1041 (2015)

Study of water droplet evaporation in a mini-channel with gas flow

Evgeniy Orlik[1,2*] and *Ekaterina* Isachenko[1,3]

[1]Institute of Thermophysics SB RAS, 630090, Novosibirsk, Russia
[2]Novosibirsk State Agricultural University, 630039, Novosibirsk, Russia
[3]Novosibirsk State University, 630090, Novosibirsk, Russia

Abstract. A series of experiments was conducted to study the evaporation of single water droplets with volume of 80~150 μl on the polished stainless steel substrate in temperature range of the substrate surface of 25-80°C in 6 mm high mini-channel with gas flow and the Reynolds number of the gas flow within 0-2500. New data on the water drop evaporation depending on the temperature of gas were obtained. The shadow method was the main method of measurement. As a result of experiments the experimental dependences for evaporation were obtained, and the influence of the gas flow rate and temperature on the water drop evaporation was analyzed.

1 Introduction

Further development of the microelectronic and power technology, which is widely used in the creation of modern ecological transport, in aircraft and space industry, is hampered by a serious technical problem of removal of high local heat fluxes, emitted on the chip and electronic elements. By far the most effective method of heat removal is based on the use of the phase transition and is directly related to the liquid evaporation. Spray cooling by the flow of droplets shows high efficiency of heat removal [1], but the very process of liquid drop evaporation is still not fully understood. Droplet evaporation has been in detail studied both numerically and experimentally, but mostly in open atmosphere. Contact angle of wetting and its hysteresis are the dominant parameters of evaporation [2-4]. Experimental studies show that the maximum heat removal is realized from the contact line [5], whose thickness is about tens microns [6]. In [7] it is shown that in the evaporation of volatile liquids, it is necessary to consider thermocapillary effects leading to the deformation of the interface and contact line. The influence of gas flow on the droplet evaporation is known, however all related studies were conducted in the open air with insufficient controlled conditions [8]. Present paper is devoted to the detailed investigation of the evaporation of water droplet placed on a heated substrate in a flat mini-channel with air flow. The experimental results obtained may be used to develop a new method of heat remvoal on the basis of near-wall drop motion.

* Corresponding author: evgeniy@gmail.com

DOI: 10.1051/matecconf/20179201002

2 Experimental setup and methods

2.1 Experimental setup

The scheme of experimental setup is shown in figure 1. It includes a channel with height varying from 4 mm to 20 mm, removable substrates, control of the substrate temperature, and the air supply system. The side walls and the top cover are equipped with optical windows for visualization. The aim of this work is to study the influence of substrate temperature and Reynolds number of the air flow on the drop evaporation rate for a channel with 6 mm height. The working fluid is ultrapure water, obtained with a Milli-Q system. The drop was placed into the channel with a syringe. The research was carried out at the initial drop volume of 60-150 mm. All experiments were performed on a substrate of stainless steel with polished surface. The Reynolds number of the gas flow varied from 0 to 2000. The temperature in the air flow was maintained equal by means of a system of thermoregulation. Heating of the substrate was carried out by Peltier elements installed in contact with the substrate. The substrate temperature during the experiment was regulated with a precision of up to 0.20 using the controller the PR-59.

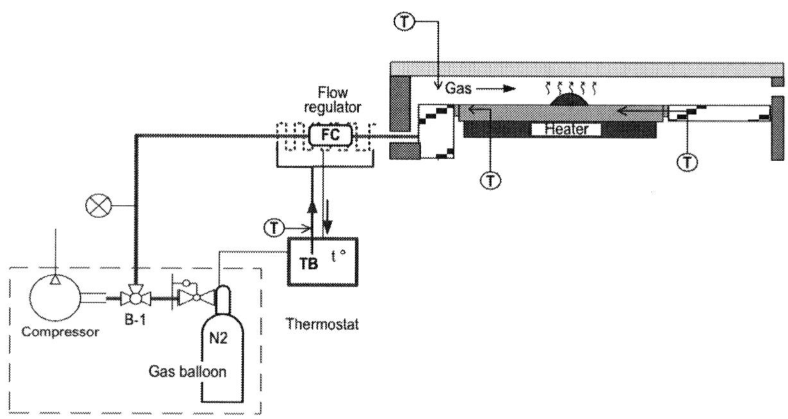

Fig. 1. Scheme of the experimental setup.

2.2 Methods

To register the contour of a sitting drop on a solid substrate we used the shadow method. Its principle is based on the fact that the physical object is illuminated by a parallel light beam, and its shadow is recorded by the camera, as shown in figure 2. Optical equipment allows obtaining images with a resolution of 6 µm/pixel (see figure 4). The obtained images were processed by different methods with the help of software (the Drop Shape Analysis by KRÜSS). This software has multiple methods for stydying the drop contour. This work used the method of height/width to calculate the height of the drop and the diameter of its base, and the elliptic method for calculating the droplet volume during the experiment.

The first stage was the estimation of the error of calculating the drop volume with the elliptic method (see Fig. 2b) and 3). With this view using the rotary device DSA-100 the angle of the substrate inclination to the horizon was changed. As a result, under the action of gravity the drop became asymmetric, thereby the process of the droplet blow-off the in the gas flow was simulated as shown in figure 2 b). From the obtained results it may be concluded that the difference between the advancing and receding angles of the droplet is less than 10^0, and the error of volume measurement is 5 %.

DOI: 10.1051/matecconf/20179201002

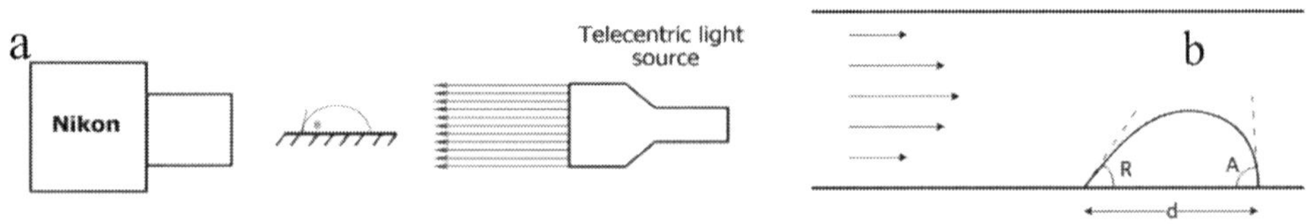

Fig. 2. a) The shadow method, b) The scheme of the drop.

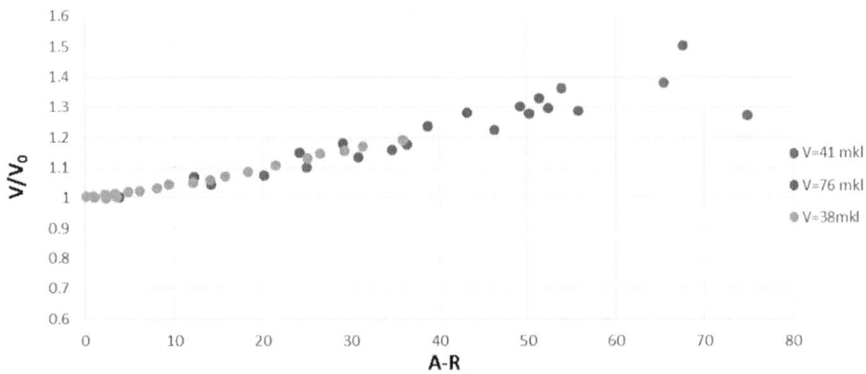

Fig. 3. Estimating the error of the drop volume calculation by the elliptic method in the software "The drop shape analysis", V is the current volume of the drop, V_0 is the drop volume at the substrate inclination by 0^0, A – advancing angle, and R – receding angle.

3 Results

Figure 4 shows photos of the process of drop evaporation at temperatures of the substrate 50 and 70^0 and two different flow rates. Registration of images was conducted with 10 second interval between frames.

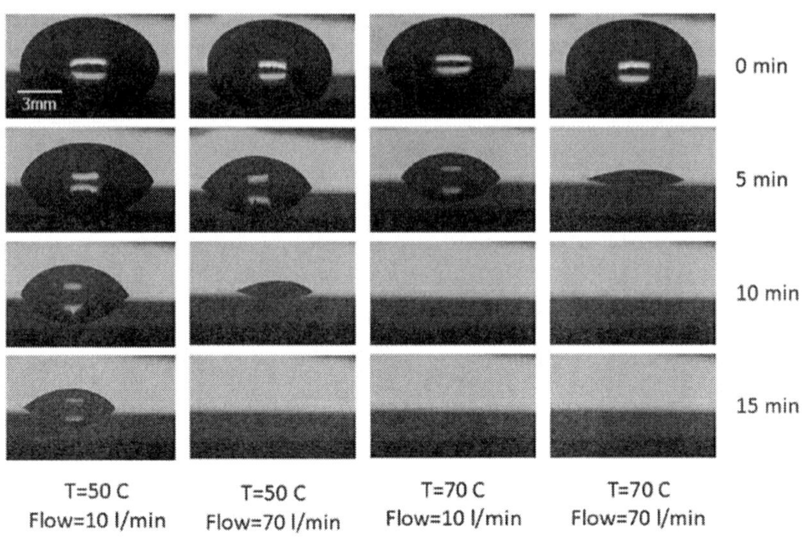

Fig. 4. Droplet evaporation depending on the substrate temperature and gas flow rate.

Evolution of droplet volume has been done in figure 5. It seems that life time of droplet can be used for generalization of evolution of droplet volume. Air temperature has a little effect for evaporation as obvious because low thermal conductivity of air, life time is almost the same for green and blue symbols.

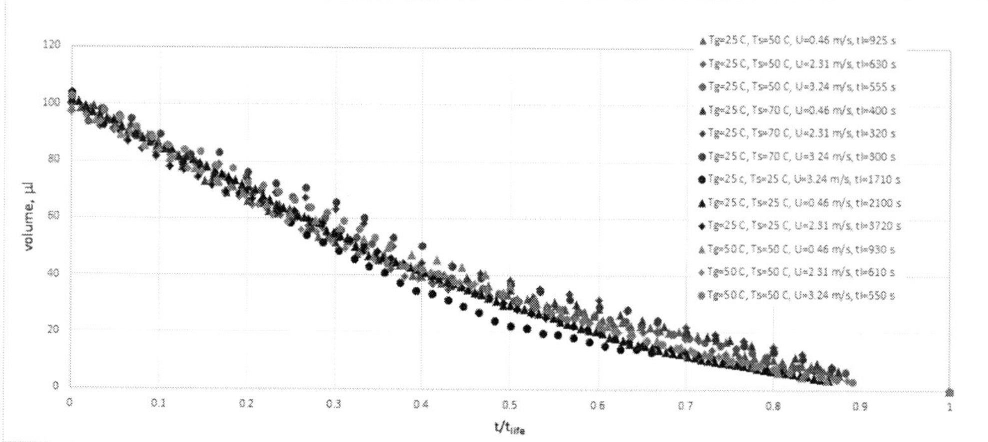

Fig. 5. Droplet volume versus t/t_{life}, subscripts: l – life time, s – substrate, g – gas. V_{init}~100μ l.

Figure 6 shows data for the evaporation rate J related to the droplet surface area depending on the droplet size, substrate temperature and velocity of the gas flow in mini-channel. The curves have a local minimum, whereas it is missing in the evaporation of a droplet in the surrounding atmosphere [2, 3]. In the initial moment of time the drop takes up to half of the channel height and the gas flow has a significant impact on evaporation. Since the gas velocity in a flat channel has a parabolic profile, the velocity of drop blowing decreases with reduction in the drop height; this explains the local minimum. The same effect is present for the evaporation of droplet blowing by air in the open atmosphere, where a characteristic scale is the boundary layer thickness relative to the droplet height [2, 3]. Following conclusion can be drown that evaporation rate J is mostly affected by gas velocity for droplets with diameter d higher than 6mm (see in figure 6) and temperature substrate 50°C and less, because blue and black symbols have the same values. Further, when the drop becomes flat (see figure 5), that is its diameter several times exceeds the height of the drop, the surface temperature of the drop becomes close to the temperature of the substrate, leading to increased evaporation. We can notice that size of droplet before disappearing depends from temperature; one of the explanations of this fact is that surface tension is reducing with increasing of temperature.

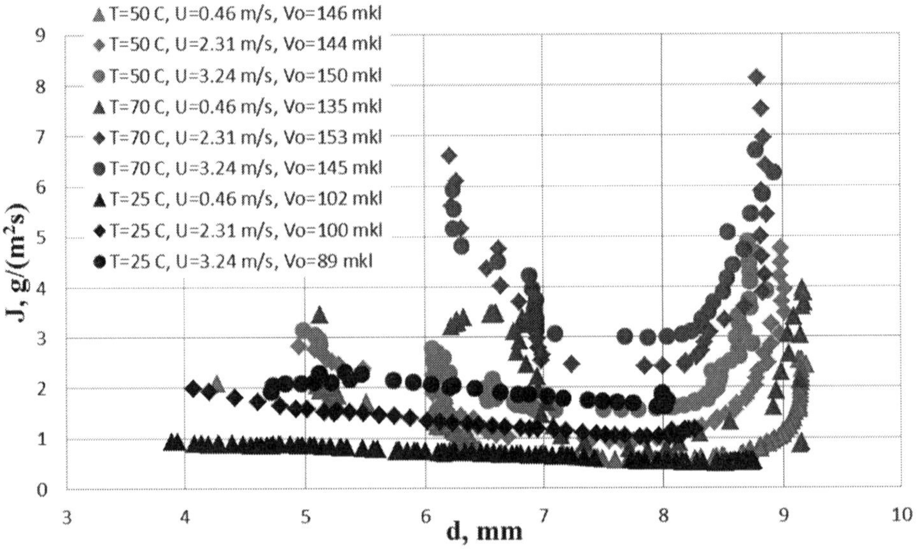

Fig. 6. The evaporation rate versus the droplet diameter. T_{gas}=25°C.

4 Conclusion

In this paper, the evaporation of a single droplet in the 6 mm channel was studied in detail; for this purpose the images of the drop were recorded with an interval of 10 sec and analyzed. The influence of the substrate temperature, the gas flow rate and temperature on the evaporation of water drops was determined experimentally. We can draw the following conclusions: the evaporation rate is significantly determined by the substrate temperature. It is found local minimum on evaporation rate curves. The gas flow rate also affects the evaporation, and this effect is maximal at the initial moment of time. Air temperature has a little effect to droplet evaporation.

The authors express their gratitude to the Russian Science Foundation for the support of this work (project No. 14-19-01755).

References

1. J. Kim, Int. J. Heat Fluid Flow **28**, 753 (2006)
2. E.Ya. Gatapova, A.A. Semenov, D.V. Zaitsev, O.A. Kabov, Colloids Surf., A **441**, 776 (2014)
3. A.A. Semenov, D.V. Feoktistov, D.V. Zaitsev, G.V. Kuznetsov, O.A. Kabov, Thermophys. Aeromech. **22**, 771 (2015)
4. O.A. Kabov, D.V. Zaitsev, Doklady Physics **58**, 292 (2013)
5. V. Cheverda, I. Marchuc, A.Karachevsky, A. Orlik and O. Kabov, Thermophys. Aeromech. **23,** 415 (2016)
6. O.A. Kabov, D.V. Zaitsev, D.P. Kirichenko, V.S. Ajaev, Nanosc. Microsc. Therm. (to be published)
7. Y. Tsoumpas, S. Dehaeck, A. Rednikov, P. Colinet, Langmuir **31**, 13334 (2015)
8. L. Bin, R. Bennacer, A. Bouvet, Appl. Therm. Eng. **31**, 3792 (2011)

Research of power fuel low-temperature vortex combustion in industrial boiler based on numerical modelling

K.Y. Orlova[1,*], *B.V.* Lebedev[1]

[1]Tomsk polytechnic university, 634050 Tomsk, Russia

Abstract. The goal of the presented research is to perform numerical modelling of fuel low-temperature vortex combustion in once-through industrial steam boiler. Full size and scaled-down furnace model created with FIRE 3D software and was used for the research. All geometrical features were observed. The baseline information for the low-temperature vortex furnace process are velocity and temperature of low, upper and burner blast, air-fuel ratio, fuel consumption, coal dust size range. The obtained results are: temperature and velocity three dimensional fields, furnace gases and solid fuel ash particles concentration.

1 Introduction

Low-temperature vortex (LTV) combustion is a technology of combustion concept in furnace volume. A horizontal vortex presence is a principal difference of this technology from the conventional flame combustion. A horizontal vortex has a horizontal axis of symmetry parallel to furnace front wall due to additional blast through the ash hopper opening neck. The main vortex is forced to appear in bottom part of the furnace volume; smaller vortex appearance is possible in upper and mid part of the furnace volume what depends on boiler design.

2 Object of study

Aerodynamics of this technology increases active burning area which in its turn let to decrease maximum temperature in the furnace and improve environmental performance due to de-crease of nitrogen oxides output. The technology makes possible to use granular fuel grind with different burning characteristics. Low-temperature vortex technology provides more stable burning and deflagration. The technology might be carried on in conventional chamber furnace by its upgrade during general maintenance.

We chose P-49 boiler furnace of the Nazarovo power station for the research. A chosen furnace placing presumes: straight-flow burners' presence on front wall, double-nozzled bottom blast, two row of tertiary blast, and aerodynamic overhang on back wall.

[*] Corresponding author: kseniashcherbakova@gmail.com

LTV furnace numerical model was made by means of FIRE-3D software which is based on modern approach and methods of mathematical formulation of complex physicochemical processes [1]. The researched boiler furnace is symmetric in width, all burners and air feeders are located in furnace fore-and-aft plane therefore the aerodynamic is plain as well (parallel to side wall). Hence, it is reasonable to use only part of the furnace volume for the modeling instead of the full volume. Figure 1 shows longitudinal and cross section of the model.

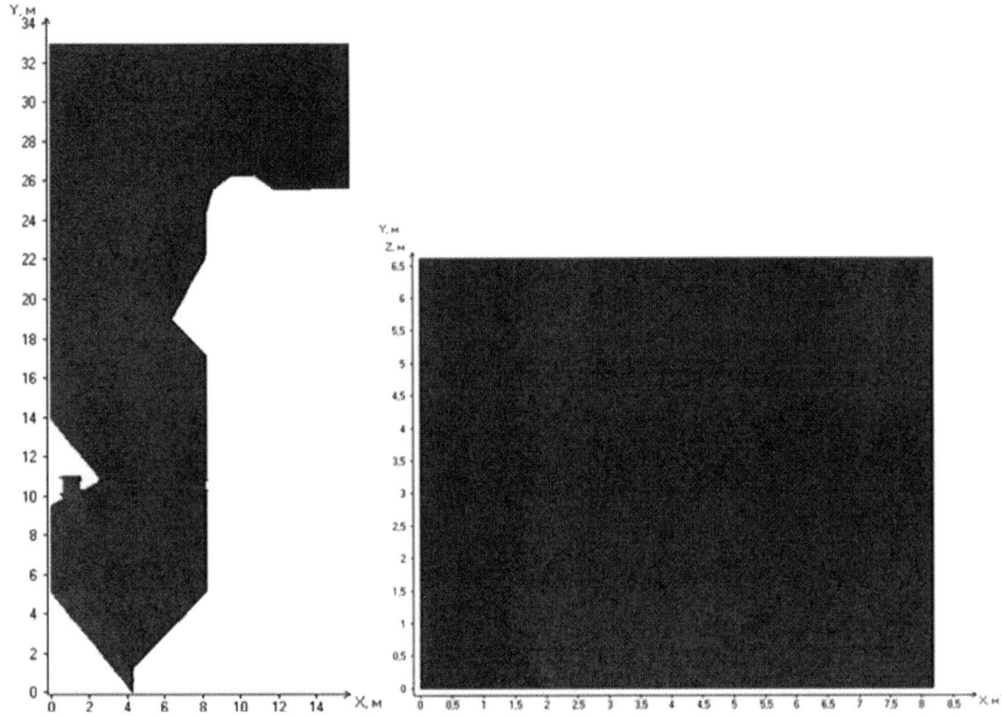

Fig. 1. Longitudinal (XY) and cross section (XZ) of the reduced size furnace.

The obtained model is used for processes research inside the furnace when Nazarovo brown coal is burned. Centrally directed pulverizing system is used for the boiler. That is why dry coal with the following characteristics is used as a base coal dust.

3 Coal dust size range

A coal dust size analysis results (Fig.2) show that researched coal dust consist mainly of 90 and 200 μm (micrometers) coal dust particles, and then smaller fractions. Particles with size more than 500 μm take not more than 6 % of coal dust mass. Nominal condition fuel consumption is 41 kg/s per boiler or 3.417 kg/s per burner.

Presence of 5 coal dust fractions let us to model fuel dispersiveness with high accuracy. Fuel dust size increase – more than 55 % of particles with size over 200 μm – improve vortex area aerodynamic due to better two streams interaction. First stream formed by fuel-air mixture; second stream consist of tertiary blast hot air. Streams are opposed one another and form force couple which creates vortex movement in the furnace bottom area.

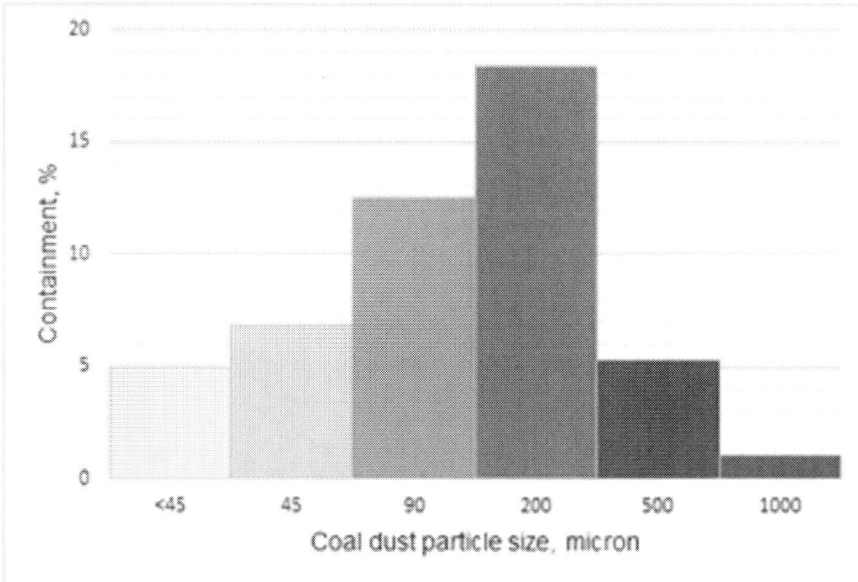

Fig. 2. Coal dust size analysis results.

4 Air regimes

Three different air regimes were calculated in order to re-search how bottom blast velocity influence. Burner blast re-mains constant – 50 m/s. Existing LTV boiler numerical model is used for the calculations [2].

Table 1 shows bottom blast velocity and models air regimes.

Table 1. Air regimes for evaluation of bottom blast velocity influence.

Parameters	Regimes		
	a	b	c
Primary blast, ω_1, m/s	50.00	50.00	50.00
Secondary blast, ω_2, m/s	50.00	50.00	50.00
Bottom blast, ω_3, m/s	30.00	40.00	50.00
excess air, α	1.22	1.20	1.20

The research resulted with the following: all three regimes show air blast process of burner blast by bottom blast. An upward current is formed in mid area of the furnace. When it interacts with the aerodynamic overhang the current is brought to rest, enters upper part of the furnace with lower speed, and leaves through convection shaft exit window.

High oxygen concentration is corresponds to places of air input in the furnace. There is almost no oxygen in area below burners' awning.

5 Temperature and oxygen distribution

During the numerical modelling we obtained new data and present it in graphs. Graphs show mean value of temperature and oxygen containment through furnace height, and their change when bottom blast velocity is changed.

DOI: 10.1051/matecconf/20179201003

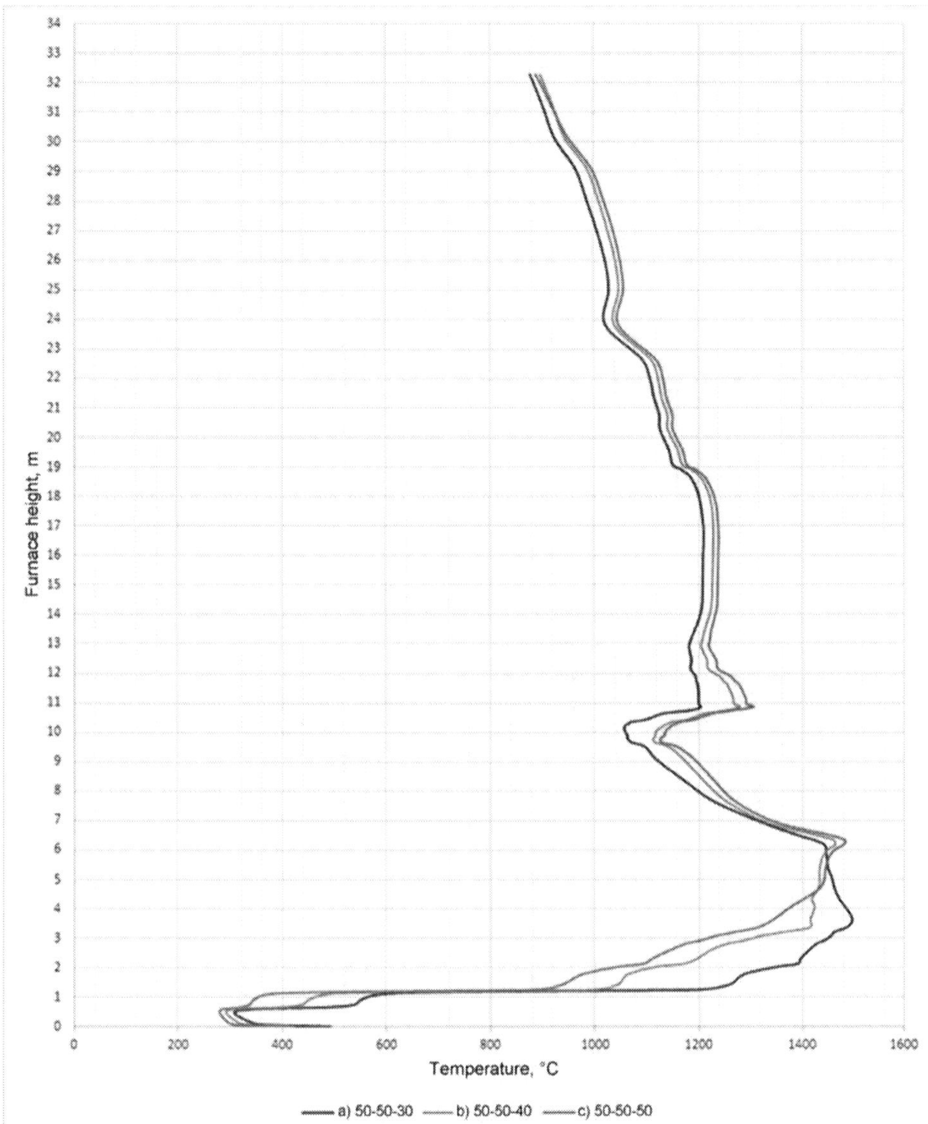

Fig. 3. Mean value of temperature through furnace height.

Temperature above burners' awning increase when bottom blast velocity is increased. This happens due to burning of particles which were thrown away from the vortex when bottom blast is increased. At the same time in the bottom area of the furnace gases temperature is lower because of cold air input through ash hopper opening neck. Oxygen containment through furnace height depends of velocity in the following way: the more bottom blast velocity, the more oxygen concentration because of the excess air presence. The excess air does not take part in combustion reaction on one condition – excess air was the same for all regimes.

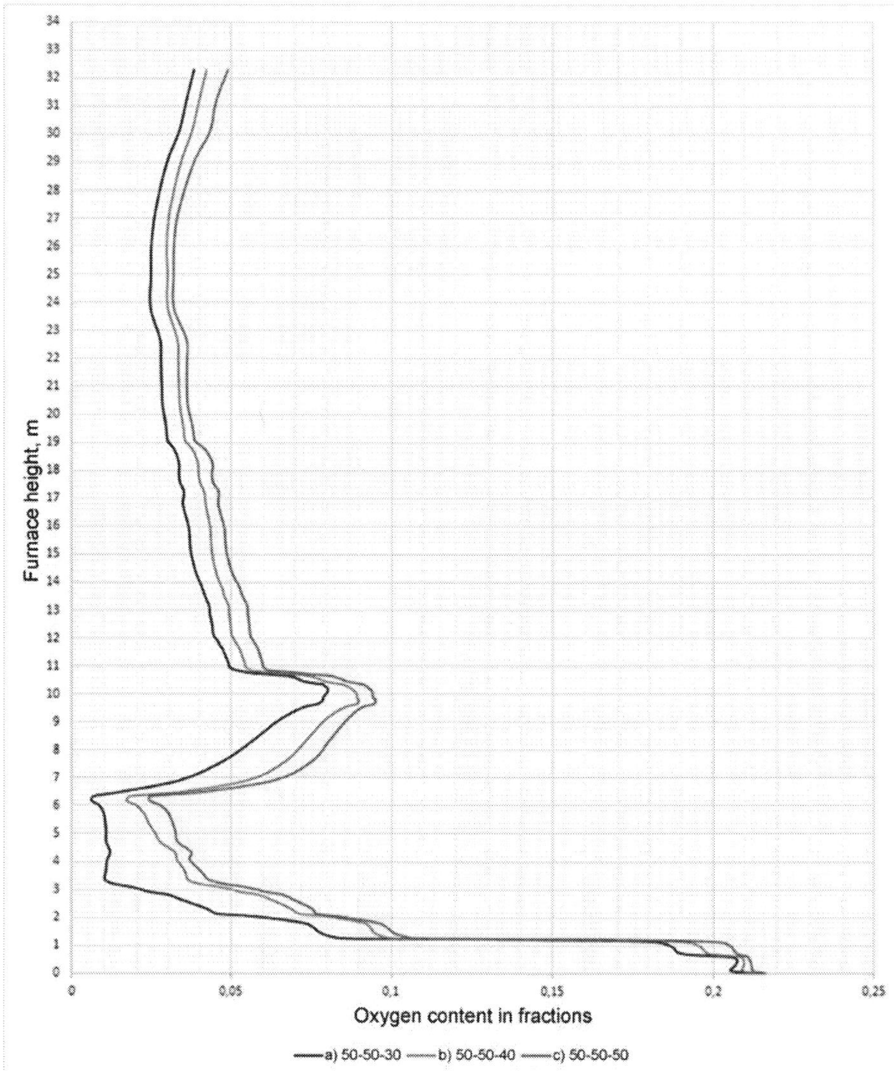

Fig. 4. Mean value of oxygen concentration through furnace height.

6. Conclusion

The conducted research shows how fuel size range affects burning processes of coombe brown coal in low-temperature vortex with horizontal rotation axis.

References

1. A.M. Bubenchikov, A.V. Starchenko, Numerical modeling of aerodispersive mixings' dymanic and combustion in channels, **221-234** (1998), ISBN 5-7511-1004-8 [In Russian].
2. P.S. Gergelizhiu, K.Y. Scherbakova, B.V. Lebedev, MATEC Web of Conf. **23**, 01013 (2015)

Microwave resonator method of dynamic measurement of mass of the samples of gasifying solid fuel

Victor V. Perov[1], and *Vladimir*. E. Zarko[1,*]

[1]Voevodsky Institute of Chemical Kinetics and Combustion, 630090 Novosibirsk, Russia

Abstract. The paper describes novel contactless method of determining instantaneous gasification rate of solid materials under intensive blowing the reacting (gasified) surface. Method is based on dynamic measurement of the mass of dielectric substance within specially designed microwave resonator. The attenuation of passing through the resonator signal is proportional to the actual value of the sample mass. Before firing experiments the setup is calibrated using samples of studied material with different channel radius. With the proper choice of data acquisition system one can achieve the web sample thickness resolution around few microns and time resolution higher than 1 kHz. The examples of preliminary tests with paraffin samples blown by air jets with temperatures of 350-1500 K and speeds of 375-700 m/s are presented.

1 Problem statement

Existing microwave techniques are intended for exploring the linear burning rate of energetic material (EM) samples [1]. They are based on registration of amplitude and phase of the microwave signal passing through a sample and reflected from the burning surface of solid fuel at the expense of sharp change of dielectric properties on the interface of between the fuel and air. Known techniques do not allow investigating the dynamic gasification processes of bored samples of solid fuel at intensive blowing the channel surface.

2 Microwave sensor

To achieve the specified goal the microwave resonator method of dynamic measurement of mass of gasifying solid fuel samples has been developed based on the measurement of the attenuation of a microwave signal passing through the resonator sensor [2] (Fig. 1) loaded with investigated sample.

* Corresponding author: zarko@kinetics.nsc.ru

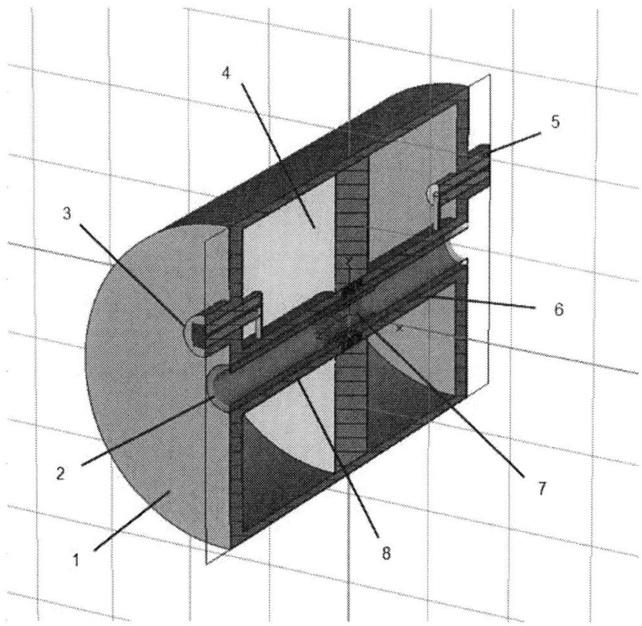

Fig 1. The microwave sensor.
1 – case; 2 – protective tube; 3 – input port; 4 – centering disk; 5 – output port; 6 – output resonator; 7 – sample; 8 - input resonator.

The sensor is executed in the form of system of two connected coaxial resonators excited on antiphase type of fluctuations through one of two ports. Another port is intended for passing the microwave signal to the receiver of the network analyzer. The working area of the sensor is formed by a gap between coaxial resonators. For achievement of the highest uniformity of distribution of high-frequency electric field in the volume of studied sample the length of working area is chosen equal to its diameter.

When carrying out researches, the resonant characteristic of the sensor is previously determined and change of transitional attenuation S21 between sensor's ports on the chosen working frequency is controlled. Insertion of a sample of solid fuel into working area of the sensor affects the form of the resonant characteristic. The first change of the characteristic is connected with the shift of resonant frequency of an antiphase type of the fluctuations, which is caused by increase in magnitude of connection between coaxial resonators at the expense of insertion of dielectric material. If the EM formulation includes the boron powder possessing considerable losses on ultrahigh frequencies, the resonant characteristic S21 (f) except shift also extends because of deterioration of the quality (Q-factor) of the system of connected resonators on the antiphase type of fluctuations.

3 Sensor calibration

Before firing experiments the sensor is calibrated by means of samples of studied material having different channel radius (Fig. 2). Let's note a high linearity of the calibration characteristic S21 (r) in the field of radiuses of 4-8 mm owed to a special design of the sensor and an optimum selection of working frequency that allows minimizing the number of calibration samples up to one sample of studied material. In this case the second point in the calibration characteristic is determined in the absence of EM material in the working area of sensor.

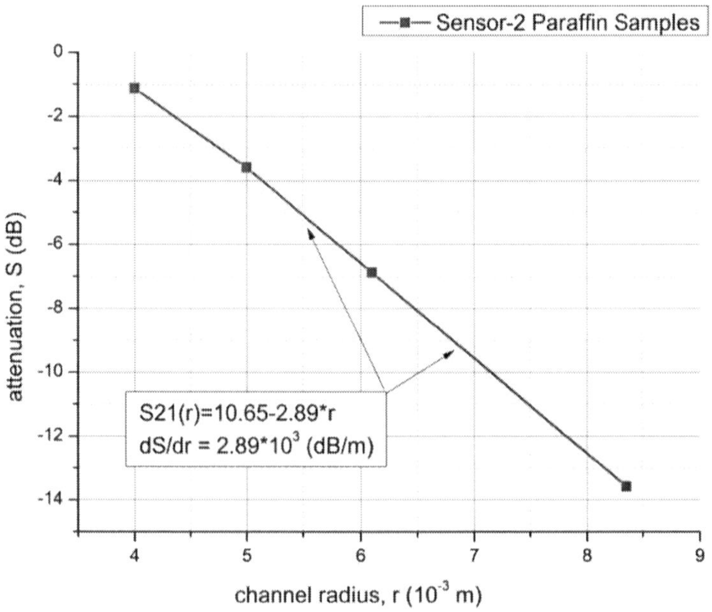

Fig 2. Dependency of the passing microwave signal attenuation on the working frequency versus the channel radius of calibrated samples made of P-2 paraffin.

Calibration curve S (r) in the case of paraffin samples is characterized by an inclination angle of dS/dr equaled $2.89 \cdot 10^3$ dB/m. Applying this result together with the results of registration of gasification of paraffin samples in the form of S (t) (Fig. 3), we can obtain the required dynamic dr/dt parameters for each concrete experiment.

4 Experiments

The microwave resonator method of dynamic mass measurement of solid fuel samples has been illustrated [3] on an example of research of gasification process of the paraffin samples blown by a high speed jets of hot air (Fig. 3, 4).

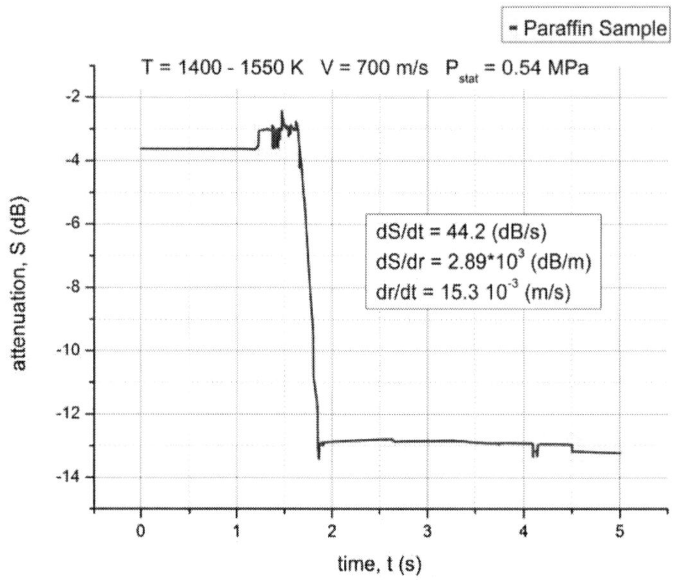

Fig. 3. Dependency S (t) upon paraffin gasification by the air jet with parameters: jet speed V=700 m/s, temperature T=1300 K, static pressure P_{stat}=0.54 MPa.

Duration of gasification process is 0.23 s. Speed of change of the channel radius in the middle of the process is $15 \cdot 10^{-3}$ m/s.

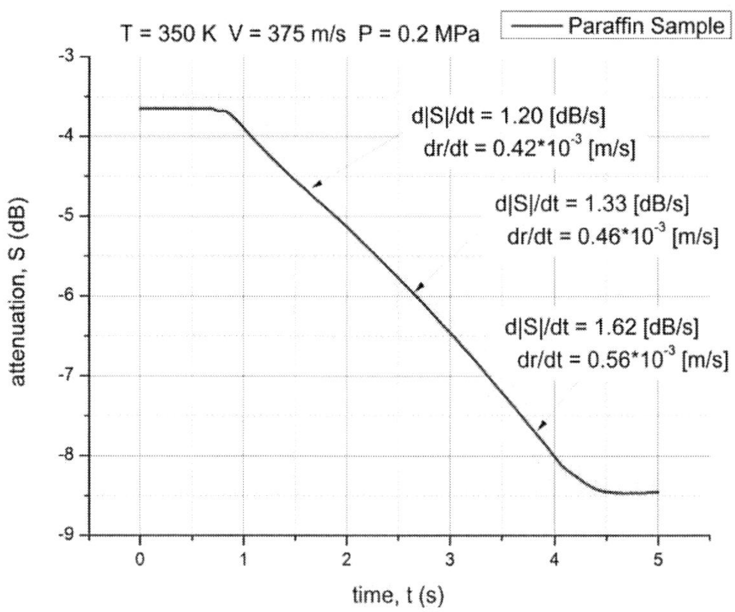

Fig. 4. Experimental dependency S (t) in the case of paraffin P2 sample gasification by the air jet with the speed of 375 m/s, temperature 350 K and gas static pressure 0.2 MPa.

Duration of gasification process is 3.7 s. Speed of the channel radius changing during gasification process increases from $0.42 \cdot 10^{-3}$ m/s to $0.56 \cdot 10^{-3}$ m/s.

As a material of samples the P-2 type paraffin has been chosen. The rate of gasification equals to $15.1 \cdot 10^{-3}$ m/s at blowing by air jet with temperature of 1300K and speed of 700

DOI: 10.1051/matecconf/20179201004

m/s and it is about $0.5 \cdot 10^{-3}$ m/s at blowing by air jet with temperature of 350 K and speed of 375 m/s.

The sample studied is fixed on a sensor axis by two tubes. Internal diameter of tubes is equal to initial diameter of the channel in a studied sample that allows providing uniform velocity profile of a gas stream at the sample entrance.

The registration area of the gasification process is determined by the length of the sensor working zone that allows excluding side areas of a studied sample from consideration and reliably determining the result caused by the gasification process of a sample central region.

5 Conclusions

Thus, the method described allows measuring the instantaneous gasification rate of the samples of dielectric materials under intensive gas blowing with the space resolution about few microns and time resolution better than 1 kHz. The method can be used to obtain the dependencies of gasification rate on the parameters of gas blowing and to study different dynamic gasification regimes at varying in time conditions of solids gasification.

This work was partially supported by the Russian Foundation for Basic Research (project number 16-29-01029).

References

1. D.A.Yagodnikov, A.V. Sergeyev, V. V. Koziche, Combust. Explos. Shock Waves **50**, 2 (2014)
2. V. V. Perov, V.E. Zarko, A.S. Zhukov, Combust. Explos. Shock Waves **50**, 6 (2014)
3. V.V. Perov, V.E. Zarko, V.I. Zvegintsev, D.G. Nalivaichenko, *Proceedings of 56th Israel Annual Conference on Aerospace Sciences* (WeL2T5.3. Haifa, Israel, 2016)

Heat transfer in the FC-72 fluid rivulet flowing down a vertical heated foil

Tatyana Ponomarenko [1], and *Vyacheslav* Cheverda [2,3,*]

[1]Novosibirsk State Technical University, 630073 Novosibirsk, Russia
[2]S.S.Kutateladze Institute of Thermophysics, SB RAS 630090, Novosibirsk, Russia
[3]National Research Tomsk Polytechnic University, 634050 Tomsk, Russia

Abstract. The paper deals with the experimental study of heat transfer in FC-72 fluid revulet flowing down the vertical heated foil with the length of 80, width of 35 mm, and thickness of 25 microns, made of constantan. Employing infrared thermography, we have obtained data on temperature distribution on the foil from the opposite side of the rivulet flow. Further we will calculate the heat flux on the foil surface. It is experimentally revealed that the rivulet width reduces with increase in thermal power due to evaporation, while it increases with increasing fluid flow rate. The FC-72 fluid has good wettability and low latent heat of evaporation. Thus this fluid is recommended for use in cooling systems.

1 Introduction

As of today, one of the main objectives is the design of a highly efficient cooling system. Every year, with each new processor model, the number of transistors increases approximately twice. In modern computer microprocessors, increase in the number of transistors improves many parameters of which processor performance is the main one. However this leads to increase in thermal power that must be removed from the processor. The maximum heat generation in the microprocessor is approximately 500 W/cm^2. Though, microelectronics is only a small part of devices which require a compact and efficient cooling system.

A promising cooling system is a system based on use of rivulet flow, which is a particular case of the film flow between two dividing lines bounding the flow from both sides. The dividing line is a contact line between three phases. Small area between the three-phase contact lines forming a micro-region is characterized by high heat transfer coefficient [1]. Thus, the rivulet flow has high cooling quality.

Maps representing varieties of rivulet flow regimes along an inclined plane were studied in detail in [2, 3]. The increase in liquid flow rate resulted in formation of drip, straight, and winding flow modes, which downstream was transformed into a film flow [2, 4, 5].

Velocity field in a fluid rivulet flowing over a cylinder is obtained in [6] using the PIV (Particle Image Velocimetry) technique.

* Corresponding author: t.evans2010@yandex.kz

Measurements of the local thickness of the rivulets having a wave structure and flowing down a vertical plate were conducted in [7] using LIF method (Laser Induced Fluorescence method). The authors obtained maps representing varieties of wave generation patterns in rivulet flow.

The heat transfer in the water rivulet flowing down a vertical heated foil was studied in [8]. Based on the images obtained from the IR scanner, the authors solved the Cauchy problem for elliptic equation to calculate heat flux on the surface with flowing rivulet. It was proved that in the area of micro-region located near the wetting contact line, heat transfer rate is higher than the average in the rivulet that causes a local temperature reduction in this area.

The main objective of the present study is to determine the heat flux in the area of the three phase contact line of the FC-72 fluid flowing down over a heated surface using data obtained through infrared thermography and solving methods of ill-conditioned problems.

In contrast to the similar experiments already conducted in [8], where highly purified water obtained through a Milli-Q system was used as a test liquid, in this work we use FC-72 fluid designed specifically for cooling supercomputers. Fluorinert Electronic Liquid FC-72 is a fluorocarbon dielectric liquid. It is colorless, transparent, and is thermally and chemically stable cooling agent, compatible with sensitive materials, nonflammable, practically non-toxic, and leaving no residue after evaporation. A special combination of FC-72 properties is ideal for use of this fluid in many areas of electronics, including immersion cooling and film cooling, being currently actively developed.

2 Experimental setup

In the course of the study, the experiments were conducted on a vertical foil with a length l = 80 mm, width w = 35 mm and thickness h = 25 μm made of constantan. The liquid on the foil was fed at a given flow rate within the range 0.1-5 ml/min through a tube with a special nozzle using the Cole-Parmer EW-74905-54 syringe pump. Foil was connected to a source of DC power TTi QPX 1200L through the brass electrode holders. Foil heating was carried out using the power source. Thermal output on the foil was varied within the range from 0.09 to 1.47 watts. The observations were carried out using Nikon D 7000 camera and infrared (IR) 570 M Titanium scanner. The foil was coated with a black graphite paint with an emissivity factor close to 1.

Wetting angle of the FC-72 fluid was measured by a gas (air) bubble technique and was 10±0.4° that indicates good wettability in comparison with satisfactory wettability of water which is 63°.

The experimental setup and the image of the testbench are shown in Figs. 1 a and b.

DOI: 10.1051/matecconf/20179201005

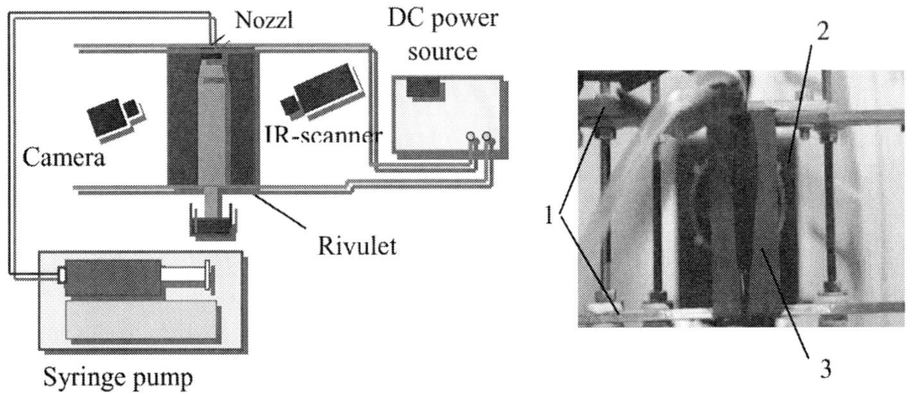

Fig. 1. a – experimental setup; b – FC-72 fluid rivulet flowing down the heated foil at Q = 1 ml/min, P = 0.09 W: 1 – brass electrodes, 2 – IR camera, 3 – foil with flowing rivulet.

3 Experimental results

An infrared scanner was used to measure the temperature distribution over the foil surface from the opposite side of the flowing rivulet (Fig. 2 a-e).

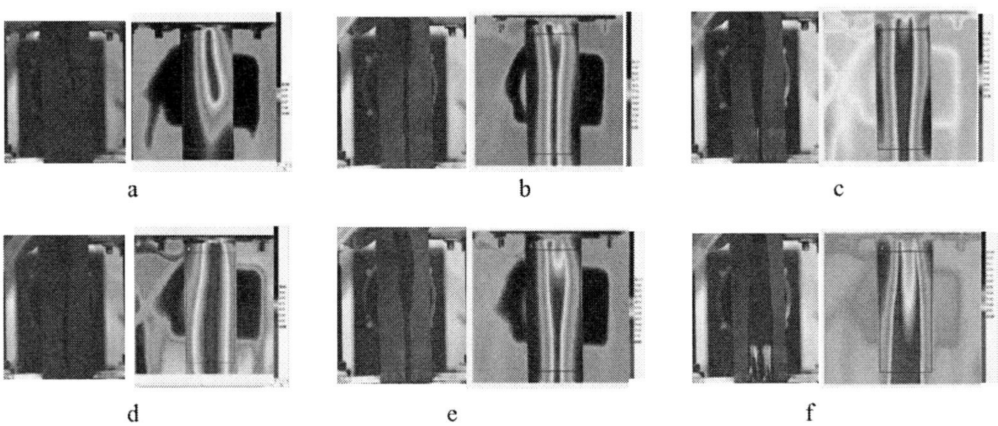

Fig. 2. Images of FC-72 fluid rivulet flowing over a heated foil, and the thermograms from the opposite side of the rivulet:
a – Q = 0.1 ml/min q_{av} = 15.049 W/m^2; b – Q = 0.2 ml/min, q_{av} = 15.049 W/m^2;
c – Q = 0.5 ml/min, q_{av} = 97.45 W/m^2; d – Q = 1 ml/min, q_{av} = 97.45 W/m^2;
e – Q = 2 ml/min, q_{av} = 237.857 W/m^2; f – Q = 5 ml/min, q_{av} = 237.857 W/m^2.

In the future, to calculate evaporation heat flux of a fluid near the contact line, we will use a special method to solve the Cauchy problem for an elliptic equation. In a similar study [8], which was conducted using water, the calculation of the heat flux from the foil was carried out using images from the IR scanner without taking into account heat fluxes in the foil. Alternatively, Cauchy problem describing the thermal conductivity in the foil was solved with due consideration of heat fluxes in the foil. It was shown that the maximum heat flux density takes place in the region near the contact line.

The FC-72 rivulet width was measured from photographs and ranged from 1.4 to 25 mm. Based on the experimental results we plotted dependences of the rivulet width versus applied thermal power at fixed flow rates, as well as dependences of the rivulet width on fluid flow rate at certain thermal power (Figs. 3 a and b).

As is obvious from figures, increasing fluid flow rate leads to increase in the rivulet width, while increasing thermal power leads to reduction in the rivulet width due to the intense evaporation of the FC-72 fluid. At that, latent heat of evaporation of FC-72 fluid is less than that of water.

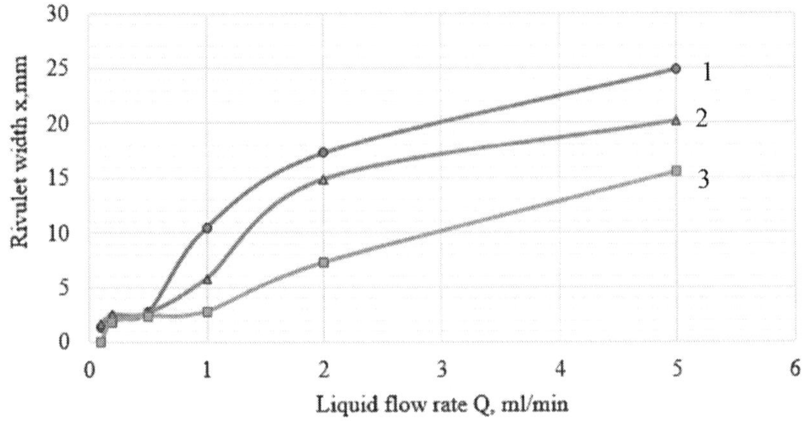

a

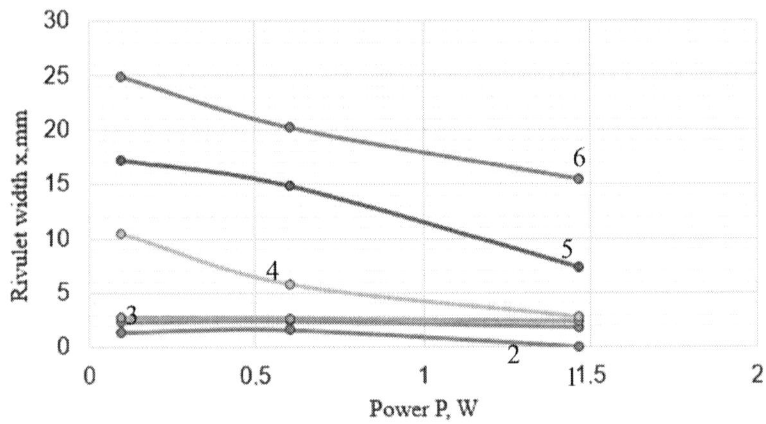

b

Fig. 3: a – the rivulet width versus FC-72 fluid flow rate:
1 – thermal power P=0.09 W; 2 – P=0.6 W; 3 – P=1.47 W;
b – the rivulet width versus thermal power at constant flow rates: 1 – Q = 0.1 ml/min;
2 – Q = 0.2 ml/min; 3 – Q = 0.5 ml/min; 4 – Q = 1 ml/min; 5 – Q = 2 ml/min;
6 – Q = 5 ml/min.

4 CONCLUSIONS

The experimental study on heat transfer in FC-72 fluid rivulet flowing down a vertical heated foil was carried out. It was established experimentally that with the increase in thermal power the width of the rivulet decreases because of evaporation. With increasing fluid flow rate the width of the rivulet increases. Good wettability (maximum contact angle is 10°) and low latent heat of evaporation of the FC-72 fluid facilitates its use when designing effective cooling systems.

This study was performed in the framework of the Russian Science Foundation grant (Agreement No. 14-19-01755).

References

1. P. Stephan, C. Brandt, Heat Transfer Eng. **25**, 78 (2004)
2. P. Schmuki, M. Laso, J. Fluid Mech. **215**, 125 (1990)
3. N. Le Grand-Piteira, A. Daerr, L. Limat, Phys. Rev. Lett. **96**, 254503-1 (2006)
4. C.W Hirt, Los Alamos: Flow Sci. **7** (1998)
5. G.W. Su, J.T. Geller, K. Pruess, F. Wen, Water Resour. Res. **35**, 1019 (1999)
6. S.V. Alekseenko, V.A. Antipin, A.V. Bobylev, D.M Markovich, Exp. Fluids **43**, 197 (2007)
7. S.V. Alekseenko, A.V. Bobylev, S.M. Kharlamov, D.M Markovich, *Proc. of 14th Int. Symp. on Applications of Laser Techniques to Fluid Mechanics* (Lisbon, Portugal, 2008)
8. V.V. Cheverda, I.V. Marchuk, A.L. Karchevsky, E.V. Orlik, O.A. Kabov, Thermophys. Aerodyn. **23**, 3 (2016)

Experimental study of thermosyphon operation when cooling the condensation part by drop irrigation

Konstantin O. Ponomarev[1,*], *Evgeniya* G. Orlova[1], *Atlant* E. Nurpeiis[1]

[1]National Research Tomsk Polytechnic University, 634050, Tomsk, Russia

Abstract. The influence of drop irrigation of the condensation part on the thermosyphon operational mode was studied experimentally. The temperature and pressure fall was found in the internal space of the upper chamber after applying the liquid on the upper wall of the thermosyphon. The temperature difference along the height of vapor channel increases and the effective resistance decreases. When complete evaporation of liquid droplets from thermosyphon, it returns to the quasi-stationary mode of operation.

1 Introduction

Two-phase thermosyphons are simple in construction, autonomous, and effective as thermal engineering devices capable of transferring significant heat flows at the minimum temperature differences [1]. Due to these advantages thermosyphons have low operating and capital costs; they are widely used in oil refining, petrochemical and gas industries [2-3], solar energy [4-5], microelectronics and other fields [6, 7]. The operational mode of thermosyphon [8-10] is known to depend on the composition and filling volume of coolant, geometric parameters, inclination angle, material of construction, heat load [11-13], conditions of cooling the condensation part [12]. In this condition, conjugated processes of conduction, convection and phase transitions in thin liquid films or droplets [13-20] are currently poorly studied in order to design energy-efficient heat-removing equipment. To ensure the conditions for safe and efficient operation of these devices, in addition to summarizing the available information, it is necessary to obtain partial dependences of certain factors on the thermosyphon operational mode.

The purpose of this work is to study experimentally the operation of two-phase thermosyphon when cooling the condensation part by drop irrigation.

2 Experimental technique

Experimental studies were carried out on the setup. The scheme of this equipment is presented in Fig. 1.

* Corresponding author: kop.tpu@gmail.com

The experimental setup consists of the following parts: the test cell 1 has the metal substrate 2, through which heat is transferred from the heating spiral 3 (connected to the power source 4) to the bottom of the lower chamber of the thermosyphon 5. Heat and mass transfer with the change in the aggregate state of the working liquid occurs in the lower chamber of the thermosyphon. The vapors of liquid 7 formed by boiling in the lower chamber pass through the vapor channel to the upper chamber. The condensate 8 flows into the evaporation zone along the walls of the vapor channel.

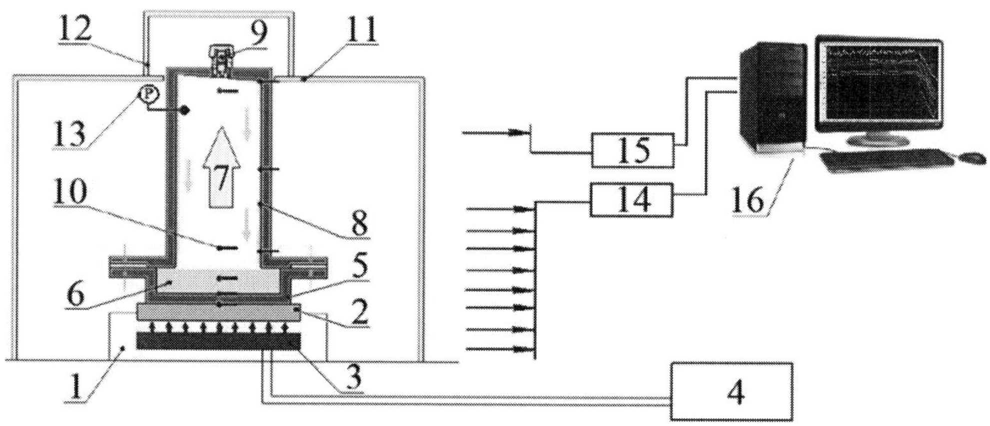

Fig. 1. The scheme of the experimental setup: 1 – test cell; 2 – metal substrate; 3 – heating spiral; 4 – power source; 5 – thermosyphon; 6 – working liquid; 7 – vapors of liquid; 8 – condensate; 9 – valve; 10 – thermocouples; 11 – glass box; 12 – cover for top opening; 13 – pressure sensor; 14 – analog-to-digital converter NI 9214; 15 – analog-to-digital converter NI USB-6000; 16 – personal computer.

The upper wall has a threaded opening for the valve 9, which is designed for pressure control. Eight thermocouples 10 record the temperature: the outer and inner bottom wall of the thermosyphon, working liquid, vapors in the lower and upper part of the vapor channel, condensate on top, center and bottom part of the side wall. To provide constant conditions of the heat exchange, thermosyphon is placed in the glass box 11 with the cover 12 for top opening.

The distilled water was used as the working liquid. After reaching the quasi-stationary mode of the thermosyphon the glass cover was removed from the condensation part, and ten drops of distilled water were uniformly placed on the upper wall of the thermosyphon by Lenpipette dispenser. The volume of a drop was 50 microliters. When complete evaporation of the liquid from the surface, condensation part was covered by the glass cover, and the load increased. The heating power was 6; 7; 9; 11.6 and 14.8 W.

3 Results and discussion

The obtained results of cooling the condensation part are shown in Fig. 2. The temperature and pressure fall was found in the internal space of the upper chamber after applying the liquid on the upper wall of the thermosyphon. At this moment, a sharp increase in temperature difference occurs (Fig. 2).

After evaporation of the applied liquid from the surface there is an increase in the temperature inside the thermosyphon and a decrease in the temperature difference along the height of the vapor channel up to values corresponding to mode without irrigation. So thermosyphon operation returns to the quasi-stationary mode. The irrigation of the condensation part of the thermosyphon does not influence the specific heat flow.

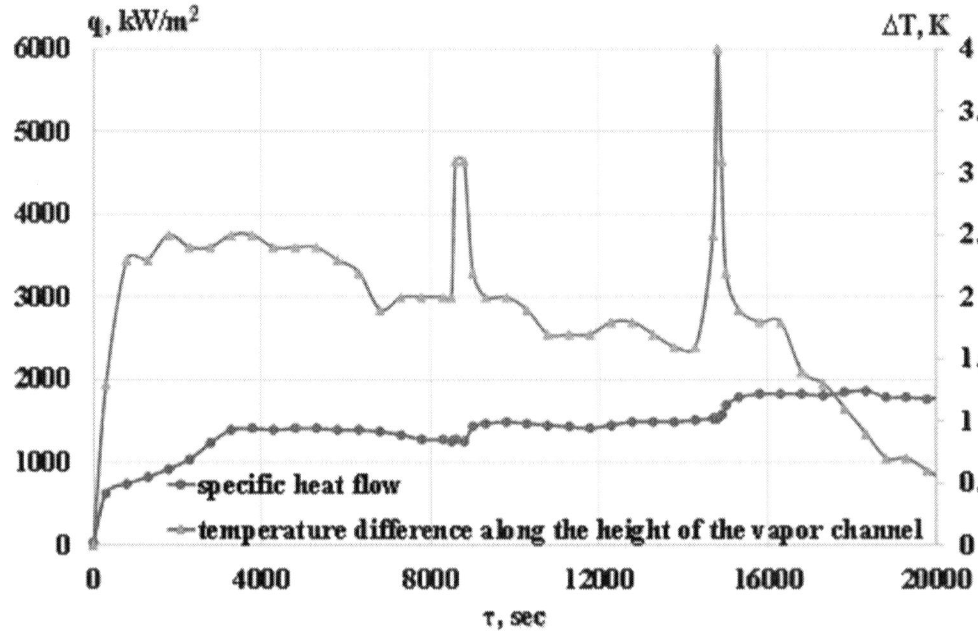

Fig. 2. Change in the specific heat flow and temperature difference along the height of the vapor channel.

The effective thermal conductivity is found to increase with increasing the specific heat flow (Fig.3). When irrigating the condensation part, the effective thermal conductivity decreases.

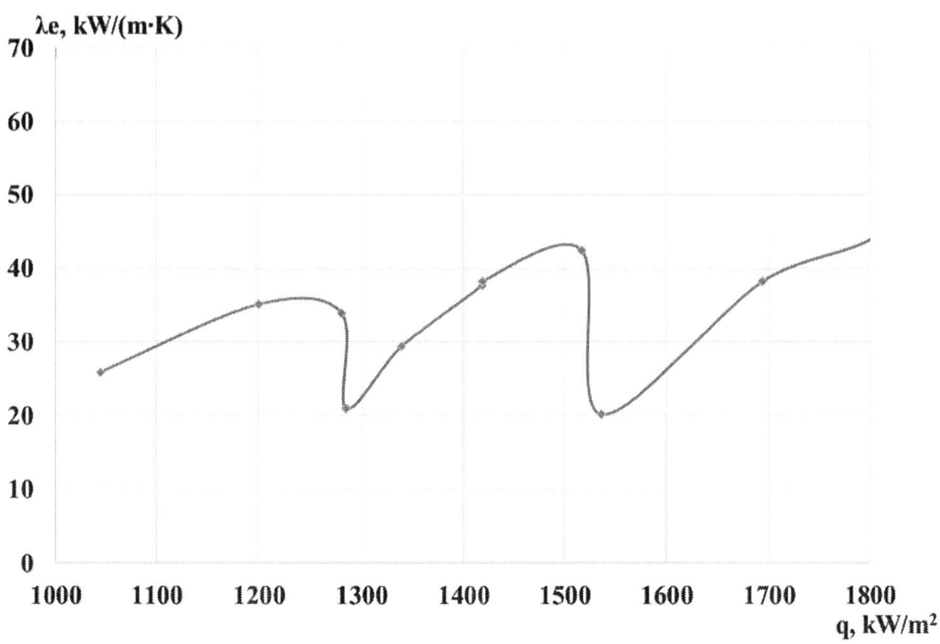

Fig. 3. Dependence of the effective thermal conductivity on the specific heat flow when cooling the condensation part.

It is connected to the fact that when evaporating the liquid drops from the top wall of the thermosyphon, heat equaled to the latent heat of vaporization is assimilated. The temperature in the upper part of the thermosyphon dramatically decreases, as a result, the temperature difference along the height of the vapor channel increases. This leads to an increase in the effective resistance. Since the effective thermal conductivity and resistance are inversely proportional, the latter one decreases.

When complete evaporation of liquid drops, the surface temperature and effective thermal conductivity increases, the temperature difference inside the thermosyphon decreases, and it returns to the quasi-stationary mode of operation.

The reported research was supported by Russian Federation President Grant for state support of the Russian Federation leading scientific schools SS-7538.2016.8 (No. 14.Y31.16.7538-SS).

References

1. V.Yu. Kravets, V.A. Chernobaj, A.A. Nikitenko, B. Golamreza, EEJET **2**, 40 (2011)
2. W. Angelo, M.H. Mantelli, F.H. Milanez, *Proceedings of the 14th International Heat Pipe Conference (14th IHPC)* (Florianopolis, Brazil, 2007)
3. G.V. Kuznetsov, M. Al-Ani, M.A. Sheremet, Tomsk State University Journal of Mathematics and Mechanics, **93** (2011)
4. S.A. Kalogirou, Prog. Energy Combust. Sci. **30**, 231 (2004)
5. M. Esen, H. Esen, Sol. Energy **79**, 459 (2005)
6. J.M. Gordon, K.C. Ng, H.T. Chua, A. Chakraborty, Int. J. Refrig. **25**, 1025 (2002)
7. A. K. Nayak, D. Lathouwers, T. H. J. J. Van der Hagen, F. Schrauwen, P. Molenaar, A. Rogers, Appl. Therm. Eng. **26**, 644 (2006)
8. A. Amiri, R. Sadri, M. Shanbedi, G. Ahmadi, B.T. Chew, S.N. Kazi, M. Dahari, Energy Convers. Manage. **92**, 322 (2015)
9. C.C. Chien, C.K. Kung, C.C. Chang, W.S. Lee, C.S. Jwo, S.L. Chen, Energy **36**, 415 (2015)
10. N.C. Chami, A. Zoughaib, Energy Build **42**, 1267 (2010)
11. S. Jaisankar, T.K. Radhakrishnan, K.N. Sheeba, S. Suresh, Energy Convers. Manage **50**, 2638 (2009)
12. H. Mirshahi, M. Rahimi, Iran J. Chem. Eng. **6**, 15 (2009)
13. G.V. Kuznetsov, D.V. Feoktistov, E.G. Orlova, K.A. Batishcheva, Colloid J. **78**, 335 (2016)
14. D. Zaitsev, D. Rodionov, O. Kabov, Tech. Phys. Lett. **35**, 680 (2009)
15. N.V. Demyanovich, V.I. Maksimov, T.A. Nagornova, Modern problems of science and education, 1 (2014)
16. D. Zaitsev, D. Kirichenko, O. Kabov, Tech. Phys. Lett. **41**, 551 (2015).
17. G. Kuznetsov, D. Feoktistov, E. Orlova, J. Eng. Phys. Thermophys. **89**, 317 (2016)
18. G. Kuznetsov, D. Feoktistov, E. Orlova, Thermophys. Aeromech. **23**, 17 (2016)
19. V. Nakoryakov, S. Elistratov, S. Misyura, J. Eng. Phys. Thermophys. **20**, 338 (2011)
20. V. Lebedev, V. Lemanov, S. Misyura, V. Terekhov, Fluid Dyn. **28**, 624 (1993)

Simulation of the processes of heat- and the mass transfer in the rooms of public building with the natural ventilation

Maria Prorokova[1,*], *Vyacheslav* Bukhmirov[1]

[1]State Educational Institution of Higher Professional Education Ivanovo State Power University named after V.I. Lenin, 1153003 Ivanovo, Russia

Abstract. In the article the mathematical model of the processes of heat exchange and mass exchange in the room of building with the natural ventilation is shown. The verification of mathematical model is performed via the comparison of the results of calculation in ANSYS Fluent with the data of experiment. Experiment was conducted in the room of educational institution. In the experiment were measured the temperature of air, air speed and moisture content in air. A low relative error in the calculation with the use of a mathematical model makes its use for predicting the parameters of microclimate after the introduction of the energy-saving measures possible

1 The mathematical model of the processes of heat exchange and mass exchange in the room

The prediction of the parameters of microclimate in the rooms of habitable, public and office buildings is urgent task. Solution of this problem will make it possible to estimate influence on the microclimate of the rooms of factors, connected with the energy-saving measures [1].

For the solution of this problem in Ivanovo State Power University was used the method of mathematical simulation. The thermal, humid and air regime of room was described by such parameters as the temperature of air ($T_{в}$, ^{0}C), air speed (w, m/s), moisture content (d, $kg/kg_{d.a.}$) and the concentration of carbon dioxide (CO_2, ppm). For calculating these parameters the system of differential equations, which contains the equation of the conservation of energy, pulse and quantity of substance, and also the integral-differential equation of the transfer of the radiant energy, was solved [2-4].

The standard k-ε model of turbulence was used for calculating the turbulent properties of air in the room [5]. The realization of the mathematical model of the processes of heat-mass transfer is executed in ANSYS Fluent [6]. The solution of the equations of the transfer in ANSYS Fluent is based on the method of final volumes.

* Corresponding author: prorokova_mv@list.ru

2 The verification of the mathematical model

The verification of the mathematical model of microclimate in the room was executed by the comparison of the results of calculation in ANSYS Fluent with the data of experiment. In the experiment in the room were measured the temperature, air speed and moisture content in air. Measurements are taken on height 0.1, 0.6 and 1.7 m from the floor (for the rooms, in which the people the large part of the time sit) [7].

During the first stage calculation in ANSYS Fluent is built the geometric model of experimental room (figure 1) and with the aid of the grid editor Meshing to the calculated region it is superimposed Cartesian multilevel grid (method CutCell). The automatic generation of adapted grid made it possible to obtain grid from 1488278 the elements and 1817190 the units with the quality of the elements average by the volume of calculated region 0.98. The exterior view of calculated grid is represented in the figure 2.

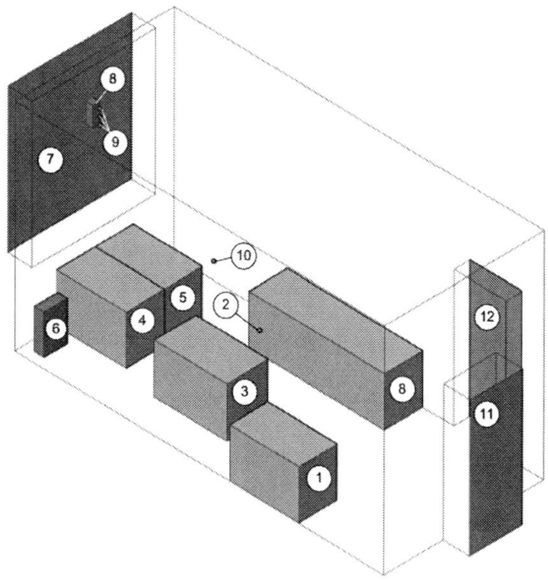

Fig. 1. The geometric model of the experimental room: 1, 3, 4, 5, 8 – table; 2, 10 – the sensors, which measure the parameters of air; 6 – the source of the heat (oil cooler); 7 – window; 8 – the source of the heat (heat-fan); 9 – sensor for measuring the parameters of surrounding air; 11, 12 - the door.

In the calculation it is accepted:

– air in the room is of four component mixture of nitrogen (N_2), of oxygen (O_2), of carbon dioxide (CO_2) and of water vapour (H_2O);

– air is subordinated by ideal gas law;

– air in the room is diathermy;

– air it enters evenly only on the perimeter of window.

For describing the boundary conditions of task the results of the experiment were used: the mean temperature of the surfaces of room, the temperature of air, the moisture content in air, air velocity and the composition of surrounding air, the temperature of the surface of man, composition and the temperature of air, which inhales and breathes out men, power of the source of heat (oil cooler).

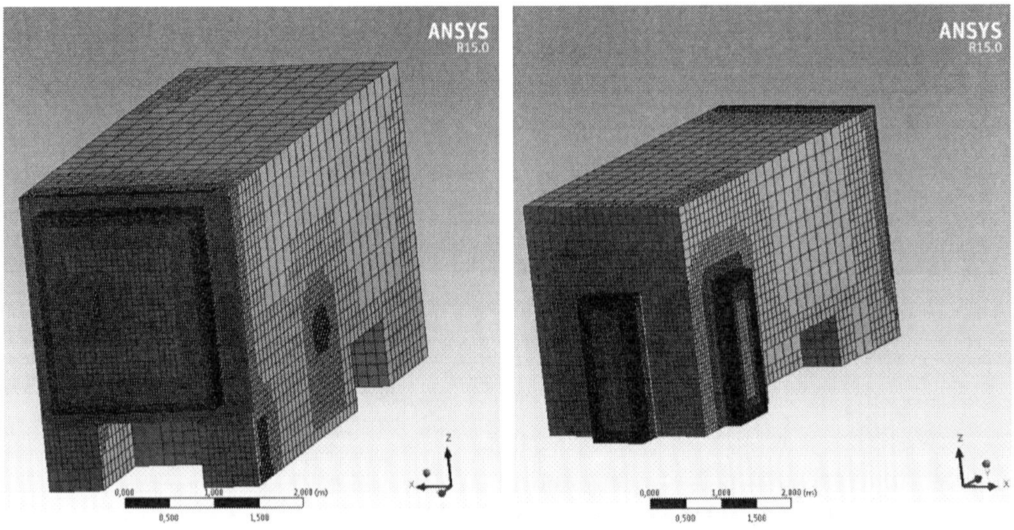

Fig. 1. Computational grid.

With the mathematical simulation of microclimate in the experimental room the influence of air composition on the basic parameters of microclimate was investigated. Calculations are executed for two air compositions:

– air in the room and tributary, and breathed out by man consists of the mixture of two gases: of nitrogen (N_2) and oxygen (O_2);

– air in the room is the four-component mixture, which consists of nitrogen (N_2), of oxygen (O_2), of carbon dioxide (CO_2) and of water vapour (H_2O).

Problem is solved for steady state of heat exchange and mass exchange with the application of the method Pseudo Transient.

The results of calculation and experiment are given in the figures 3, 4 and 5.

In the tables 1 and 2 is given an relative error in calculated temperature and moisture content in air in ANSYS Fluent (in comparison with the experiment).

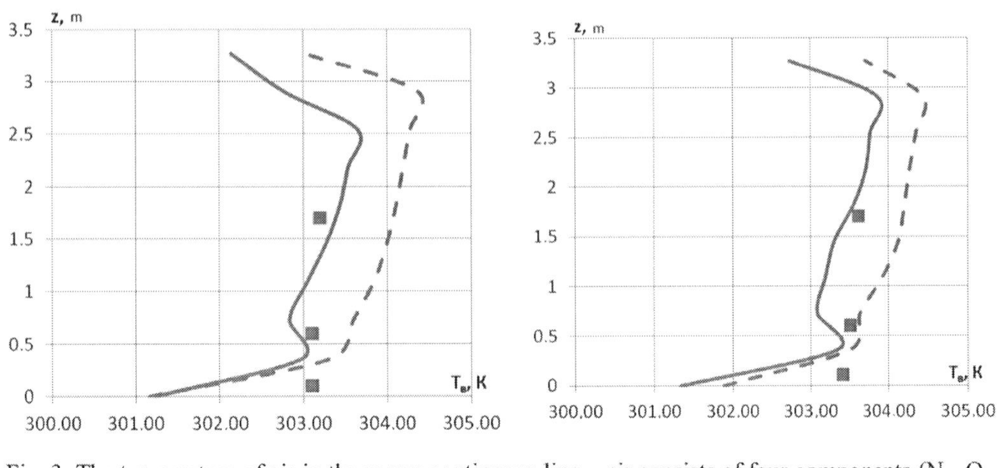

Fig. 3. The temperature of air in the room: continuous line – air consists of four components (N_2, O_2, CO_2 and H_2O); dashed line – air consists of two components (N_2 and O_2); point – the experiment.

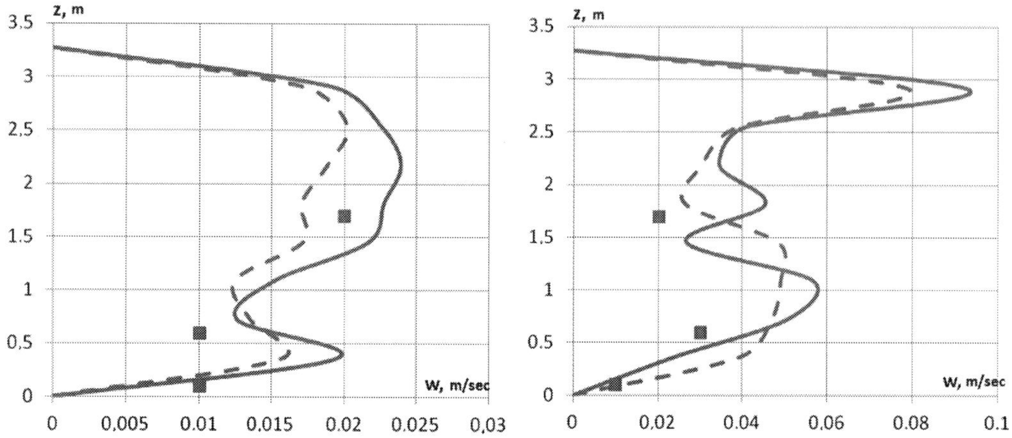

Fig. 4. The velocity of air in the room: continuous line – air consists of four components (N_2, O_2, CO_2 and H_2O); dashed line – air consists of two components (N_2 and O_2); point – the experiment.

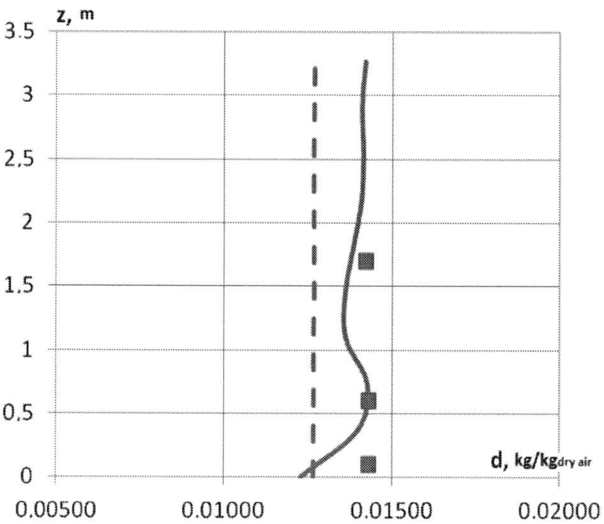

Fig. 5. Moisture content of the air in the room: continuous line – air consists of four components (N_2, O_2, CO_2 and H_2O); dashed line – air consists of two components (N_2 and O_2); point – the experiment.

Table 1. An error in the calculated temperature of air in ANSYS Fluent.

№ the point	Height above floor level, m	$T_в$, K (experiment)	Air consists of two components		Air consists of four components	
			$T_в$, K (calculation)	Relative error, %	$T_в$, K (calculation)	Relative error, %
1	0.1	303.1	301.94	3.85	302.18	3.06
	0.6	303.1	303.67	1.89	303.05	0.17
	1.7	303.2	303.95	2.48	303.37	0.56
2	0.1	303.4	302.05	4.44	302.50	2.96
	0.6	303.5	303.80	0.98	303.35	0.49
	1.7	303.6	304.47	2.84	303.33	0.88

DOI: 10.1051/matecconf/20179201007

Table 2. An error in the calculated moisture content of the air in ANSYS Fluent.

№ the point	Height above floor level, m	d, $kg/kg_{d.a.}$ (experiment)	Air consists of two components		Air consists of four components	
			d, $kg/kg_{d.a.}$ (calculation)	Relative error, %	d, $kg/kg_{d.a.}$ (calculation)	Relative error, %
1	0.1	0.0143	0.0127	11.19	0.0130	9.09
	0.6	0.0143	0.0127	11.19	0.0142	0.70
	1.7	0.0142	0.0127	10.56	0.0151	6.34

The analysis of figures 3, 4 and 5 and tables 1 and 2 shows that the calculation of water vapour and carbon dioxide in air increases the accuracy of the mathematical simulation of microclimate. A difference in the calculated and experimental values of the air speed and moisture content is caused by both the error in the numerical calculation and by error in the experimental determination of these parameters.

3 Conclusion

Is proposed and realized in ANSYS Fluent the mathematical model of the processes of heat transfer and mass transfer in the room of building, which considers water vapour and carbon dioxide in air. Is proven the authenticity of mathematical model with the aid of the comparison of the results of calculation with the experimental data.

References

1. V.V. Bukhmirov, M.V. Prorokova, Vestnik IGEU **4** (2015) [in Russian]
2. B. Gebkhart, Y. Dzhaluriya, R. Mekhadzhan, B. Sammakiya, *Svobodnokonvektivnye techeniya, teplo- i massoobmen* (Mir, Moscow, 1991) [in Russian]
3. Yu.A. Tabunshchikov, M.A. Brodach, AVOK, 2002 [in Russian]
4. A.M. Grimitlin, T.A. Datsyuk, D.M. Denisikhina, AVOK Severo-Zapad, 2013 [in Russian]
5. I.A. Belov, S.A. Isaev, *Modelirovanie turbulentnykh techeniy* (Balt. gos. tekhn. un-t, 2001) [in Russian]
6. I. Weinhold, J. Parry, The Three Waves of Commercial CFD. URL: www.mentor.com [12.07.2016]
7. GOST 30494-2011 *Zdaniya zhilye i obshchestvennye. Parametry mikroklimata v pomeshchenii* (2011)

Theoretical and numerical characterization of the thermal conductivity of carbon ceramic material from realistic microscale imaging

Pavel V. Prosuntsov[1,*], and *Nikita* Y. Taraskin[1]

[1]Bauman Moscow State Technical University, 105005 Moscow, Russia

Abstract. The paper is concerned with the theoretical and numerical characterization of the thermal conductivity of porous carbon ceramic materials. The material structure models are created based on 3D computer tomography (CT) images. The temperature functions of the molecular and effective thermal conductivity for the porous carbon-ceramic material were obtained. It was investigated, whether a simplified model could be used to estimate the thermal conductivity. The thermal conductivity data obtained with the computer tomography method and with the proposed simplified model are compared.

1 Subject of research

Carbon-ceramic composite materials (CCCM) hold a prominent position in the space rocket engineering. They are used to make the spacecraft frames temperature-, impact- erosion- and chemically resistant, protecting them from the high-speed gas flow [1]. State-of-the-art CCCM [1] retain their efficiency at temperatures up to 2000 K and possess sufficiently high mechanical properties. Despite this CCCM cannot be employed as the mass efficient thermal protection due to their high density (up to 2000 kg/m^3) and high thermal conductivity (up to 60 W/(m·K)) [2]. Thus, porous CCCM with a relatively low density and a far smaller thermal conductivity appear to have a great potential [3, 4].

However, the data on the porous CCCM properties, especially with regard to their thermal conductivity coefficient – the key operation parameter, are practically non-existent. Experiments to estimate the thermal conductivity coefficient of such high-temperature CCCM are highly challenging. Moreover, the experiments can determine only the so-called effective thermal conductivity coefficient, which estimates the total heat transfer including both conductive heat transfer along the reinforcement and radiative heat transfer in the pores. The conductive heat transfer is described by means of the molecular thermal conductivity coefficient [5]. Therefore, at this stage of research it appears reasonable to perform the theoretical and numerical estimation of the porous CCCM thermal conductivity separating the heat transfer constituents.

* Corresponding author: pavel.prosuntsov@mail.ru

DOI: 10.1051/matecconf/20179201008

The probable characteristics of porous CCCM with the woven and non-woven reinforcement are theoretically validated in [4]. The heat conductivity analysis of the porous CCCM (with 1000-2000 kg/m^3 density) accounting for the radiative heat transfer in the pores is conducted in [5].

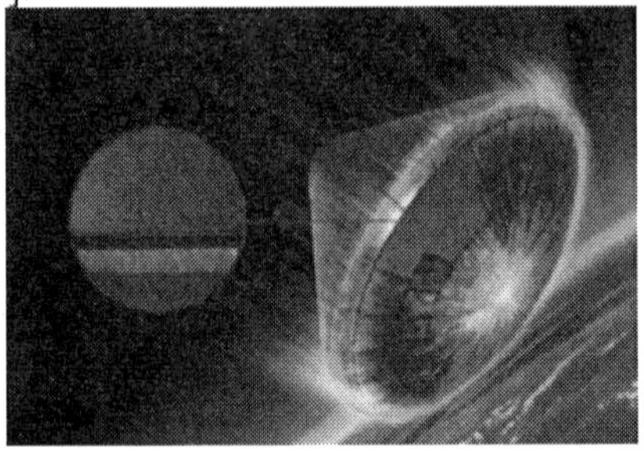

Fig 1. Illustration of a multifunctional HOST heatshield concept for an Earth entry application and a close-up of the carbon-carbon outer layer [2].

The porous CCCM thermal conductivity was estimated based on the numerical simulation of the heat transfer process in the representative volume element. This method employed relatively simple structure models, in the form of intercrossing cylindrical fibres. It was assumed that all the fibres had equal diameter. The study in [5] demonstrated the evident relationship of the effective and molecular thermal conductivity from the temperature and porosity.

2 Porous CCCM structure investigation

The reference data regarding the porous CCCM structure were obtained based on the sample microstructure analysis using the Brucker Skyscan 1172 micro-CT scanner. This device uses a multitude of X-Ray images at various angles to create a 3D model of the sample. Figs. 2-4 present 3D structure models of the porous CCCM with 27-29 % porosity.

Fig 2. 3D structure model of a porous CCCM sample, obtained with X-ray micro-CT scanning.

Both the original sample model (Fig. 2) and the separate areas (Figs. 3-4) are presented. It is evident that the material structure is significantly different from the models used in [5].

Fig 3. 3D structure model of a porous CCCM sample area, obtained with X-ray micro-CT scanning.

Fig. 4. 3D structure model of a porous CCCM sample area, obtained with X-ray micro-CT scanning.

3 Determining the thermal conductivity of the porous CCCM by means of X-ray micro-CT scanning method

In order to determine the effective thermal conductivity coefficient of the porous CCCM an approach similar to [5] was used. First, a finite-element model of a representative volume element was created in the Siemens NX software package (Fig. 5). This model comprised a porous material zone and two contacting plates. The outer surfaces were assumed to have different temperatures. The side surfaces of the plates were assumed to be thermally insulated. The thermal contact between the plates and the porous material fibers was assumed ideal. Additional mirror reflecting surfaces (zero thickness; reflectance = 1) were located on the side surfaces of the porous material. The surfaces of the fibers and plates were assumed to be grey, diffusely reflecting, with the emissivity factor equalling 0.75. The thermal conductivity of the monolythic CCCM, which was used for calculation, is presented in Fig.6.

The heat transfer simulation in this representative volume element (Fig. 7) resulted in estimating the value of the total heat flux through the porous material and calculating its effective heat transfer coefficient. The resulting temperature function of the molecular and effective heat transfer coefficient of the porous CCCM is presented in Fig.8 (graphs 1 and 2) and corresponds qualitatively to the data in [4].

Fig. 5. Finite-element model of the representative volume element of the porous CCCM.

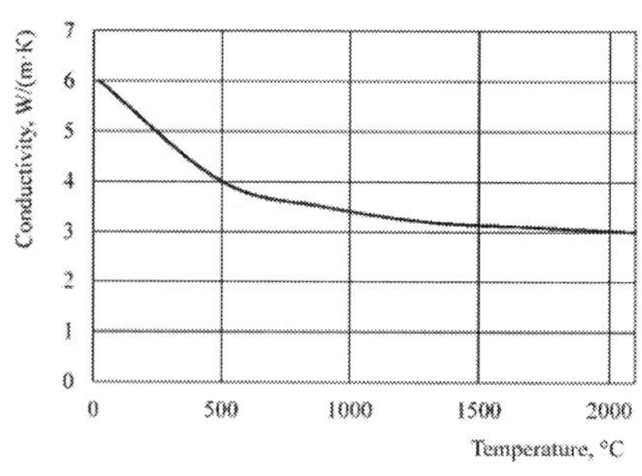

Fig. 6. Thermal conductivity of the monolithic CCCM as a function of temperature.

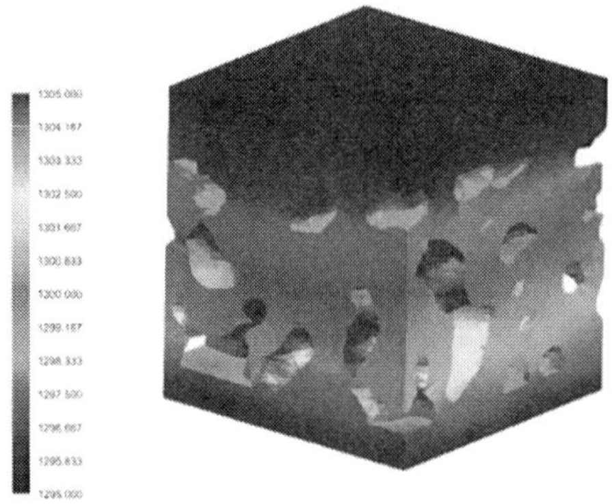

Fig. 7. The thermal state of the representative volume element of the porous CCCM, °C.

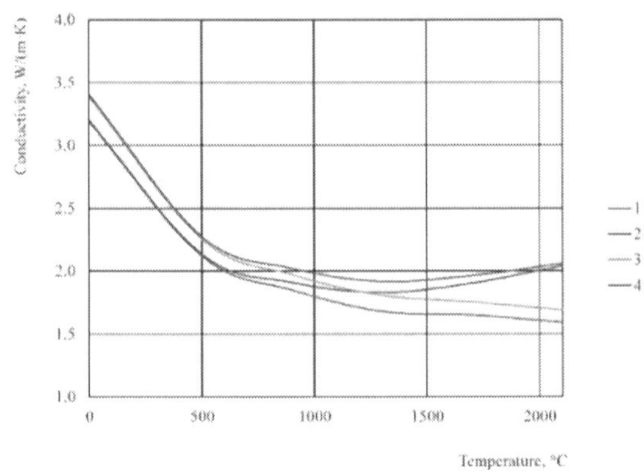

Fig. 8. Molecular (1, 3) and effective (2, 4) thermal conductivity of the porous CCCM, obtained with the model from the micro-CT scanning (1, 2) and a simplified model (3, 4).

4 Estimating the thermal conductivity of the porous CCCM with the simplified model

Simulating the radiative-conductive heat exchange using the CT data model involves time- and labour-consuming calculations. This is true for the finite-element model stage, and especially true for the calculation stage due to the high complexity of the representative volume element shape.

All of the above brings up the question of creating a simplified structure model that will adequately reproduce its characterisitics and will significantly reduce the time- and labour-intensity of the simulation, which is critical at the initial calculation stages.

The MSC.Digimat software package was used to create a structure model comprising the monolithic material and ellipsoid pores with 2:1 semiaxes ratio. Fig. 9 presents the geometric and finite-element simplified structure models; Fig. 10 presents the thermal state of the representative volume element.

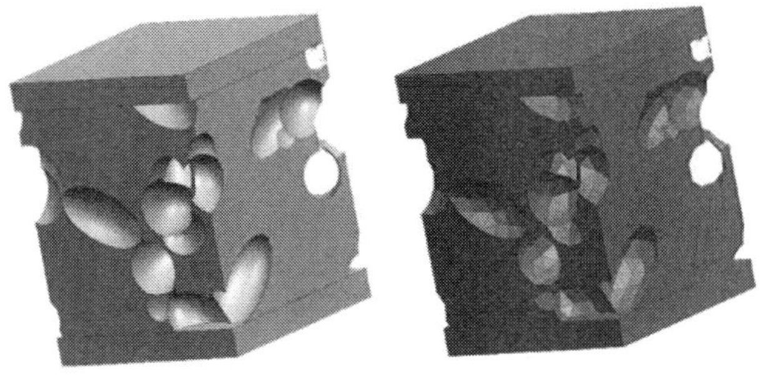

Fig. 9. Finite-element model of the representative volume element of the porous CCCM.

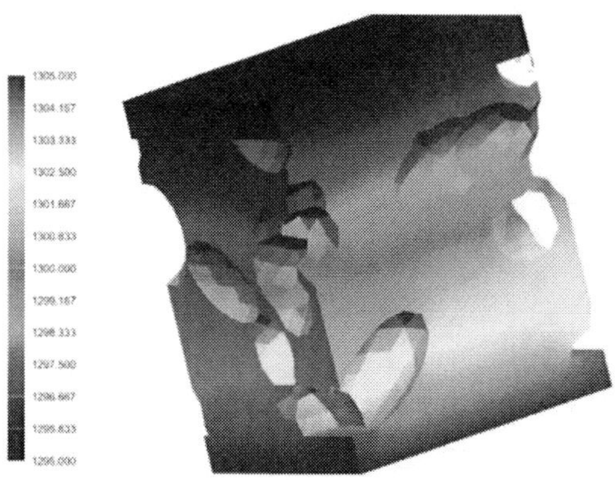

Fig. 10. Thermal state of the representative volume element of the porous CCCM, °C.

After that, as in Part 3, the temperature functions of molecular and effective thermal conductivity coefficients were determined for the simplified model (Fig. 8, graphs 3 and 4). It is evident that for the both models, the values of the thermal conductivity coefficient differ by no more than 4 %, while the calculation time decreases almost by a factor of 10. Consequently, the simplified structure model of the porous CCCM can be successfully used for the thermal conductivity calculations of this class materials.

Some results of this paper were obtained as part of the grant agreement No. 14.577.21.0099 by the Russian Federation Ministry of Education and Science. Unique identifier for Applied Scientific Research – RFMEFI57714X0008.

References

1. P. Narottam, *Bansal Handbook of Ceramic Composites* (Kluwer academic publishers, Boston, 2005)
2. S. P. Walker, et. al. *20th AIAA International Space Planes and Hypersonic Systems and Technologies Conference* (Scotland, UK, 2015)
3. B. Laub, M.J. Wright, E. Venkatapathy, *6th International Planetary Probe Workshop, Atlanta, Georgia Short Course on Extreme Environments Technologies* 06/21-22 (2008)
4. S.V. Reznik, P.V. Prosuntsov, K.V. Mikhailovskii, J. Eng. Phys. Thermophys. **88**, 3 (2015)
5. P.V. Prosuntsov, N.Y. Taraskin, Matec Web of Conf. **72**, 010929 (2016)

Experimental investigation of pool boiling of water-based AL_2O_3 nanofluid on a copper cylinder

Maxim Pryazhnikov[1,2,*], *Konstantin* Meshkov[1], *Andrey* Minakov[1,2], and *Alexander* Lobasov[1]

[1]Siberian Federal University, 660041 Krasnoyarsk, Russia
[2]Institute of Thermophysics SB RAS, 630090 Novosibirsk, Russia

Abstract. Saturated boiling of nanofluid on a copper cylinder is experimentally studied. The studied nanofluid were prepared using distilled water and Al_2O_3 nanoparticles. The volume concentration of the nanoparticles was equal 3 %. Cylinder diameter was equal 25 mm. The time dependence on the excess temperature at different boiling regimes were obtained. It shown increase of heat transfer coefficient under boiling of nanofluid.

1 Introduction

Studies of the last two decades have shown that nanofluid have unusual transfer properties. In particular, small adds of nanoparticles into carrying liquid can considerably increase its heat conductivity and viscosity [1, 2]. This stimulated many thermophysical applications of nanofluid, particularly, aimed at intensification of heat exchange. It has been found that nanofluid have really enhanced coefficients of heat transfer (see, for example, [3, 4] and literature cited therein). The further intention to increase this coefficient stimulated the study of heat transfer of nanofluid during boiling. These works have been performed quite intensely during the last decade. Nevertheless, the obtained results are rather contradictory. For example, in [5] it is noted that adding of nanoparticles does not change heat transport considerably, with an even decrease in the heat transfer coefficient during boiling been seen in [6]. Conversely, in [7, 8], this coefficient rose.

Saturated boiling of nanofluid on a cylindrical heater is experimentally studied by researchers in their works [9, 10]. The studied nanofluid were prepared using distilled water and iron oxide (II, III) nanoparticles or diamond nanoparticles. The volume concentration of the nanoparticles was changed from 0.05 to 1.0 %, their diameters varied from 10 to 100 nm, and the heater diameter was increased from 0.1 to 0.3 mm. It is established that the critical heat flux density in boiling nanofluid depends on the size and material of nanoparticles and on the heater diameter. The critical heat flux density increases with increasing size of nanoparticles and decreases with increasing diameter of the heater.

* Corresponding author: arrivent@yandex.ru

In this work, saturated boiling of the nanofluid were prepared using distilled water and Al_2O_3 nanoparticles was experimentally studied. The present work was aimed at studying the influence of nanoparticles on rate of cooling and heat transfer coefficient under boiling of nanofluid on a copper cylinder.

2 Experimental apparatus and procedure

The study of boiling process occurred on the copper cylinder (diameter is 25 mm).

Method of procedure. The cylinder is heated in the furnace to temperature of 350-500 °C. After that it is immersed within studied fluid (distilled water or nanofluid) to the depth of 15-60 mm (see figure 1). Since the surface of the cylinder has a high temperature, when immersed in water formed a vapor film. The studied fluid was heated up saturation point to not spend the warm on heating fluid. Cooldown period occurs at a constant coolant temperature. The value of heat transfer coefficient is constant in the film boiling; that is the boundary condition for steady cooling regimen. Cylinder temperature is measured using thermal couple which mounted into it. The decline rate of the cylinder and the heat transfer coefficient are meant on the score of evidence from the research.

Fig. 1. The scheme of the experimental setup.

Study of boiling was drown using the distilled water, water-based nanofluid and Al_2O_3 nanoparticles. The volume concentration of the nanoparticles in water was equal 3 %. Average particle size of Al_2O_3 nanoparticles is 36 nm. Preparation of nanofluid was carried out based on standard two step process. After adding to the base fluid the required amount of nanopowder, the nanofluid was first thoroughly mixed mechanically, and then was placed for the half-hour into an ultrasonic disperser Sapphire to destruct conglomerates of particles. The nanoparticles were purchased from "Advanced Powder Technologies" company LLC (APT) (Tomsk).

3 Results and discussion

The series of five experiments was carry out for distilled water and the nanofluid containing Al_2O_3 nanoparticles.

Boiling regimes in distilled water at different times was shown in figure 2. Unfortunately, addition of nanoparticles leads of decrease in transparency of nanofluid. For this reason the photo of boiling of Al_2O_3 nanofluid are less informative, but the process of boiling are similar.

The dependence of excess temperature θ (the temperature difference between the temperature of cylinder surface and the saturation temperature of the fluid) on time, obtained in the experiment, were shown in figure 3.

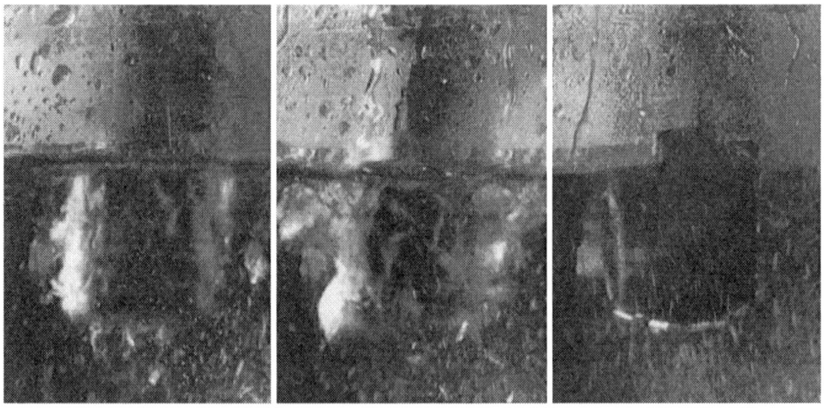

Fig. 2. Boiling regimes at different times. Left to right: film boiling, transition boiling, nucleate boiling.

The average (for all replicates) dependences of excess temperature θ on time for the distilled water (dotted line) and Al_2O_3 nanofluid (solid line) were shown in figure 3. In this figure there were boiling conditions, that realized in these experiments and shown in figure 1-2. That is change slope of a curve indicates the changes.

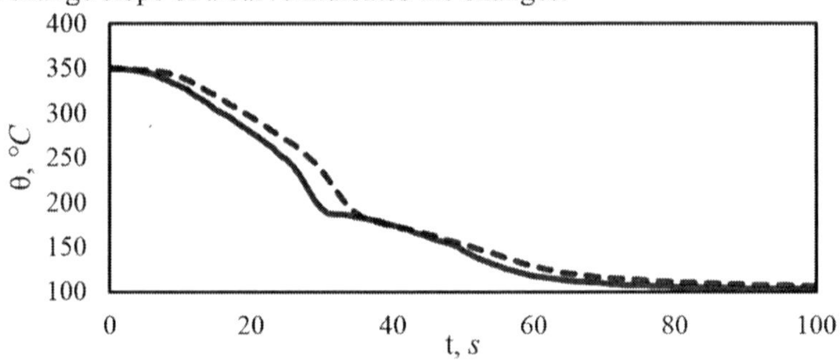

Fig. 3. The dependence of excess temperature on time.

It can be observed that the boiling of distilled water and the nanofluid are identical in order of merit, but they differ quantitatively because it is necessary to notice that:

a) a transition region from film boiling to transition boiling (see figure 4) – set transition in the nanofluid occurs in about 5 seconds earlier then in distilled water and excess temperature θ is at about 10-15 °C higher in the first instance, which is equivalent of increase of heat transfer coefficient of the nanofluid on 7.5-10 % in comparison to heat transfer coefficient of distilled water.

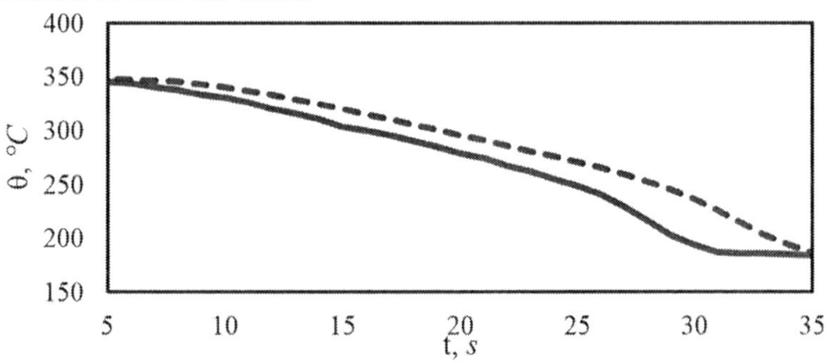

Fig. 4. The dependence of excess temperature on time in the transition region from film boiling to transition boiling.

b) a transition region from transition boiling to nucleate boiling (see figure 5) – set transition in the nanofluid occurs in about 25-30 seconds earlier then in distilled water, which is equivalent of increase in heat transfer coefficient of the nanofluid on 20-40 % in comparison to heat transfer coefficient of distilled water.

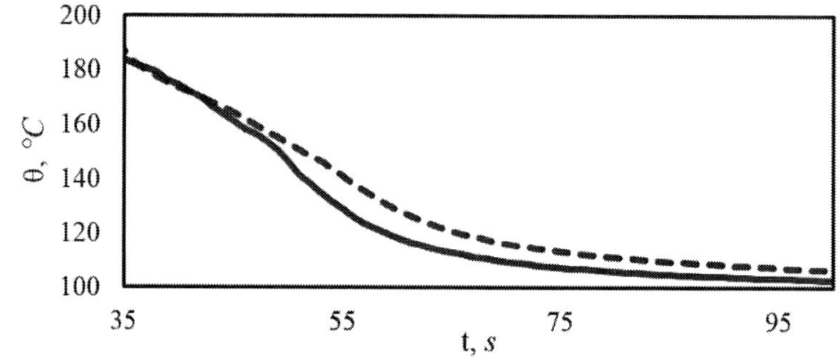

Fig. 5. The dependence of excess temperature on time in the transition region from transition boiling to nucleate boiling. transition boiling to nucleate boiling.

4 Conclusions

Based on the experimental results, it may be concluded that using of nanofluid, connected to phase transition, is effectually. Using nanofluids increases the heat transfer coefficient, this can increase the energy efficiency of setups in which the main process is boiling.

The current work is performed at partial support of the projects funded by the Russian Foundation for Basic Research and Krasnoyarsk Regional Fund for Support of Scientific and Scientific-Technical Activities (Contract No. 16-48-243042\16).

References

1. S.K. Das, S.U.S. Choi, W. Yu, T. Pradeep, *Nanofluid: Science and Tecnology* (Wiley-Interscience, New Jersey, 2007)
2. X.Q. Wang, A.S. Mujumbar, Int. J. Therm. Sci. **46,** 1 (2007)
3. V.I. Terekhov, S.V. Kalinina, and V.V. Lemanov, Thermophys. Aeromech. **17**, 157 (2010)
4. D.V. Guzei, A.V. Minakov, V.Y. Rudyak, A.A. Dekterev, Tech. Phys. Lett. **40**, 34 (2014)
5. S.M. Kwark, R. Kumar, G. Moreno, J. Yoo, S.M. You, Int. J. Heat Mass Tran. **53**, 972 (2009)
6. S.K. Das, G. Prakash Narayan, A.K. Baby, Int. J. Heat Mass Tran. **51,** 1099 (2008)
7. J.S. Coursey, J. Kim, Int. J. Heat Fluid Fl. **29**, 1577 (2008)
8. M.H. Shi, M.Q. Shuai, Z.Q. Chen, Q. Li, Y. Xuan, J. Enhans. Heat Transf. **14**, 223 (2007)
9. A.V. Minakov, A.S. Lobasov, V.Y. Rudyak, D.V. Guzei, M.I. Pryazhnikov, Tech. Phys. Lett. **40**, 562 (2014)
10. M.I. Pryazhnikov, A.V. Minakov, V.Y. Rudyak, Tech. Phys. Lett. **41**, 891 (2015)

New Approach to Simulation of Heat State of Compartments from Lattice Composite Shells for Space Engineering Products

Alexander F. Razin[1], *Mikhail* N. Slitkov,[1] and *Anatoly* N. Garashchenko[1,*]

[1]Central Research Institute for Special Machinery, JSC, Khotkovo, Moscow Region, Russia

Abstract. A new approach to the simulation of the heat state of the compartment of lattice polymer composite materials (PCM), not providing for the use of known commercial software packages, has been proposed. The simulation has been performed using the PCM interstage of the Proton rocket as an example with due account of aerodynamic heating, solar radiation and acting of jets of auxiliary propulsion units. At the first stage of numerical analysis, a problem of unsteady heat conduction in the system "skin-air gap-heat insulation" has been solved. An effect of changing a pressure inside a compartment on thermal conductivity of heat insulation was taken into account. The effective thermal conductivity in gaps was used. An effect of a temperature of equipment on a value of radiant heat flux was also taken into account. At the second stage, the heat state of the system "skin-rib" was analyzed. A mathematical model in the form of a system of nonlinear equations for heat balance of control elements on which a rib and a skin section were partitioned, including an information about a temperature of heat insulation received at the first stage of the simulation, was used.

1 Introduction

The creation of up-to-date rocket and space technology provides for an analysis of temperature fields of both separated structural elements and a product as a whole in connection with high level and variety of heat loads on majority of units and their blocks. The determination of temperature fields (or even if characteristic temperatures of separate sections of the most heated surfaces) under actual operational conditions is practically impossible.

Therefore, studies in this field show that the mathematical simulation is the main instrument for that analysis. At the same time, the simulation of heat transfer processes proceeding in complex, as a rule multiply connected geometrically and physically, fields of products of rocket and space technology is an independent problem difficult to solve.

* Corresponding author: a.n.gar@mail.ru

Known methods and approaches do not provide the possibility of its solution. Main difficulties of numerical solution of heat transfer problems for structures, for example, of the PCM lattice compartment type (Fig. 1) lie in substantial non-one-dimensionality of heat transfer conditions, non-uniformity of heat fluxes on boundaries of a solution range (outer and inner contours of a product), multiple lamination of actual structures, breaks of transport factors on these boundaries, substantial contribution of radiation and convection to forming the temperature fields of a specific object.

Fig. 1. Appearance of the compartment from the outside and on the inside.

To solve this kind of nonlinear boundary problems of mathematical physics, not only numerical methods [1] but and specific methodical and algorithmic techniques [2-5] of obtaining reliable theoretic consequences, for example, when studying natural convection processes in areas with local power sources [2], igniting of structurally non-uniform substances [3], natural convection in closed areas with radiant heat sources [4], igniting of fuels under local heating [5] have been developed. Solutions of singular-disturbed problems [2-5] are obtained by finite difference method for irregular and non-uniform, basically, difference nets. With this, to overcome one or several above-listed problems, the use of very small steps of a difference net in time (up to 10^{-7} s) was the most reliable technique. However, using this efficient, but taking long time for calculations, technique when solving heat transfer problems for products of rocket and space technology is impossible in connection with their large sizes and complex configuration, high heat loads and long operation time. These reasons caused the objective necessity of developing a new approach to the simulation of heat transfer processes in large-sized and non-uniform property materials of separate elements of structures for long time, various heat loading conditions and high heat fluxes to heating surfaces.

The objective of the present paper is to analyze unsteady temperature fields in PCM structural elements made in the form of carbon reinforced plastic lattice structures that are distinguished by high load-bearing capacity at minimum mass [6, 7]. Lately, such structures began to be widely used as rocket compartments in space technology. They are applied, for example, when making the interstage (upper) and the skirt of the tail compartment for the stage II of the Proton rocket, compartments for the Angara rocket, etc.

2 Statement of the problem

The new approach was used to the mathematical simulation of heat transfer in PCM lattice structures with complex configuration. Heating of the outer side of a compartment at the expense of convection and heat transfer on inner surfaces from equipment located in an object at the expense of radiation and natural convection were considered. Besides, at the end of the system work, the action of jets of auxiliary propulsion units (PU) on the compartment surface was taken into account.

The mathematical simulation was carried out for a compartment in the form of a lattice shell (outside diameter of 4100 mm, length of 3731 mm) formed by a set of spiral and annular ribs, with a skin. Its appearance is shown in Figure 1 and the layout of equipment blocks (II & III) is presented in Fig. 2. The thickness of a skin was 1.5 mm, of a spiral rib was 6 mm and of an annular rib was 4 mm. The inner surface of the compartment is covered by heat-insulating material. There is a gap between the material and the skin equal to the height of ribs (23 mm). The skin consists of layers of different PCMs and outer heat-protective coating (OHPC) whose thickness depends on the skin section. The outer surface (OHPC surface) is covered with white enamel. It is subject to aerodynamic heating.

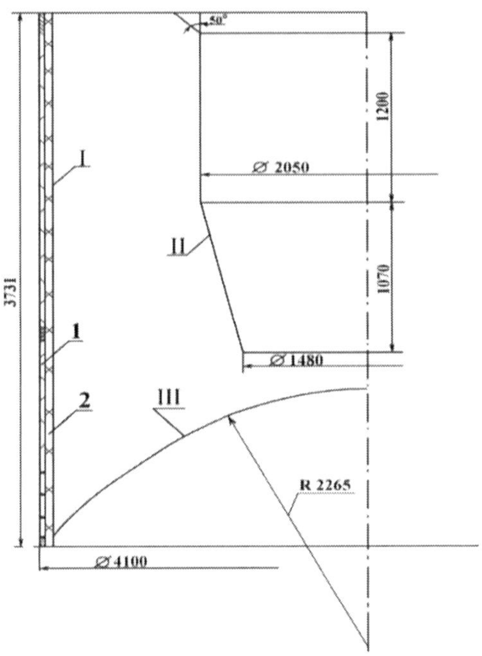

Fig. 2. Layout of equipment in inner volume of the compartment: 1 – shell; 2 – heat insulation; II, III – equipment blocks.

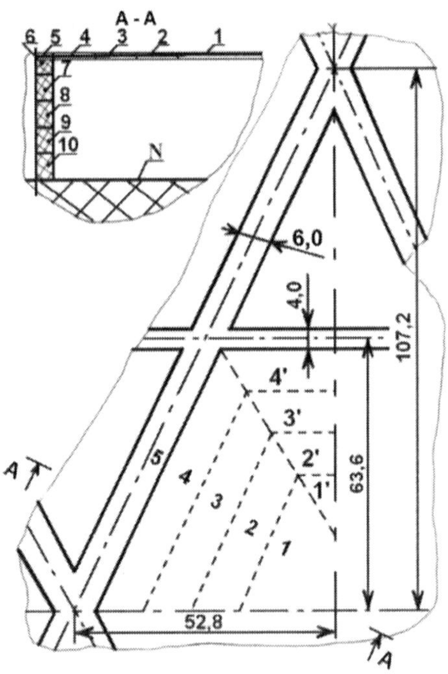

Fig. 3. Figure 3. Diagram of solution area for a skin and ribs (1, 2, …, 10 are element numbers).

The ambient temperature $T_e(t)$ and coefficient of convective heat exchange $\alpha_{ei}(t)$ are known. From a certain moment of time, the heat flux density $q_e(t)$ on this boundary is given. In addition, at the end of work, the outer surface is subject by the action of jets of auxiliary propulsion units.

At the first stage of simulating, the heat state of the most warmed-up element of the compartment, i.e. a skin, was determined. The mathematical model of the process is a system of nonlinear equations of unsteady heat conduction in multilayered cylinder (skin-air gap-heat insulation). It includes boundary conditions on the inner surface that take into account convective heat exchange (up to a moment of pressure drop down to zero) and resulting radiant heat flux. The latter was calculated by method of "radiant balance" in a closed system of bodies [8]. The heat conduction equation to be used has the form:

$$c_m \rho_m \frac{\partial T}{\partial t} = \frac{1}{r}\frac{\partial}{\partial r}(\lambda_m r \frac{\partial T}{\partial r}), \quad t > 0,\ r_0 < r < r_n, \quad r_m = r_0 + \sum_{m=1}^{m} \delta_m, \quad m = 1,2...,n,$$

where T, r, t are temperature, radial coordinate and time;

c, ρ, λ are specific heat capacity, density and thermal conductivity;

m is number of a layer of multilayered cylinder.

This model has features consisting in the account of an influence of the pressure change inside a compartment on thermal conductivity of heat insulation and the use of effective thermal conductivity in a gap between a skin and heat insulation with due account of re-emission. The natural-convective heat transfer coefficient and the air temperature in inner volume of a compartment were parameters of the mathematical model for calculating of which the nonlinear differential heat-balance equation was used.

The OHPC thicknesses were selected so that the skin temperature during the all process of heating (including the action of propulsion unit jets) did not exceed the level of beginning thermal destruction of a binder (~210°C ... 220°C). In places of affecting PU, values of OHPC thicknesses were determined with due account of removal values (δ_d) which were estimated from the equation:

$$\delta_d = \frac{Q}{\rho H_{ef}} = \frac{1}{\rho\ H_{ef}} \int_0^{t^*} \alpha (T_e - T_d) dt,$$

where Q is an integral of heat flux density;

H_{ef} & T_d are effective enthalpy and OHPC destruction temperature.

3. Results and discussion

A high number of estimation values of the OHPC thickness as well as non-uniformity of heat flux to the outer surface during the work caused the necessary to analyze the heat state of a skin on many its sections. The results of simulating the temperature fields of different sections are obtained in the form of time variation of temperatures on boundaries of skin material layers and heat insulation surface. As an example, Fig. 4 presents temperature values for one of the sections.

Temperature fields of a skin were used at the second stage of simulating when analyzing the heat state of spiral and annular ribs of a compartment. For this, a mathematical model in the form of a system of nonlinear equations of heat balance for control volumes of elements into which a rib and a skin section being adjacent to it and dissected by adiabatic planes as is shown in Fig. 2 were partitioned. With such partitioning, it was adopted that a temperature in all volume of an element is the same (temperature field is uniform). When stating a problem, it is assumed that outer surfaces of elements (from 1 to 5) are affected by heat flux analogous to one to be taken into account before on the outer surface of a skin. With this, the skin sections whose temperatures are the most high and were calculated formerly (see Fig. 4) were considered. Each of the skin elements under consideration (besides an element 5) participates in the process of radiant heat exchange with elements of a rib (6...10) and the heat-insulating material surface (N) whose temperature is adopted known from the results of calculations conducted before.

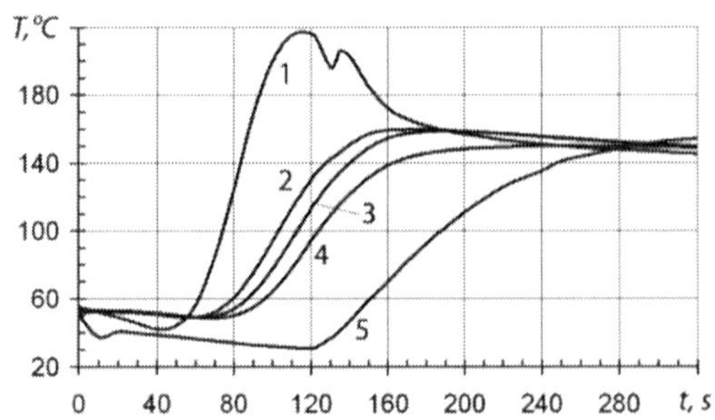

Fig. 4. Temperature changes in elements of system "skin - air gap - heat insulation": 1- OHPC outer surface; 2 - OHPC-skin boundary; 3 - inner surface of a skin; 4 - outer surface of heat insulation; 5 - inner surface of heat insulation.

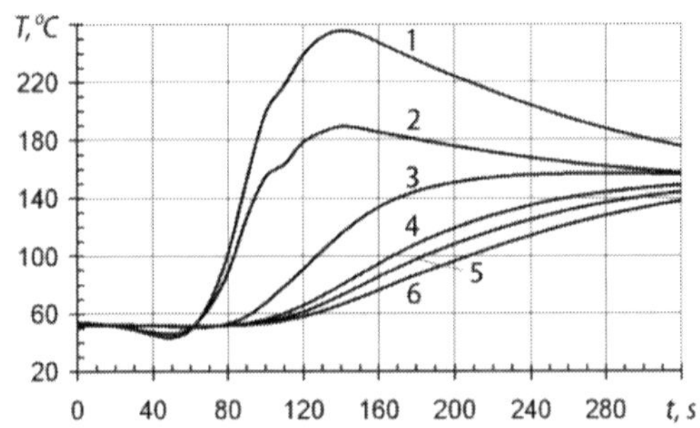

Fig. 5. Temperature changes in annular rib with skin sections being adjacent to it: 1 - skin element (No. 1); 2 - skin element at a rib (No. 5); 3, 4, 5 - rib elements Nos. 8, 9, 10 (respectively).

It is also assumed that the heat transfer from element to element is carried out only by heat conduction. The heat exchange between elements within mutual "visibility" occurs at the expense of radiation. A system of equations is stated for skin sections as well as for sections on a rib. For sections of a skin between ribs, this system has the following form:

$$C_i \frac{dT_i}{dt} = q_i F_i - \sum_{j=m+1}^{j=n} K_{i-j}(T_i - T_j) - K_{i-j+1}(T_i - T_j) - K_{i-N}(T_i - T_N). \quad (1)$$

For section of a skin on a rib:

$$C_m \frac{dT_m}{dt} = q_m F_m - K_{m-j}(T_m - T_j). \quad (2)$$

For sections of a rib:

$$C_j \frac{dT_j}{dt} = q_i F_i - \sum_{j=m+1}^{j=n} K_{i-j}(T_i - T_j) - K_{j-j+1}(T_j - T_{j+1}) - K_{j-N}(T_j - T_N). \quad (3)$$

The following symbols have been adopted: C is heat capacity of element to be determined as $C = \rho c V$ (V is element volume); F is surface area of element; ε is emissive power; indices: $i = 1,\ldots,$ m is symbol of element on a skin, $j = m+1,\ldots,$ n is symbol of element on a rib (in this case n=5, m=5); e is environment; w is outer surface of a skin; N is heat-insulation surface.

Density of the resulting heat flux on the outer surface of skin elements as well as radiant heat transfer from i-th skin section to j-th rib section are calculated from the formulae [8]:

$$q_i = \alpha_e(T_e - T_i) + q_e - \varepsilon_N \sigma T_i^4; \quad (4)$$

$$K_{i-j} = F_{i-j} \sigma \varepsilon_i \varepsilon_j \varphi_{i-j} \frac{T_i^4 + T_j^4}{T_i - T_j}, \quad (5)$$

where $K_{i\text{-}j}$ is parameter of heat transfer by effective heat conduction from heater to wall, taking into account re-emission of surfaces and natural convection in plane slit [8];

$F_{i\text{-}j}$ is area of mutual irradiation of two elements;

φ_{i-j} is angular coefficient.

Typical results of temperature computations for the annular rib (which is more warmed up with comparison to the spiral rib) with skin sections being adjacent to it are presented in Fig. 5 in the form of temperature-time relations for each element. In this case, a cross-section with formerly determined temperature of a skin was considered (see Fig. 4). From results of simulating it follows, in particular, that maximum temperature for a rib (139 °C) is reached by the end of heat effect.

The truth of numerical temperature values was verified by comparison for the heat state of a skin obtained by traditional and above-described methods for element under No. 1 (see Fig. 3) which is subject to the influence of ribs to a least extent.

The similar approach was applied when simulating the time variation of temperatures of elements in the tail compartment of the Proton rocket. The load-bearing structure of its compartment case consists of similar lattice shell formed by a family of spiral and annular ribs and a skin. The skirt case has a cylindrical shape 2390 mm in length and an outside diameter of 4100 mm.

For the interstage and tail compartments of the rocket, similar simulating was the important part of their design and development. The serviceability of the compartments has been confirmed by the results of flight tests. Besides, the considered method was successfully used while simulating the heat state of the Angara rocket compartments.

4 Conclusion

A new approach to the simulation of the heat state of structural elements of a polymer composite lattice compartment has been proposed. This approach was used when designing and developing of various products with a complex configuration under conditions of an intensive heat effect. The fidelity of calculations has been confirmed by the results of flights of the rockets with compartments under consideration.

References

1. V.L. Strakhov, A.N. Garashchenko, G.V. Kuznetsov, V.P. Rudzinskii, Combust. Explos. Shock Waves **37**, 2 (2001)
2. G.V. Kuznetsov, M.A. Sheremet, Int. J. Heat Mass Tran. **54**, 260 (2011)
3. O.V. Vysokomornaya, G.V. Kuznetsov, P.A. Strizhak, Russ. J. Phys. Chem. B **5**, 4 (2011)
4. G.V. Kuznetsov, N.I. Kurilenko, V.I. Maksimov, G.Ya. Mamontov, T.A. Nagornova, J. Eng. Phys. Thermophys. **86**, 3 (2013)
5. D.O. Glushkov, G.V. Kuznetsov, P.A. Strizhak, Russ. J. Phys. Chem. B, **5**, 6 (2011)
6. V.V. Vasiliev, A.F. Razin, Compos. Struct. **40**, 1-2 (2006)
7. V.V. Vasiliev, V.A. Barynin, A.F. Razin, Compos. Struct. **94**, 3 (2012)
8. E.M. Sparrow, R.D. Cess, *Radiation Heat Transfer* (Hemispere Publishing Corporation, USA, 1978)

Interaction particles from the surface of the curved pipeline

Michail Vasilevsky[1], *Aleksandr* Razva[1,*], *Aleksandr* Naumkin[1], *Dmitriy* Zaharov[1] and *Eugene* Zykov[1]

[1]National research Tomsk polytechnic university, 634050 Tomsk, Russia

Abstract. The interaction of the agglomerated fine dust particles from the surface of the rotary pyleprovoda, given deposit formation evaluation. The interaction of large particles to the surface of the rotary pyleprovoda. The analysis of hydrodynamic phenomena in the means of protection against wear.

1 Introduction

The disperse material coming to the dust flue consists of sets of primary particles, the aggregated particles, and represents an aerosol together with the bearing environment. The agglomerates are formed by coagulation turbulent, with flow disrupting deposits from surfaces with suction materials, particle segregation as a result of various transportation devices in [1]. Rotating flow in curved channel leads to the centrifugal inertial force acting across the flow, which in turn changes the conditions of gas motion. There are secondary flow of the first and the second kind [1], whose impact on the separation of the particles is shown, depending on the shape of the channel, the concentration and size of particles [1]. Curvilinear motion of the dust-laden stream is accompanied by interaction with the surface of the particles, if their speed due to centrifugal force is much greater than the turbulent transfer rate near the surface [1]. In [2] shows the calculated distribution model particle concentrations that are not deposited on the wall due to turbulent eddies. It is assumed that the total flux of particles due to centrifugal forces and turbulent diffusion transfer should be zero.

The agglomerates are composed of many particles, so have a lot of large compared to the mass of the individual particles. Depending on the ratio of the radial and tangential velocity of sinter approaching it to the surface is carried out at different angles (Figure 1).

Large particles (greater than 20 microns) are abrasives. The intensity of abrasive wear of the apparatus depends on the hardness, size, and particle density. Causes of wear: impact action of abrasive particles, particles of metal cutting, fatigue accumulation in the surface layer of the wear material, heat and fatigue failure after repeated collision particles. As a result of repeated plastic deformation of the material causing the residual metal lattice distortion, the process of the fracture surface.

* Corresponding author: razva@mail.ru

2 The calculated ratio of the interaction with the surface of the agglomerate

The hade defines the nature of its interaction with a surface. Agglomerate density is several times lower compared to the density of the particulate material of which it consists. Therefore ricocheting hardly possible. Depending on the angle and speed of interaction of interaction will agglomerate or blurred, and again dispersed to coarsen, or vice versa, will condense and form deposits. If deposits are strong, they do not break the flow. Figure 1 shows the motion trajectory of the agglomerate in the channel conversion circuit agglomerate A (dashed line) in the sealed clot B. Agglomerates composed of fine particles under the influence of centrifugal force moves relative to the periphery of the gas and the interaction with the sealing wall extends stage deformation aftereffect.

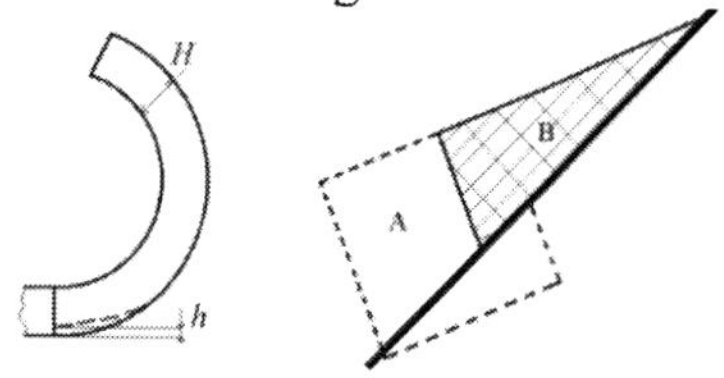

Fig. 1. Trajectories of the movement of agglomerate in the port and the scheme of transformation of agglomerate A (dotted line) in the condensed clot B.

Changing the amount of movement is equal to the impulse force generated in the formed contact surface. Furthermore, it can be used the energy change rate equation relating the kinetic energy and rate of change of energy of the peripheral forces applied to the deformable surface. All these data allow us to predict the status and parameters of sinter, if you know the parameters of the particle layer deposition.

The particle speed relative to the gas stream in a curvilinear radial direction can be determined from the dimensionless equation [1]

$$\frac{\overline{W_\phi^{\,2}}}{\bar{r}} = \Psi\left(\mathrm{Re}_\delta\right)\frac{\mathrm{Re}_\delta^{\,2}}{24 Stk \cdot \mathrm{Re}_{\delta,w}}, \tag{1}$$

$$\mathrm{Re}_{\delta,w} = \frac{W_0\delta}{\nu},\quad Stk = \frac{\delta^2}{18\nu}\frac{\rho_\delta}{\rho}\frac{W_0}{R_c},\quad Stk\cdot\mathrm{Re}_{\delta,w} = \frac{\rho_\delta}{\rho}\frac{W_0^2}{R_c}\frac{\delta^3}{18\nu^2},\quad \mathrm{Re}_\delta = \frac{\Delta U\delta}{\nu}, \tag{2}$$

where ρ, ρ_δ – gas and particle density, δ – particle diameter, ν – kinematic speed gas, ΔU – particles relative to the gas speed in the radial direction, W_0 – gas speed at the inlet of the apparatus, $\overline{W_\phi} = \frac{W_\phi}{W_0}$ – circumferential dimensionless particle speed, R_c – base value turning radius.

Agglomerates near a wall pass the site with the reducing stream speed, and on the wall the speed of a stream is equal to zero. District speed of gas at distance from a wall [3]:

$$W_\phi(y) = \sqrt{\frac{\tau_w}{\rho}}(9.5 + 1.32\ln\eta),\quad \eta = \frac{y}{\nu}\sqrt{\frac{\tau_W}{\rho}},\qquad \tau_W = 0.5C_f\rho W_\Delta^2/2, \tag{3}$$

where $C_f \approx 0.006$, W_Δ – the value of the circumferential speed of the wall on the border zone, $W_\Delta = 0.95W_0$.

Assume that half the wall of intensive speed $\Delta \approx 0.03R$ drop zone in the agglomerate does not affect the flow of force is interested, and this zone sinter passes through inertia. We will estimate the value of finite speed of approach of the agglomerate to the wall. Let $x = R-y$, speed of approach to the wall zone ΔU_Δ, speed the fall of the wall in the normal direction ΔU_{st}.

The equation of a decelerated motion of agglomerate in the radial direction has an appearance:

$$m\frac{d\Delta U}{dt}=\xi\frac{\pi\delta^2}{4}\frac{\Delta U^2}{2}\rho;\quad m=\frac{\pi\delta^3}{6}\rho_\delta;\quad \Delta U=\frac{dx}{dt};\quad \frac{d\Delta U}{dt}=\frac{d\Delta U^2}{2dx},\tag{4}$$

After integration, we will receive the following expression

$$\frac{\Delta U_{cm}}{\Delta U_\Delta}=\sqrt{\exp\left(-\frac{3}{2}\xi\frac{\rho}{\rho_\delta}\frac{\Delta}{2\delta}\right)}\tag{5}$$

Let's consider the simplest process of sealing of agglomerate at its normal contact interaction with a barrier. Speed of blow of V_0 is directed on a cylinder shaft perpendicular to the barrier plane. Process of blow is followed by compression of the disperse environment with small change of cross-sectional area. This can happen in the case of interaction of the particles within the cylinder, in which each contact is inelastic, and the particles are attracted to each other autogezionny forces. This is typical of a size less than 10 microns. The initial length L_1 of the cylinder, the length L_2 in the compressed state (figure 2).

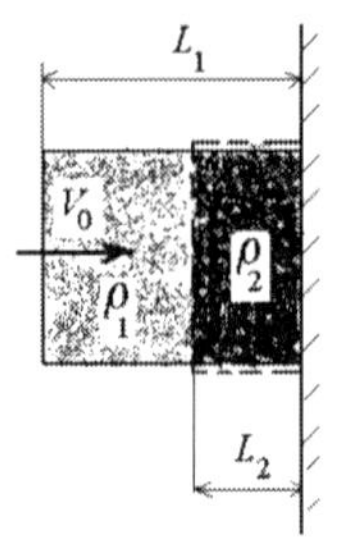

Fig. 2. Schematic of the interaction with the surface of the cylindrical sinter at a right angle.

At blow the disperse environment from volume with a length of L_1-L_2 passes into volume with L_2 length. Let's put that before blow and after blow firmness of the environment in volumes is distributed evenly. The condition of preservation of mass of the disperse environment in position of cylinders 1 and 2 gives the following ratio:

$$\rho 1L1 = \rho 2L2,\tag{6}$$

The equation of impulses at interaction of agglomerate with a wall has an appearance

$$\rho_1 L_1\frac{dV}{dt}=-\sigma_s;\quad dt=\frac{dl}{V_0},\tag{7}$$

Where t – variable time, l – the variable length, V – variable speed, σ_s – contact piece tension on a wall.

The last equation can be written in difference form:

$$\rho_1 L_1\frac{\Delta V_0^2}{\Delta l}=-\sigma_s,\tag{8}$$

Sharing and taking into account factors that $\Delta l= L_1–L_2$, using equation (3), we obtain the relation between the density, speed and voltage:

$$\sigma_s = V_0^2 \frac{\rho_1}{1-\frac{\rho_1}{\rho_2}}, \tag{9}$$

Firmness of agglomerate defines its strength indicators which, in turn, are defined by autogezionny internal forces. From equation (4), that the contact voltage, which are under the above assumptions and the voltage in the agglomerate at the end of the contact interaction, determined by the density and speed to agglomerate and the density of impact after impact. However, it is clear that tension in agglomerate has to decrease to a running end therefore tension, average on length, has to be less, than tension at a butt.

Represent the distribution of stress over the length of a volume with a length L_2:

$$\sigma = \sigma_s \left(1 - \frac{l}{L_2}\right)^k, \tag{5}$$

After the termination of influence of all loadings (the after-effect period) agglomerate in a new state appears in balance. Agglomerate is affected by the carrying-away aerodynamic forces from a stream, gravity, on the other side of - the wall force attracting due to adhesion and an autogeziya of particles. In a new state at a free butt there is no pressure force whereas adhesive forces act on a contact butt.

The formula (3) and formulas (4) and (5) allow to carry out calculation of contact tension and to estimate distribution of tension in agglomerate during interaction it with a wall. Calculation is carried out for two values of the relations of firmness $\rho 1/\rho 2=0,7$ and $\rho 1/\rho 2=0,4$, $\rho 1=300$ by kg/m^3. In a formula (2) $\Delta Uc_m \equiv V_0$.

Calculation of tension in the agglomerate on a formula (5) it shows the rapid reduction of tension to a running end.

Thus, information on firmness of deposits can be also information on the speed of approach of agglomerate to a surface and about its weight. Estimates show that tension in interactions of agglomerate with a surface can reach the values comparable to atmospheric pressure that is the reason of formation of strong deposits.

3 Hydrodynamic effects in anti-devices

It was found that the intensity of abrasive wear is proportional to the third or fourth degree of speed of particles at the time of blow [4]. For reduction of wear of a wall of the dust flue by its surfaces, carry out pockets or fins (Figure 3) [5].

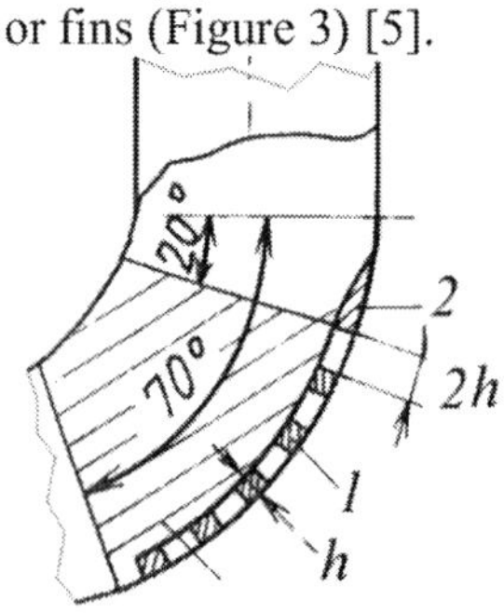

Fig. 3. The circuit surface protection rotary dust flue wear: 1– rib height h; 2– configuration of the first rib to the unstressed entering the oncoming flow.

When passing a stream in a branch pipe of a particle of ashes deviate to its external ribbed wall. The large-scale whirlwinds arising near a concave are split up by edges for smaller vortex educations which scale is proportional to the cell size between edges. In lack of large-scale turbulent transfer the particles weighed in a stream on the whole can't reach a wall of the rotary gas flue. Speed of those particles that hit against a gas flue wall, considerably decreases in the presence of edges owing to braking of a stream, formation of aerodynamically shadow zones and stagnant currents of small scale in intercostal space [5].

The intensity of the induced motion of the gas in the intercostal region can be estimated [6], where it obtained the hydrodynamic flow pattern of a viscous liquid in an open cavity in conditions of forced convection and studied the influence of the model parameters on the character of the movement. After contact with the fine particles in the cavity between the ribs and variations are possible deposition cavity. In this case, you can use the results of [7], in which the numerical simulation of heat transfer and hydrodynamics in motion of a viscous incompressible non-isothermal liquid in an open rectangular cavity in view of the melting process of its walls. With intensive heat removal motion of the gas in the cavity is calculated using the results of [8], which studied the effect of model parameters on the nature of the flow and the cooling cavity.

Results of tests of rotary dust flues with fins showed essential reduction of wear of surfaces at invariable hydraulic resistance [5]. Test results from the rotary dust pipe fins showed a significant reduction in wear surface at a constant flow resistance [5]. [9] The calculation of the abrasive wear of pipelines and pneumatic transport systems of TPP ash coal dust, [10] provides recommendations for optimizing the parameters of piping bends his knees pneumatic ash handling systems and installations pulverization TPP.

4 Conclusion

To improve the reliability of dust systems must take into account the properties of dust particle flows, with the possible prediction of possible deposits on turning dust pipe areas, and at high content of large abrasive particles to take measures for the protection of surfaces from wear.

References

1. M. V. Vasilevsky, *Dust removal of gases inertial devices* **248** (Tomsk, TPU, 2008)
2. M. V. Vasilevsky, E.G. Zykov, and A. S. Razva, Theor. Found. Chem. Eng. **45**, 321 (2011)
3. A. N. Shtym, *Aerodynamics of cyclonic vortex cameras* (Far Eastern University, Vladivostok, 1985)
4. J. A. Tadolder, *On wear of metals at high jet speeds* 83 (Works of the Tallinn polytechnic institute, 1975)
5. V. A. Reznik, N. N. Prokofiev, E. L. Alexandrov, *Turning the boiler dust pipeline* (patent Russian № 2059932 10.05.1996)
6. G. V. Kuznetsov, A. V. Krainov, J. Appl. Mech. Techn. Phys. **42**, 851 (2001)
7. G. V. Kuznetsov, A. V. Krainov, J. High Temp. **41,** 252 (2003)
8. G. V. Kuznetsov, A. V. Krainov, EPJ Web of Conf. **76,** 01023 (2014)
9. V. Y. Putilov, I. V. Putilova, Therm. Eng. **9**, 60 (2003)
10. V. Y. Putilov, I. V. Putilova, Energy saving and water treatment, **3**, 67 (2004)

Comparative analysis of the honeycomb and thin-shell space antenna reflectors

Sergey V. Reznik[1,*], and *Andrey* D. Novikov[1]

[1]Bauman Moscow State Technical University, 105005 Moscow, Russia

Abstract. Parabolic three-layered reflectors from polymer composite materials with the aluminium honeycomb fillers became widely used in space communication systems in the past decades. There are technological possibilities for creating reflectors in the form of thin-walled ribbed shell with the lower linear density than that of the three-layered structures. The paper presents the results of the temperature and stress-strain analysis for the two types of structures, which could help to select the variant with the best performance characteristics.

1 Introduction

The state-of-the-art space antenna reflectors are essentially parabolic shells with various types of reinforcement that enable the required stiffness. The most frequent design layout is a three-layered shell with the honeycomb reinforcement, less frequent is a thin shell with the reinforcing ribs over the convex surface [1]. The common features for all structures are, firstly, the requirement for the high shape and dimensional stability under condition of varying temperature as a result of the spacecraft entering the Earth shadow and, secondly, low linear density (the mass in relation to the surface area). The design of the reflectors with the honeycomb filler meets the most stringent requirements with the regard to the shape and dimensional stability, while having a relatively small linear density in the range of 3.5-5.0 kg/m^2 [2]. Sources [3] and [4] presented the results of simulating the thermal and stress-strain behavior of several reflector variants with the concave shell ribbed reinforcement for the geostationary Earth orbit (GEO) conditions, which proved the possibility of the designs with the small linear density and high stiffness [4]. After the analysis of various reinforcement patterns [5], the "five-point star" pattern was singled out, as being superior to the other variants with respect to stiffness and linear density.

This research aims to compare the characteristics of the space antenna reflector layouts with the honeycomb filling and ribbed reinforcement. In order to achieve this aim, the following objectives were fulfilled consecutively:

- Geometric models were developed;
- Thermal physical characterisctics of the honeycomb filling were determined;
- Temperature and stress-strain state at GEO was determined.

* Corresponding author: sreznik@bmstu.ru

2 Geometric models development

The comparative analysis involved building geometric models of the design with the honeycomb filling (Fig. 1) and the ribbed reinforcement (Fig. 2). The shells in the both variants were made from carbon fibre composite.

The temperature and stress-strain simulation was performed using the finite-element method in the Siemens NX PLM software package. The carbon fibre composite was assumed to have the following thermal physical and mechanical properties: thermal conductivity coefficient 1000 J/(kg·K); density 1550 kg/m^3; emissivity 0.85; absorptivity 0.735; CLTE $5.27 \cdot 10^{-7}$, Young's modulus 140 GPa, Poisson ratio 0.3.

In the honeycomb filled design each carbon fibre composite shell was 0.6 mm thick, the honeycomb layer was 25 mm thick.

Fig. 1. Three-layered construction with the honeycomb filling.

The ribbed reinforcement design was essentially a smooth parabolic shell 0.4 mm thick with flat ribs, 90 mm high and 0.6 mm thick.

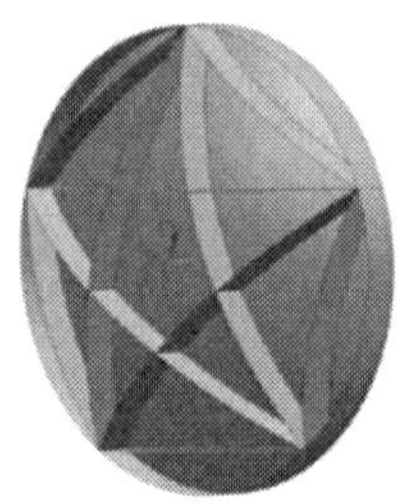

Fig. 2. Ribbed reinforcement design.

3 Honeycomb filler characterization

For the finite-element analysis of the thermal state, the information about the thermal physical and mechanical properties of the honeycomb fillers is essential. The unified reference data for the honeycomb materials are not available, which necessitates conducting thermal physical and mechanical characterization for each design layout.

The mechanical characteristics of the honeycomb materials were determined by menas of the e-Xstreme Digimat 6.0. The input data included the AMr-2 alloy properties, in particular, Young's modulus 71 GPa, density 2680 kg/m^3; the honeycomb characteristics: hexagonal pattern, 0.015 mm wall thickness, 25 mm filler height. The mechanical properties values for the honeycomb material are given in Tables 1 and 2.

Table 1. Young's modulus of the honeycomb material.

Axis	Young's modulus, MPa
x	0.035
y	0.035
z	484.989

Table 2. Poisson's ratios and shear modulus of the honeycomb material.

Plane	Poisson's ratio	Shear modulus, MPa
xy	0.99	0.0087
yz	0.33	91.1633
zx	0.33	91.1633

The thermal physical properties in the longitudinal direction were determined using a 3D model of the honeycomb elementary cell, with 0.015 mm wall thickness and 25 mm height. Hexagonal polygonal bodies 0.6 mm thick were added above and below it to simulate the carbon fibre shell. After that a finite-element model was created and the boundary conditions were specified in the form of temperature and radiative heat transfer inside the cell.

The Siemens NX Nastran was used to conduct the finite-element analysis for 15 calculating cases simulating various operating temperatures at 20 °C interval: from minus 120 °C to plus 160 °C. The heat flux density through the elementary honeycomb cell was determined in accordance with Fourier law:

$$q = -\lambda_{эф} \frac{T_1 - T_2}{\delta_\Sigma}; \tag{1}$$

where q is the heat flux density; λ_{eff} is the effective heat conductivity, T_1 is the temperature on the upper shell surface, T_2 is the temperature on the lower shell surface.

Temperatures T_1, T_2 for each case were specified with a 2 °C difference. For example, for the 60 °C case the temperature T_1 constituted 59 °C, T_2 – 61 °C. After the heat density flux was determined, the honeycomb material thermal conductivity was estimated

$$\delta_\Sigma = \delta_1 + \delta_2 + \delta_3; \tag{2}$$

$$\frac{\delta_\Sigma}{\lambda_{эф}} = \frac{\delta_1}{\lambda_1} + \frac{\delta_2}{\lambda_2} + \frac{\delta_3}{\lambda_3}. \tag{3}$$

where λ_1, λ_3 are the shell thermal conductivities; λ_2 is the filler thermal conductivity.

The estimation of the thermal efficiency in the transverse direction employed the layout similar to that for the longitudinal direction. The thermal load conditions were identical to the estimation of the longitudinal thermal conductivity.

The thermal conductivity of the honeycomb filler in the longitudinal and transverse directions is presented in Fig. 3.

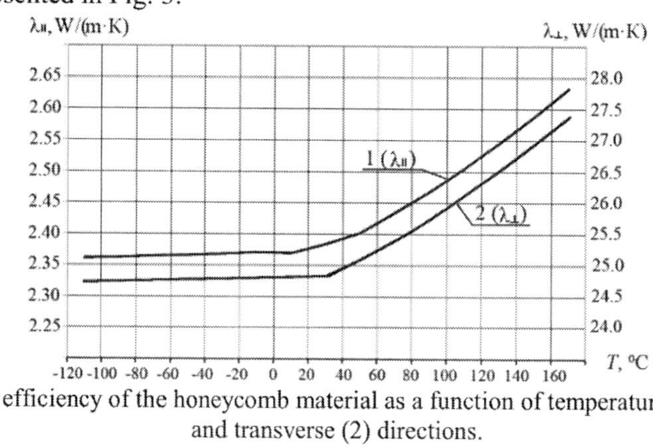

Fig. 3. Thermal efficiency of the honeycomb material as a function of temperature, longitudinal (1) and transverse (2) directions.

4 Determining the stress-strain behavior of the honeycomb filler structure

The temperature gradient was determined by means of Space Systems Thermal solver from the Siemens NX software package. The research involved analyzing 24 moments of operation in the GEO. The largest gradient coincided with the moment when the reflector axis of rotation was at 150 ° relative to the Earth-Sun axis. The thermal calculation data were used as the reference data for stress-strain calculation. The simulation results are presented in Fig. 4.

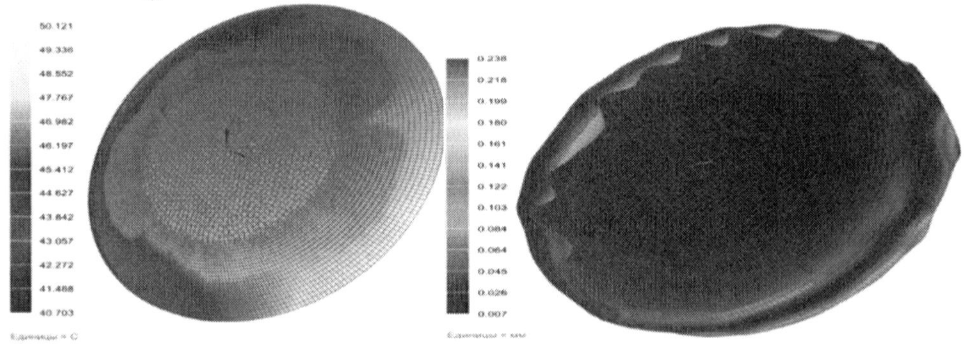

Fig. 4. Thermal (left) and stress-strain (right) behavior of the three-layered structure at the moment, when the reflector axis of rotation was at 150° angle the Earth-Sun axis at GEO.

5 Determining the stress-strain behavior of the ribbed reinforcement structure

To estimate the stress-strain behavior of the ribbed reinforcement structure, 24 moments of the work at GEO were analyzed. The largest temperature gradient coincided with the moment when the reflector axis of rotation was at 120° relative to the Earth-Sun axis. The thermal calculation data were used as the reference data for stress-strain calculation. The simulation results are presented in Fig. 5.

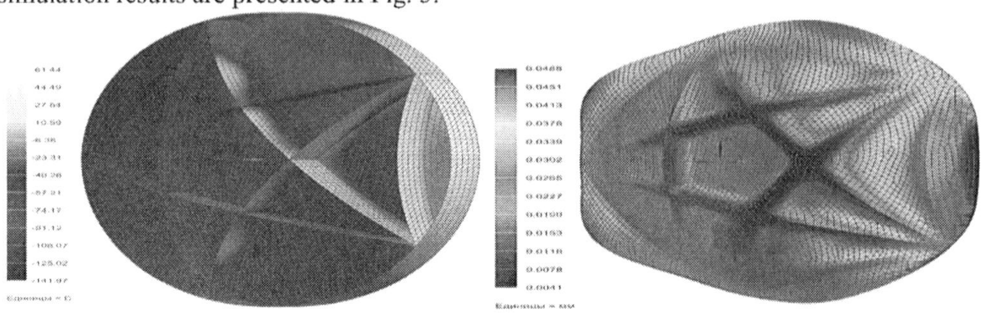

Fig. 5. Thermal (left) and stress-strain (right) behavior of the ribbed reinforcement structure at the moment, when the reflector axis of rotation was at 120° angle to the Earth-Sun axis at GEO.

6 Analysis

Table 3 presents the comparison of the design layout characteristics under investigation.

Table 3. Comparative characteristics of the design layouts.

Layout	**Temperature gradient, °C**	**Maximum displacement, mm**	**Mass, kg**
Three-layered structure	9.42	0.238	2.6
Ribbed reinforcement structure	184.54	0.049	1.224

As is evident from Table 3 the honeycomb filler structure is characterized by a more uniform distribution and a lower temperature gradient. The three-layered structure had 0.238 maximum displacement, while the mean displacement across the surface is in the range of 0.04 to 0.06 mm. The maximum displacement for the ribbed reinforcement structure constituted 0.049. However, the ribbed reinforcement layout has a clear advantage with the mass twice as low as that of the honeycomb filler layout. Therefore, the ribbed reinforcement layout is selected for the further development.

Some results of this work were obtained in the framework of the project №1864 under the state commission №2014/104 for the public research activity as part of the commission by the Ministry of education and science of the Russian Federation.

References

1. S.V. Reznik, P.V. Prosuntsov, A.V. Azarov, J. Eng. Phys. **88**, 674 (2015)
2. Pfeiffer, K.E. *Highly stable CFPR antenna reflectors HPS GmbH* (2007)
3. S.V. Reznik, P.V. Prosuntsov, A.V. Azarov, J. Eng. Phys. **88**, 945 (2015)
4. P.V. Prosuntsov, S.V. Reznik, K.V. Mikhailovsky, A.D. Novikov, IOP Conf. **153,** 012012 (2016)
5. S.V. Reznik, P.V. Prosuntsov, A.D. Novikov, *Proceed. XV Minsk International Heat and Mass Transfer Forum* **2**, 423 (2016)
6. *GOST 21631-76. Sheets of aluminum and aluminum alloy* (Standarts Publishing House, 1993)

Experimental investigation of the two-phase flow in a short horizontal microchannel with the height of 50 μm and width of 20 mm

Fedor Ronshin[1,*]

[1]Kutateladze Institute of Thermophysics, 630090 Novosibirsk, Russia

Abstract. The two-phase flow has been studied experimentally in a short horizontal microchannel with the height of 50 μm and width of 20 mm. The following regimes of two-phase flows have been registered: jet, bubble, stratified, annular, and churn. The regime map of two-phase flow has been plotted. This map has been compared with the regime map plotted for the channels of larger cross-section; it is shown that the height and width of a rectangular channel has a significant effect on the boundaries between flow regimes.

1 Introduction

Currently, there is rapid development of electronics of micro- and nanoscale, which promotes wide spread application of miniature heat exchangers. It is possible to use microburners, where liquid fuel combusts in minichannels [1]. Studies show that the heat exchange systems with the use of mini- and microchannels can take away the higher heat fluxes than the systems with a typical channel size larger than 1 mm or with freely falling films [2]. The surface-to-volume ratio of the flat channel increases with decreasing channel height. This feature provides high heat transfer rate in such systems.

The two-phase flows used in various systems in microelectronics, aerospace industry, transport, energy and other industries, has been actively studied in recent years. Many studies on the two-phase flow in mini- and microchannels have been published. The review of publications on two-phase flows in microchannels of different configurations is presented in [3, 4]. It is shown that many studies use round microchannels although the rectangular microchannels have more prospects for being used in the systems of thermal stabilization. In the rectangular microchannels, the flow pattern corresponds qualitatively to the regimes in tubes; despite the boundaries between the flows differ greatly. The mechanisms that influence formation of regimes in microchannels are described in [5, 6]. It is shown that the microchannel height has a significant influence on formation of the two-phase flow.

The goal of this study is investigation of characteristics of the two-phase flow and boundaries between the flow regimes in a short (with the length of 90 mm from the point of liquid inlet to the channel) rectangular microchannel with the height of 50 μm and width of

* Corresponding author: f.ronshin@gmail.com

20 mm. The results will be compared with the flow regimes in the microchannels of the larger cross-section.

2 Experimental setup and methods

The microchannel with the height of 50 mm and width of 20 mm is used as a working section. The bottom part of the working section is made of stainless steel plate, 160 mm long and 55 mm wide, processed by grinding, with a nozzle for liquid injection made at the angle of 11°. The microchannel is closed from the top by a quartz plate with previously applied antireflection coating. Between the glass and stainless steel plates, there are two constantan inserts of 50-μm height, defining the height of the channel. The contact angles have been measured before and after the experiment using the KRUSS DSA 100 installed on both surfaces. The microchannel height has been measured after the assembly at several points using the confocal technique. The average height of microchannel in the observation area is 48.7 ± 5.3 μm.

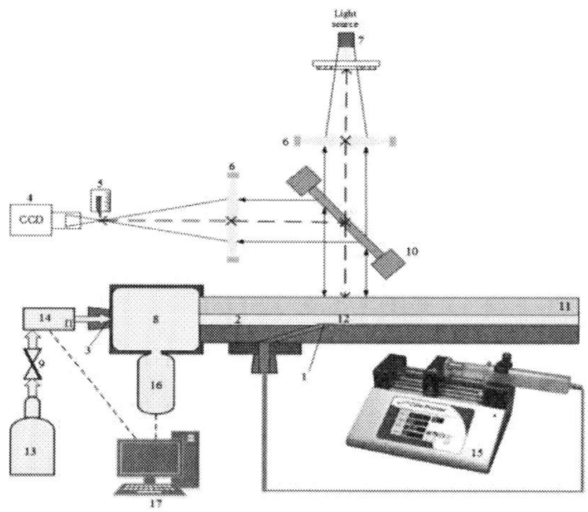

Fig. 1. Principle scheme of experimental setup: (1) liquid inlet to microchannel; (2) microchannel; (3) gas inlet to microchannel; (4) digital camera; (5) Schlieren-knife; (6) lenses; (7) light source; (8) gas chamber; (9) reducer; (10) beam splitter; (11) optical window; (12) observation area; (13) tank with gas; (14) regulator and sensor for flow rate measurements; (15) high precision syringe pump; (16) pressure sensor; (17) personal computer.

Gas mixture is fed to the central part of microchannel from tank (13). The gas flow rate is varied from 100 to 5000 ml/min and kept constant by the El-Flow flow controller of Bronkhorst Company (14). Gas is supplied to gas chamber (8), then it is fed to the microchannel through gas nozzle (2). The liquid flow rates are varied from 0.5 to 50 ml/min by high precision syringe pump Cole-Parmer EW-74905-54 (15). Liquid is introduced into the microchannel through liquid inlet (1). Superpure distilled deionized nano-filtered water, pre-cleared by Direct-Q® 3 UV installation, has been used as liquid. High-purity nitrogen has been used as gas. The distance between the gas and liquid nozzle is about 70 mm. The pressure in the gas chamber (8) is measured by the pressure sensor WIKA Type P-30 (16). The readings of this pressure sensor and current flow rate of gas are written in a file on a PC (17).

Liquid and gas interaction in the microchannel has been visualized in area (12) with the help of digital video and photo cameras by Schlieren photography technique. The Schlieren technique has been used to register and visualize surface deformations of the thin liquid film. Light from the source enters the microchannel with the gas-liquid flow through diffuser (7), lens (6), prism (10) and optical window (11). Light reflected from the gas-liquid interface is transmitted through beam splitter (10), lens (6) and camera lens filter (4). Knife (5) moved by a microscrew cuts the central part of the luminous flux. As a result, the camera captures a grayscale image, where a certain angle of the liquid-gas interface inclination corresponds to each gray level.

To visualize the two-phase flow, the high-speed camera FASTCAM-500M and digital camera Nikon D7000 in the Shlieren regime have been used. The detailed description of the Schlieren technique is presented in [7].

3 Experimental results and discussion.

The following regimes of two-phase flows have been registered in the channel of 50-μm height and 20-mm width during the experiment: jet, bubble, stratified, annular, and churn ones. The regime map is shown in Fig. 1. Inserts in this figure represent the characteristic patterns of the flow in the studied channel. Superficial velocities of liquid and gas, determined as the ratio of volumetric flow rate to cross-sectional area of the channel, have been used as the coordinates.

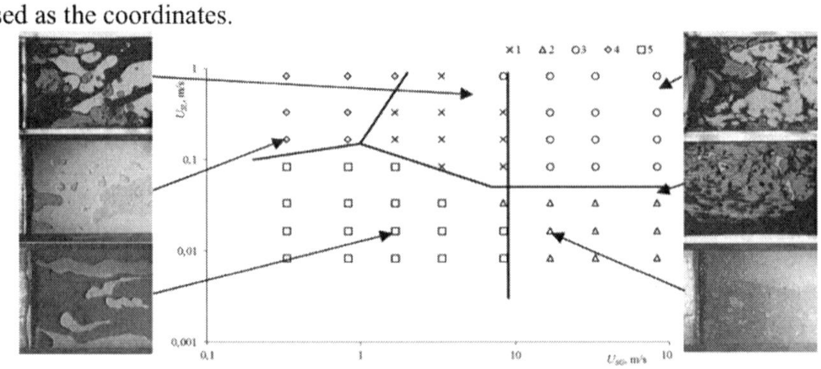

Fig. 2. The regime map of two-phase flows in microchannel with cross-section of 0.05×20 mm^2. Flow regimes: (1) churn; (2) stratified; (3) annular; (4) bubble; (5) jet.

At very low superficial liquid velocities, gas moves in the microchannel center, and the bulk of liquid moves at its periphery along the sidewalls and as jets in the center. With increasing superficial liquid velocity, it starts occupying a larger part of microchannel, and a gas flow moves in the center. There are no disturbances on the liquid surface. The stationary jet flow is observed at low superficial liquid and gas velocities, when the gas flow occupies less than a half microchannel width. The jet regime is specific for the flat mini- and microchannels. An increase in the superficial velocity of liquid increases pulsation frequency and amplitude, and leads to a loss of stability of the jet two-phase flow.

At high superficial velocities of liquid and low superficial velocities of gas, the amplitude of liquid perturbations in the lateral parts of microchannel reaches its half, forming the stable liquid bridges, and the bubble flow starts. Under this regime, liquid with many small gas bubbles moves along the channel. The size and number of bubbles change depending on the liquid and gas flow rates, but the bubble sizes are always much smaller than the channel width. There are also many small liquid droplets inside the bubbles. Formation of such droplets is caused by destruction of bridges or liquid films.

At low superficial liquid velocities and high superficial velocities of gas, the stratified regime is observed. Under this regime, a part of liquid moves along the bottom wall of the microchannel in the form of a film entrained by a gas flow. The upper wall of microchannel is dry. Gas under this regime occupies more than a half microchannel cross-section. The stratified regime is characteristic only of the non-circular microchannels because in the circular microchannels the film closes, forming the annular flow. The liquid droplets move along one of the microchannel sidewalls, leaving the traces on the liquid film. On the other side of microchannel, the film is not formed and droplets do not move. Two kinds of droplets can be distinguished: the large ones with the diameter of 1 to 3 mm, which represent the liquid bridges between the top and bottom walls of microchannel, and small droplets with the diameter less than millimeter on the microchannel wall or on the liquid film.

With increasing superficial liquid velocity, the film is formed on the upper wall of microchannel, and transition to the annular regime occurs. Under this regime, we can see the mobile liquid droplets that slide over the liquid films on the top and bottom walls of microchannel. The transition from the stratified to the annular flow is determined using the Schlieren technique. In the annular regime, liquid moves along the microchannel walls in the form of a film; in the central part, gas with droplets forms the flow core. Gas takes up much more volume than liquid. The liquid film is formed on the top wall of microchannel at a distance of several millimeters from the zone of liquid input to the microchannel. In previous experiments, in microchannels with the height of less than 200 μm, the film was formed on the top wall of microchannel directly near the liquid input into the microchannel due to front instability [5].

At superficial velocities of liquid from 0.1 to 1 m/s and superficial gas velocities from 1 to 10 m/s, the churn flow is observed. This regime is characterized by the features of the jet and bubble flows. This regime is characteristic of the vertical channels, and it occurs in the wide horizontal microchannels. The broken bridges are typical of this regime. The churn flow is caused by development of jet flow instability and increasing pulsation frequency of liquid, moving near the sidewalls of microchannel under the influence of the gas flow. Both the gas flow (with discontinuity) and gas bubbles can be observed under this regime. The churn flow is considered in detail in [8].

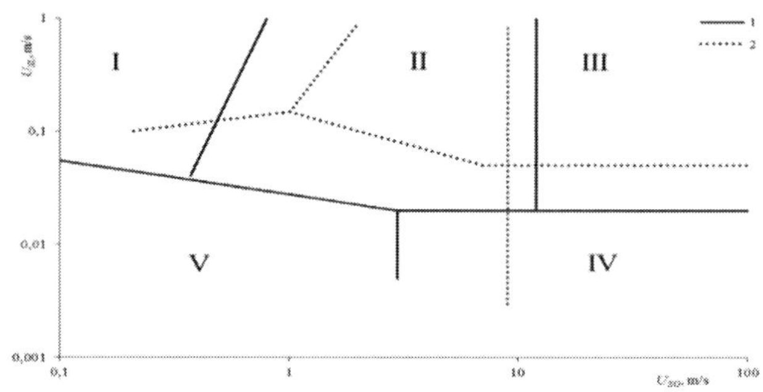

Fig. 3. The effect of microchannel width on the boundaries between the flows for the channels with different cross-sections. Flow regimes: I – bubble, II – churn, III – annular, IV – stratified, V – jet; channel cross-sections: 0.05×40 mm^2 (1), 0.05×20 mm^2 (2).

A comparative regime map plotted for microchannels with cross-section of 0.05×20 mm^2 and 0.05×40 mm^2 is shown in Fig. 3. It is seen that with an increase in the microchannel width, the zones of the churn and bubble flow increase (the boundary between the churn and jet flow moves towards the lower superficial liquid velocities), compressing significantly the zone of the jet flow. The boundary between the bubble and churn regimes moves towards the lower superficial gas velocities, narrowing the area of the bubble flow. The boundary between the jet and stratified flows moves towards the lower superficial velocities of gas with an increase in the microchannel width, and the region of stratified flow increases. The boundary between the annular and stratified regimes shifts toward the lower superficial velocities of liquid.

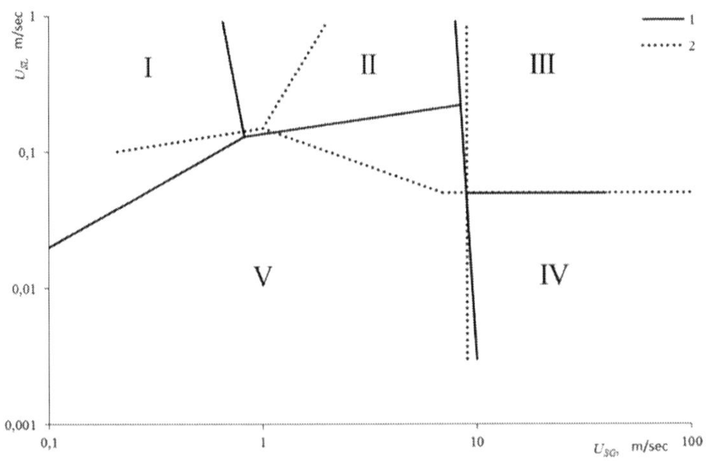

Fig. 4. The effect of microchannel height on the boundaries between the flows for the channels with different cross-sections. Flow regimes: I – bubble, II – churn, III – annular, IV – stratified, V – jet; channel cross-sections: 0.1×20 mm^2 (1), 0.05×20 mm^2 (2).

A comparative regime map plotted for microchannels with cross-section of 0.1×20 mm^2 and 0.05×20 mm^2 is shown in Fig. 4. It is seen that with an increase in the microchannel height, the boundary between the churn and jet flows moves towards the higher superficial velocities of liquid, and the boundary between the churn and bubble flows shifts towards the lower superficial velocities of gas. With an increase in the microchannel height, the boundary between the bubble and jet flows also moves towards the lower superficial velocities of liquid, narrowing the region of the jet flow. The regions of the stratified and annular flow regimes are not affected by the microchannel height.

In conclusion, it should be noted that in the channel of rectangular cross-section of 50×20 mm^2, the main regimes of the two-phase flow have been registered and the boundaries between them have been determined. The regime maps plotted for the channels with different cross-sections have been compared. It is shown that the height and width of the rectangular microchannel have a significant effect on the boundaries between the regimes.

We gratefully acknowledge the support from the Ministry of Education and Science of Russia (No. 14.604.21.0053 Project identifier RFMEFI60414X0053).

References

1. E. A. Chinnov, J. Heat Transf. **138**, 9 (2016)
2. V. V. Zamashchikov, A. A. Korzhavin, E. A. Chinnov, Combust. Explos. Shock Waves **50**, 381 (2014)
3. E. V. Rebrov, Theor. Found Chem. Eng. **44**, 355 (2010)
4. E. A. Chinnov, F. V. Ron'shin, O. A. Kabov, Thermophys. Aeromech. **22**, 265 (2015)
5. E. A. Chinnov, F. V. Ron'shin, O. A. Kabov, Int. J. Multiph. Fl. **80**, 57 (2016)
6. E. A. Chinnov, F. V. Ron'shin, O. A. Kabov, Interfacial Phenomena and Heat Transfer **3**, 243 (2015)
7. V. Cheverda, A. Glushchuk, P. Queeckers, S. B. Chikov, O.A. Kabov, Microgravity Sci. Technol. **25**, 73 (2013)
8. E. A. Chinnov, F. V. Ron'shin, O. A. Kabov, Techn. Phys. Lett. **41**, 817 (2015)

The Evaporation of Liquid Micro-Drops on the Heated Substrate

Andrey Semenov[1,2,3*]

[1]Kutateladze Institute of Thermophysics SB RAS, 630090 Novosibirsk, Russia
[2]Novosibirsk State University, 630090 Novosibirsk, Russia
[3]National Research Tomsk Polytechnic University, 634050 Tomsk, Russia

Abstract. Evaporation of a heated sessile water micro-drop was studied experimentally at the substrate temperature and surrounding atmosphere from 30 to 50 ^{0}C. The studies were performed on the float glass substrate with aluminum nanocoating of optical quality. The research has shown that the specific rate of evaporation (mass loss per unit of the drop surface area) increases with the decrease in droplet volume and at the last stage several times exceeds the initial value.

1 Introduction

Some of the effective solutions for removal of high heat flux densities are spray cooling systems [1] and systems using the stratified two-phase flow in a microchannel [2, 3]. In certain regimes in these systems the heat is removed due to intense evaporation of liquid micro-drops. In the literature, the evaporation of liquid drops on the heated plate was, as a rule, studied using sufficiently large droplets with an initial volume of about 100 μl [4-9]. This paper investigates the evaporation of liquid droplets having an initial volume of less than 1 μl.

2 Experimental setup and methods

A liquid drop is placed on the surface with a known wettability. Then, the droplet is heated up to the onset of quasistationary heat transfer between the solid substrate and the liquid drop. The used working fluid is distilled deionized nanofiltered water of Milli-Q company. The use of water with a relatively high boiling point as working fluid provides for a rather detailed investigation of the influence of temperature difference between the solid surface and the surrounding atmosphere on the intensity of evaporation and heat transfer. The temperature difference between the solid surface and the atmosphere varies from 30 to 50 ^{0}C.

The experiments on water drops evaporation in the open atmosphere were made on the experimental stand whose scheme is presented in Fig. 1. The work area was an aluminum

* Corresponding author: semenov.itp@gmail.com

plate of 15 mm thickness, on whose bottom side the heater was located. The used heater was a Peltier element with dimensions 40x40 mm2. On the upper side of the aluminum plate the investigated substrate was fixed. It consisted of a disk with a diameter of 52 mm and a thickness of 3.2 mm. Measurement of the surface temperature of the studied substrate was carried out using a flat thermocouple, fixed at a distance of 1-2 mm from the contact line of the liquid drop. The temperature under the substrate was measured by three thermocouples of K-type. Thermocouples were placed radially at different distances from the center of the disc and served to measure the temperature distribution along the substrate radius. The working surface was made of float glass with aluminum nanocoating of optical quality. The relative humidity of air in the room during the experiments was 20-30 %.

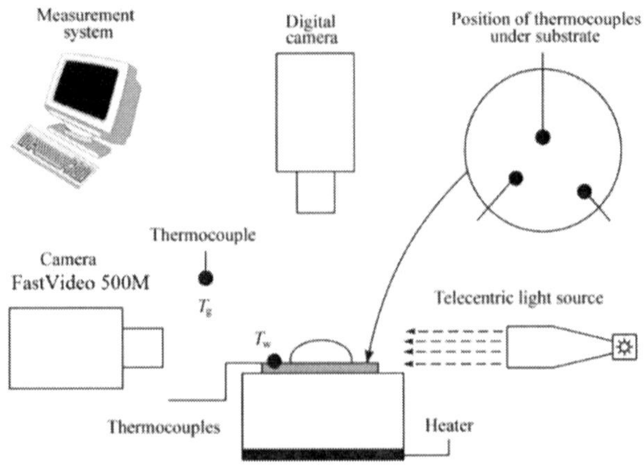

Fig. 1. Scheme of experimental stand.

The authors used two optical techniques. The first one is a shadow method, consisting of the halogen light source Edmunds Optics MI-150, connected by a fiberoptic lightguide with a system of lenses of Edmunds Optics Telecentric backlight illuminator. This system generates a plane-parallel light beam of 50 mm diameter at the output of one side of the work area. High-speed video camera FastVideo 500M with a microscope lens 10X Mitutoyo Plan Apo Infinity-Corrected Long WD, connected through the focusing bellow Novoflex Balpro TS, on the other side of the droplet. The spatial resolution of this system was 1 μm/pixel, and the frequency of shooting was up to 500 frames per second. Using the shadow method we obtained shadow images of the liquid drop profile. The images were processed by the Drop Shape Analysis software of KRUSS company. Shadow photography was used to select the drop profile by the method of Young-Laplace. Further, the following geometric parameters of the liquid drop were determined: the base diameter, maximum diameter, height, volume, contact angles (right, left, middle), the surface area, and the basal area.

The second optical technique consisted of a digital video camera Point Grey Grasshopper3 (4.1 Mpix) with a microscope lens 20X Mitutoyo Telecentric Objective. It was placed over the liquid drop and allowed visualizing the evaporation from the top and controlling the symmetry of the droplet. Spatial resolution of the system was 0.5 μm/pixel, and the frequency of shooting was up to 90 frames per second. If the drop lost its symmetry, the experiment repeated.

3 Results

The experiment began with a preliminary preparation of the investigated substrate, which consisted in a thorough cleaning of the working surface. Then the whole substrate was placed in a reservoir with distilled water for a day. Thereafter, to remove residual water from the surface the substrate was flushed with clean compressed air designed for cleaning optical components. A liquid drop of preset size was placed on the surface, heated to the desired temperature using a high-precision dispenser Lenpipet. The lens used in the shadow method was focused (depth of field was a few mm); and after that the periodic shooting was performed until complete evaporation of the liquid droplet (Fig.2). The upper chamber was switched on simultaneously with the shadow method.

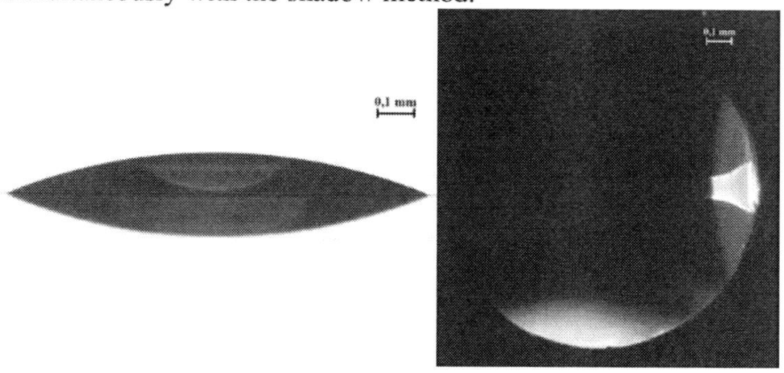

Fig. 2. Images of liquid droplets from the side and from above, V=0.06 μl, T_s=40^{0}C.

The dependence of the specific rate of liquid evaporation on time at different initial surface temperature is shown in Fig. 3. The evaporation rate was calculated as a loss in the droplet mass per unit of droplet surface area per unit of time. All built-in data were synchronized to the time of complete evaporation of liquid droplets. The research has shown that the specific rate of evaporation increases with the decrease in droplet volume and at the last stage several times exceeds the initial value. The obtained results are in good agreement with the results of [5, 6], studying the evaporation of drops with initial volume of about 100 μl.

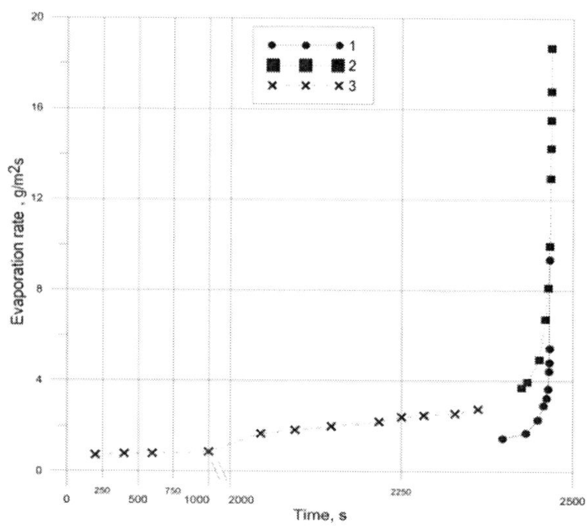

Fig. 3. The dependence of specific evaporation rate on time. 1 – data obtained with the substrate temperature T_s=30^{0}C, the initial drop volume V_0=0.22 μl; 2 - T_s=45^{0}C, V_0=95.16 μl; 3 – data obtained in the work [5] at T_s=45^{0}C, V_0=95.16 μl.

The study was supported by the grant of the Russian Science Foundation (agreement No. 14-19-01755).

References

1. J. Kim, Int. J. Heat Fluid Fl. **28**, 753 (2006)
2. O.A. Kabov, Yu.V. Lyulin, I.V. Marchuk and D.V. Zaitsev, Int. J. Heat Fluid Fl. **28**, 103 (2007)
3. D.V. Zaitsev, D.A. Rodionov and O.A. Kabov, Tech. Phys. Lett. **35**, 680 (2009)
4. O.A. Kabov and D.V. Zaitsev, Doklady Physics **58**, 292 (2013)
5. E.Ya. Gatapova, A.A. Semenov, D.V. Zaitsev, O.A. Kabov, Colloid Surf. A **441**, 776 (2014)
6. A.A. Semenov, D.V. Feoktistov, D.V. Zaitsev, G.V. Kuznetsov, and O.A. Kabov, Thermophys. Aeromech. **22**, 771 (2015)
7. I.V. Marchuk, A. Karchevsky, A. Surtaev, and O.A. Kabov, Int. J. Aerospace Eng. **2015**, 5 (2015)
8. G. V. Kuznetsov, D.V. Feoktistov, E. G. Orlova, Thermophys. Aeromech. **23**, 17 (2016)
9. G.V. Kuznetsov, D.V. Feoktistov, E.G. Orlova, K.A. Batishcheva, Colloid J. **78**, 335 (2016)

MATEC Web of Conferences **92**, 01015 (2017) DOI: 10.1051/matecconf/20179201015
Thermophysical Basis of Energy Technologies - 2016

Method of visualization on droplet dynamics using two optical systems

Anna M. Shonina[1,2,*]

[1]Kutateladze Institute of Thermophysics SB RAS, Novosibirsk, Russia
[2]Novosibirsk State University (NSU), Novosibirsk, Russia

Abstract. The methods of optical visualization of droplet dynamics are described. One of them for top view uses microscope system with CCD-camera and the second is a high-resolution shadow technique. Droplet evaporation dynamics are recorded and processed using special programs. Evolution of droplet base diameters and volumes are presented.

1 Introduction

The wettability of a surface is usually described by the contact angle, measured within the liquid at the three phase contact line. Due to the topographical peculiarities of the solid surfaces, in reality the contact angle does not exhibit a unique value. The contact angle hysteresis, which is the difference between advancing and receding contact angles, plays an important role on wetting and dewetting processes. The nanostructured surfaces allow control and significantly influence on the processes of heat transfer [1, 2] and hydrodynamics [3, 4] in a variety of applications. The roughness and the number of micro-holes can be increased. On the other hand, contact angle, capillary effects, porosity can be controlled. This leads to change in the nature of two–phase flow, hydraulic resistance, critical heat flux, condensation, evaporation, etc. One of the most actual issues is the behavior of the three phase contact line [5, 6] on surfaces with different coatings [7, 8]. The methods of optical visualizations in order to have high–resolution system is challenging issue [9, 10].

This paper presents the results of applications of two optical methods for three phase contact line dynamics during evaporation of hanging droplet.

2 Visualization systems

Scheme of experimental setup is presented in Fig. 1. The glass slides (Struers) are used as substrate. The contact angle of water droplet on glass slides has been measured using KRUSS DSA-100 (Fig. 2). And it is found to be of 16 degrees. The ultrapure water (MilliQ) is used as the working fluid. The liquid of volume of 2 μl is deposited to the glass surface using a precision syringe pump Cole Parmer. The syringe of 0.5 ml volume is used in order to make a droplet of the precise volume. An atmosphere is controlled by Testo

* Corresponding author: annamishonina@gmail.com

device and barometer. Top view visualization of evaporation dynamics has been done using Olympus BX51 microscope with different lenses and monochrome camera Point Grey. The resolution of 1 μm per pixel is provided by using 10×lens and 100 nm per pixel by 100× lens. Recording frequency is from 10 to 40 fps. Side view of droplet profile is recorded using the shadow method (Fig. 1) providing resolution of 3 μm per pixel by the present time. Contact angles are measured at different times using Advance software of KRUSS DSA-100. Dynamics of the three phase contact line during droplet evaporation on the clean glass are studied.

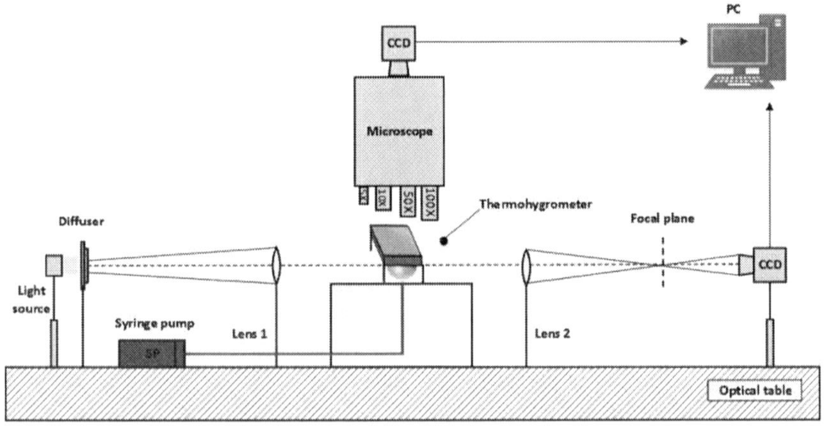

Fig. 1. Scheme of experimental setup.

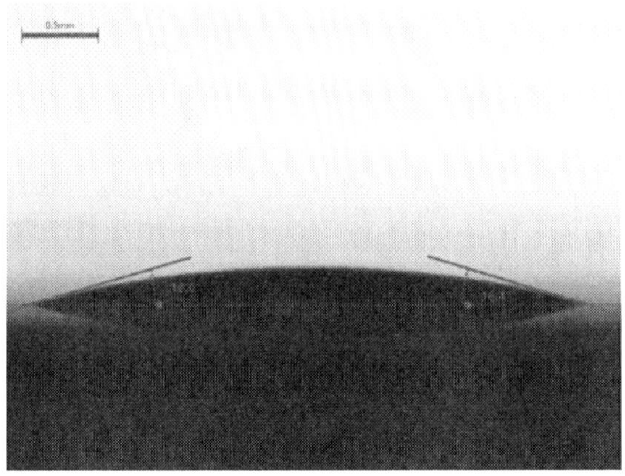

Fig. 2. Contact angle measurement by KRUSS DSA-100.

3 Results and discussions

The examples of images taken by two techniques are presented in Fig. 3. Finally, records are processed in ImageJ and Advance (KRUSS) software. The data on time, diameter, area and perimeter of droplet base, volume, velocity and contact angle are obtained using these programs. Some characteristic plots of evolutions of droplet volume via time (obtained from the side view) and of droplet diameter (obtained from the top view) are shown in Fig. 4a and Fig. 4b, respectively. Here, zero time is the time of complete droplet disappearance.

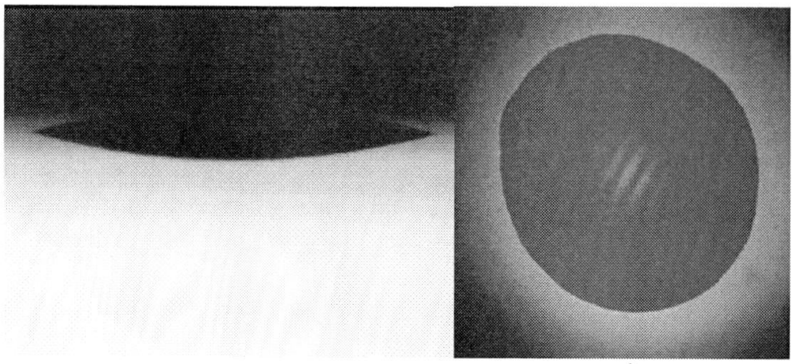

a b

Fig 3. Images of droplet on the glass substrate. Initial volume is 2μl. *a* –side view by shadow method, *b* –top view by using microscope.

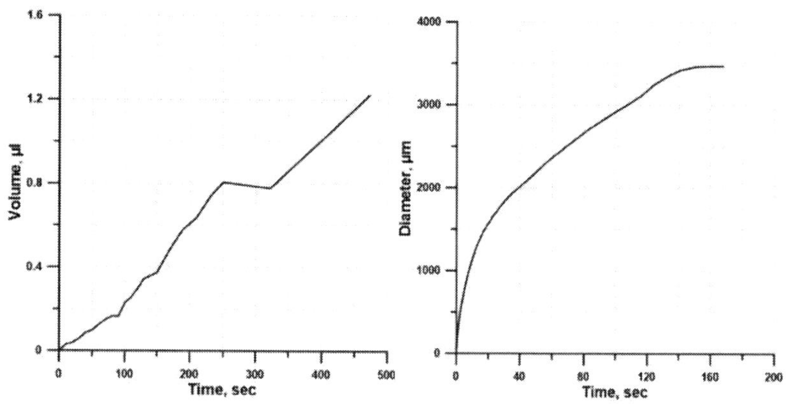

a b

Fig. 4. Volume and diameter plots. *a* –side view by shadow method, *b* –top view by using microscope.

4 Conclusion

In conclusion, two optical methods for visualization of droplet dynamics have been adopted. One of them for the top view uses microscope system with CCD-camera and the second is a high-resolution shadow technique. The top view technique can provide with the resolution of up to 100 nm per pixel. The synchronization of top and side video registrations via time is organized.

The authors gratefully acknowledge the support of this work by the Ministry of Education and Science of the Russian Federation (Agreement 14.616.21.0016, project identifier RFMEFI61614X0016).

References

1. N. Miljkovic, R. Enright, Y. Nam, K. Lopez, N. Dou, J. Sack, E.N. Wang, Nano Lett. **13**, 179 (2013)
2. C. Lin, *Droplet Impact and Evaporation on Nanotextured Surface for High Efficient Spray Cooling* in An Overview of Heat Transfer Phenomena ed. by S. N. Kazi (In Tech, Croatia, 2012)
3. E.Y. Gatapova, V.S. Ajaev, O.A. Kabov, JETP Letters **101**, 160 (2015)

4. V.S. Ajaev, E.Y. Gatapova, Kabov O.A., Adv. Colloid Interfac. **228**, 92 (2016)
5. A. Zou, A. Chanana, A. Agrawal, P.C. Wayner Jr., S.C. Maroo, Sci. Rep. **6,** 20240 (2016)
6. E.Ya. Gatapova, O.A. Kabov, Microgravity Sci. Tec. **19**, 132 (2007)
7. A. Dubov, A. Mourran, M. Moeller, O. Vinogradova, J. Chem. Phys. **141**, 074710 (2014)
8. E.Ya. Gatapova, A.A. Semenov, D.V. Zaitsev, O.A. Kabov, Colloid Surf. A **441**, 776 (2014)
9. S.S. Panchamgam, A. Chatterjee, J.L. Plawsky, P.C. Wayner, Jr., Int. J. Heat Mass Tran. **51**, 5368 (2008)
10. M.S. Hanchak, M.D. Vangsness, L.W. Byrd, J. S. Ervin, J.G. Jones, Appl. Phys. Lett. **103**, (2013)

Research of interaction of a natural and forced convection in a vortex chamber of the chemical reactor

Alexander V. Shvab[1*], *Natalia* I. Gicheva[1], and *Vladimir* N. Brendakov[2]

[1]National Research Tomsk State University, Faculty of Physics and Engineering, Department of Applied Aerodynamics, 634050 Tomsk, Russia
[2] Seversk Technological Institute - branch of State Autonomous Educational Institution of Higher Education "National Research Nuclear University "MEPhI", 636036 Seversk, Russia

Abstract. In this paper, a chemical reactor for producing refractory metals was considered. A physical and mathematical model of fluid motion and heat transfer in a vortex chamber of the chemical reactor under forced and free convection has been described and simulated. The numerical simulation was carried out in "velocity–pressure" variables by using an alternating direction implicit scheme. The velocity field and the temperature distribution in the reactor were obtained. Parametric studies on effects of the Reynolds, Prandtl and Rossbi criteria on the flow characteristics were also performed. The graphs presented show that natural convection has a significant impact on the hydrodynamics of the flow and intensifies the heat transfer. Reliability of the calculations was verified by comparing the results obtained by another method

1 Problem statement

Modeling of viscous gas dynamics and heat transfer is considered in a vortex chamber that present a cylindrical chamber (See Fig. 1).

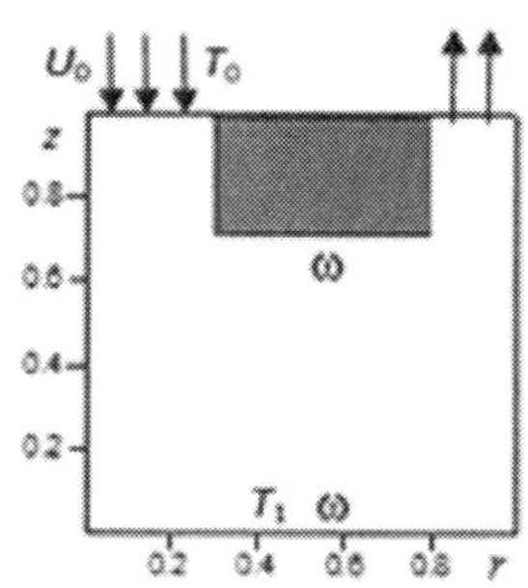

Fig.1. The computational domain of the vortex chamber.

* Corresponding author: avshvab@inbox.ru

DOI: 10.1051/matecconf/20179201016

The carrier gas flow having axial velocity U_0 and temperature T_0 enters from the pipe along the axis above, flows over the rotating disk and exits through the annular channel at the periphery of the top of the vortex chamber. The top and bottom walls of the apparatus are rotating with an angular velocity ω. Part of the bottom wall is maintained at a temperature T_1, and other walls including a small surface at the periphery of the bottom wall of the chamber are considered as heat-insulated. By virtue of relatively small velocities and temperature differences, gas is assumed incompressible and density change is taken into account only in terms that includes the force of gravity and besides density difference is proportional to the temperature difference with the opposite sign according to the Boussinesq approximation. Subject to the foregoing, non-dimensional transport equations of momentum and heat and continuity equation in cylindrical coordinates in the assumption of axial symmetry in angular direction are follows:

$$\frac{\partial u}{\partial \tau}+\frac{\partial u^2}{\partial r}+\frac{\partial uw}{\partial z}-\frac{1}{\mathrm{Re}}\left(\frac{\partial^2 u}{\partial r^2}+\frac{\partial^2 u}{\partial z^2}\right)=-\frac{\partial p}{\partial r}+\frac{1}{r\,\mathrm{Re}}\frac{\partial u}{\partial r}+\frac{v^2-u^2}{r};$$

$$\frac{\partial v}{\partial \tau}+\frac{\partial uv}{\partial r}+\frac{\partial wv}{\partial z}-\frac{1}{\mathrm{Re}}\left(\frac{\partial^2 v}{\partial r^2}+\frac{\partial^2 v}{\partial z^2}\right)=\frac{1}{r\,\mathrm{Re}}\frac{\partial v}{\partial r}-\frac{2uv}{r};$$

$$\frac{\partial w}{\partial \tau}+\frac{\partial uw}{\partial r}+\frac{\partial w^2}{\partial z}-\frac{1}{\mathrm{Re}}\left(\frac{\partial^2 w}{\partial r^2}+\frac{\partial^2 w}{\partial z^2}\right)=-\frac{\partial p}{\partial z}+\frac{1}{r\,\mathrm{Re}}\frac{\partial w}{\partial r}-\frac{uw}{r}+\frac{Gr}{\mathrm{Re}^2}\vartheta; \qquad (1)$$

$$\frac{\partial \vartheta}{\partial \tau}+\frac{\partial u\vartheta}{\partial r}+\frac{\partial w\vartheta}{\partial z}-\frac{1}{\mathrm{Pr}\,\mathrm{Re}}\left(\frac{\partial^2 \vartheta}{\partial r^2}+\frac{\partial^2 \vartheta}{\partial z^2}\right)=\frac{1}{\mathrm{Pr}\,\mathrm{Re}\,r}\frac{\partial \vartheta}{\partial r}-\frac{u\vartheta}{r};$$

$$\frac{\partial u}{\partial r}+\frac{\partial w}{\partial z}+\frac{u}{r}=0.$$

The dimensionless form of equations is obtained by using the following scales: vortex chamber radius R_0, axial velocity at the input U_0, density at the input ρ_0 and the maximum temperature difference T_1-T_0.

The system of equations obtained contains criteria of Reynolds, Prandtl and Grashof:

$$\mathrm{Re}=\frac{\rho_0 U_0 R_0}{\mu};\ \mathrm{Pr}=\frac{c_p\mu}{\lambda};\ \mathrm{Gr}=\frac{g\beta(T_1-T_0)\rho_0(\mathrm{R}_0)^3}{\mu^2}. \qquad (2)$$

The dimensionless temperature is determined as $\vartheta = (T-T_0)/(T_1-T_0)$. In addition, the following notations are introduced: u is radial velocity component, v is peripheral velocity component and w is axial velocity component.

2 Numerical calculations

The numerical simulation of a set (1) was carried out in "velocity–pressure" variables using method of pressure and velocity correction [1]. One can present a system of equations consisting of the momentum equations and the continuity equation in the vector form as follows:

$$\frac{\partial \mathbf{V}}{\partial t} = -\frac{\nabla p}{\rho} + F(\mathbf{V}); \qquad div\mathbf{V} = 0. \tag{3}$$

One can obtain the two equations using time-splitting method for the momentum equation:

$$\frac{\mathbf{V}^{+} - \mathbf{V}^{n}}{\Delta t} = -\frac{\nabla p^{n}}{\rho} + F(\mathbf{V}^{+}, \mathbf{V}^{n}); \tag{4}$$

$$\frac{\mathbf{V}^{n+1} - \mathbf{V}^{+}}{\Delta t} = -\frac{\nabla(\delta p)}{\rho}, \tag{5}$$

Here δp is the pressure correction which is equal to the difference between the pressures at new n+1st layer of time and at the nth layer known, $\mathbf{V}^{+}$ is an intermediate grid function. After scalar multiplication of the equation (5) by the gradient and taking into account the solenoidality of velocity on the n+1th layer, we have obtained the Poisson equation for the calculation of pressure correction as follows:

$$\nabla^2(\delta p) = \rho \frac{\nabla \cdot \mathbf{V}^{+}}{\Delta t}. \tag{6}$$

One can represent this equation as transient Poisson equation for the convenience of calculation:

$$\frac{\partial(\delta p)}{\partial \tau} - \nabla^2(\delta p) = -\rho \frac{\nabla \cdot \mathbf{V}^{+}}{\partial t}. \tag{7}$$

In Eq. (8) $\Delta\tau = B\Delta t$ is the time step performing the role of iteration parameter which value is chosen so that ensure the most rapid convergence of problem solution. For the difference scheme to construct, it is used a staggered grid. The set of equations obtained was being solved by using an evolutionary method of relaxation in time. For the each equation in the set obtained to solve, an alternating direction implicit scheme was used. This method is second-order accurate in time and it is unconditionally stable. Convective and diffusion terms in transport equations for a staggered grid are represented by dint of the Exponential Scheme [2] based on the control volume method.

For a unique solution to obtain, the boundary conditions in dimensionless form are set. At the input of chamber we have:

$$\frac{\partial u}{\partial z} = 0, \quad v = 0, \quad w = -1, \quad \vartheta = 0.$$

At the symmetry axis by $r = 0$ we have

$$u = 0, \quad v = 0, \quad \frac{\partial w}{\partial r} = 0, \quad \frac{\partial \vartheta}{\partial r} = 0.$$

At walls of chamber there are no-slip and heat insulation conditions:

$$u = 0, \quad v = 0, \quad w = 0, \quad \frac{\partial \vartheta}{\partial n} = 0,$$

with the exception of rotating surfaces, for which no-slip condition cause an emergence of another criterion, the inverse Rossby number $\mathrm{R\omega} = R_0\omega/U_0$. The value of peripheral velocity at the rotating surface is $v = \mathrm{R\omega}\cdot r$. We have $\vartheta = 1$ at the bottom wall by $r < 0.8$ and $\partial\vartheta/\partial z = 0$ by $0.8 < r < 1$ (heat insulation condition).
At the chamber output, the Neumann conditions are used:

$$\frac{\partial u}{\partial z} = \frac{\partial v}{\partial z} = \frac{\partial w}{\partial z} = \frac{\partial \vartheta}{\partial z} = 0.$$

3 Conclusions

The results of the numerical solution of the set (1) are presented in Fig. 2 and 3, and besides, Fig. 2 illustrates the solution by natural convection not taken into account and Fig. 3 shows the solution with natural convection by the same parameters and criteria of the flow considered. The figures presented show that a natural convection impacts significantly on the swirling flow hydrodynamics. A position of the circulation zone of the gas streamlines changes. The area of large values of peripheral velocity moves in the centerline direction. The temperature increases in a large part of the vortex chamber. This evidences the intensification of heat transfer from the bottom wall of vortex chamber.

For the calculations verification the problem considered was converted to "vortex–stream function" variables and solved by the alternating direction implicit scheme using non-staggered grid. This method gave the same results. In additional, the created model was applied for numerically solving the classical test problem of the velocity distribution in an annular channel and that of a rotating infinite disk in a stationary liquid. The study findings showed a good agreement with the exact solutions.

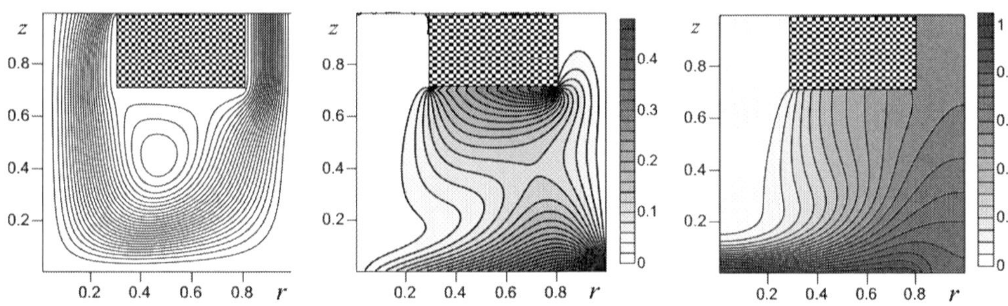

a) b) c)

Fig.2. Stream function (a), peripheral velocity (b) and temperature (c) distribution at Re=50, Rω=0.5, Pr=1, Gr=0.

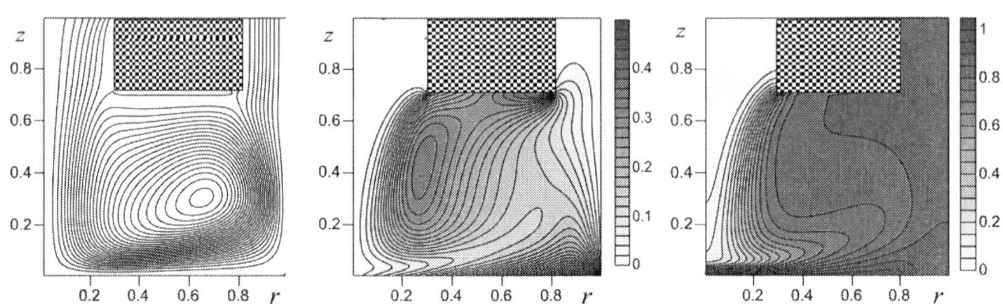

a) b) c)

Fig.3. Stream function (a), peripheral velocity (b) and temperature (c) distribution at Re=50, Rω=0.5, Pr=1, Gr=10^5.

References

1. R. Peyret, T. Taylor, *Computational Methods for Fluid Flow* (Springer-Verlag, Berlin, 1983)
2. S. Patankar, *Numerical Heat Transfer and Fluid Flow* (Hemisphere Publishing Corporation, New York, 1980)

Analyzing the thermal regime of power supply units in portable betatrons by using infrared thermography

O.S. Simonova[1,*], *S*.V. Kasianov[1], *A*.I. Moskovchenko[1], and *G*. Xingwang[2]

[1]National Research Tomsk Polytechnic University, 634050 Tomsk, Russia
[2]Beihang University, 100000 Beijing, China

Abstract. Potentials of infrared thermography in analyzing a thermal regime of the 7.5 MeV betatron power supply are discussed. Both the heating rate and thermal inertia of particular electronic components have been evaluated by processing pixel-based temperature histories. The data treatment has been performed by using the original ThermoFit Pro software to illustrate that some advanced processing algorithms, such as the Fourier transform and principle component analysis, are valuable in identifying thermal dynamics of particular power supply parts.

1 Introduction

Portable betatrons are widely used in various industrial sectors as sources of electrons and X rays, in particular, for nondestructive testing of materials, components and installations, examining luggage and cargo at airports, sea ports and heavy truck check points, performing X ray computer tomography and treating malicious tumors in medical practice [1-4]. The known advantages of betatrons are their relative simplicity and low mass compared to other types of high-energy X ray sources, as well as in-use comfort and low cost.

National Research Tomsk Polytechnic University is a unique world developer of portable betatrons which implements a whole production cycle starting from designing and finishing with commercial units for turn-key supply.

Reliable operation of betatrons depends on characteristics of a thermal regime in principle accelerator parts. Betatrons operate within a heating/cooling mode with the whole cycle time period being about 45 and 15 minutes for an emitter and a power supply respectively.

Figure 1 shows the construction of a typical betatron [1]. Each betatron is supplied with an air cooling system which helps to cool down overheated components. However, it is still a challenge to sink thermal energy from those components which are not in contact with the cooling air. Such components include a magnet guide which consists of the central poles (2) with the inserts (3) separated by dielectric gaskets, as well as the backward magnet guide

*Corresponding author: ossimonova@mail.ru

consisting of six supports which also represent a section of the power supply case (1). The magnet guides are made of transformer steel plates bonded altogether with epoxy adhesive.

Most of materials used in betatrons are of a low thermal conductivity; therefore, the heat energy sink by heat conduction is negligible. Until recently, the designing of betatrons has been fulfilled by using standard analytical approaches which can be scarcely applicable to the complicated 3D geometry of betatrons. Therefore, there is still a practical need to ensure lower heating level of betatron key components.

2 Research Questions

In this study we investigate applicability of infrared (IR) thermography for analyzing thermal regimes of the power supply unit used in a 7.5 MeV betatron (Fig. 2a).

IR thermography as a technique intended for non-contact analysis of physical objects is widely used in electrical engineering and electronics, particularly, for evaluating temperature distributions of boards and electronic components under electrical load. A review of these applications is presented in [5-8]. It is worth noting that the most of practical applications of IR thermography involve steady-state test procedures where object temperature remains constant during inspection.

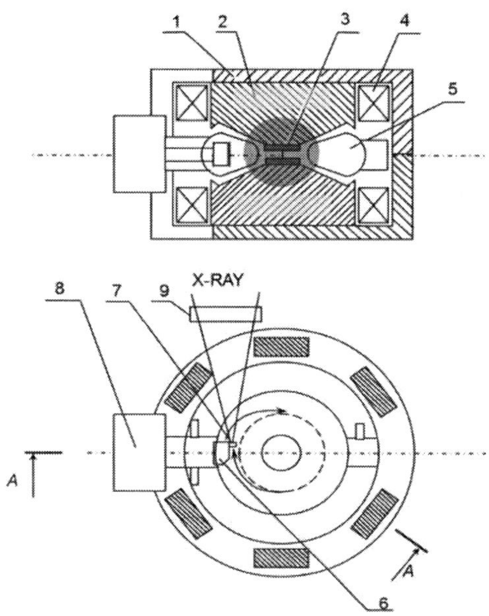

Fig.1. Betatron scheme.

1 – magnetic guide;
2 – pole;
3 – central inserts block;
4 – magnetizing winding;
5 – tube;
6 – injector;
7 – target;
8 – HV transformer;
9 – dose rate monitor.

90 – 95 °C in 20 minutes 400 Hz

60 – 65 °C

It is known that a major problem of IR/thermal diagnostics in electronics is the necessity of determining component emissivity which significantly varies depending on a type of a component thus distorting measurements of true temperature [9, 10]. In this study, we consider a relatively unexplored aspect of IR diagnostics of electronic boards and devices, namely, the analysis of temperature dynamics which appears in transition (switch on/switch off) work regimes. The concept of such approach is borrowed from active nondestructive testing of materials where hidden defects reveal themselves in surface temperature distributions due to different behavior of transient temperature in defect and non-defect areas [11]. The analysis of temporal characteristics of temperature distributions allows using modern mathematical approaches in application to IR image sequences evolving in time, such as Fourier and wavelet analysis, the technique of singular value decomposition (method of principal component analysis-PCA), non-linear fitting and many others [11].

3 Research Methods

A power supply unit of a small-size portable betatron for 7.5 MeV (Fig. 2a) was examined by using a NEC Avio TH-9100 IR imager which operates in the 7-13 μm wavelength band and ensures the temperature sensitivity of 30 mK (in the averaging mode) with the standard image format being 320×240 pixels. At the inspection side, the unit was covered with a 60 μm-thick polyethylene foil in order to simulate heat convection in a closed box.

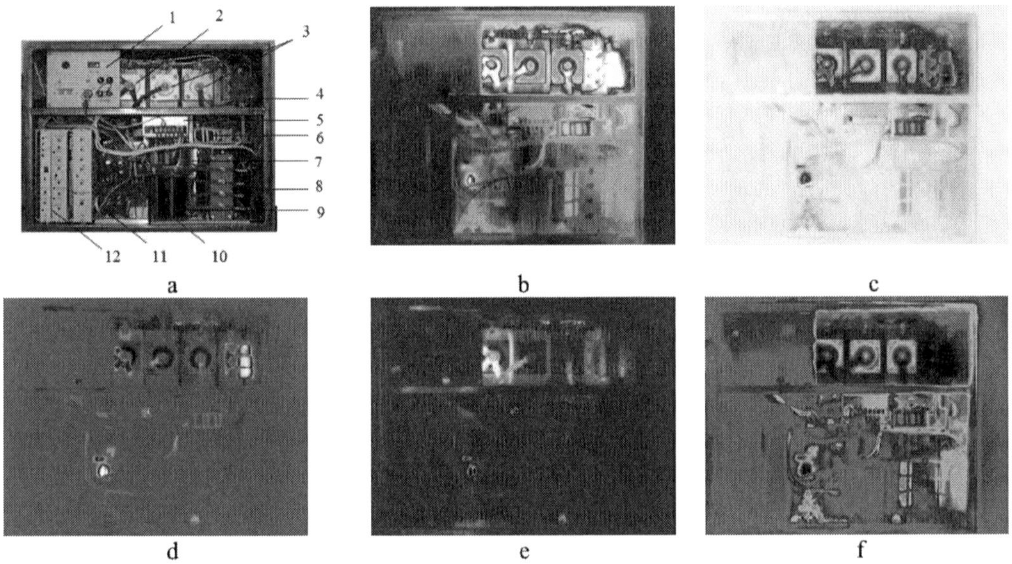

Fig. 2. Results of infrared/thermal inspection of power supply unit components after switching on a 7.5 MeV betatron:
a – 7.5 MeV betatron power supply unit (photo): 1–control panel, 2–preliminary contour diode, 3–power contour thyristor, 4–preliminary contour resistors, 5–auxiliary power supply unit, 6–damping throttle, 7–power contour capacitors, 8–filtwering circuit throttle, 9–filtering capacitors, 10–charge switch-off delay board, 11–rectifying diode module, 12–control unit; *b* – IR thermogram in 10 min after betatron switched on; *c* – after applying principle component analysis (PCA), 1st component; *d* – PCA, 2nd component; *e* – PCA, 3rd component; *f* – correlogram.

The presence of the foil distorted temperature read-outs not worse than by 1.5 % due to high transparency of polyethylene for IR radiation. The acquisition frequency was 5 Hz thus allowing the recording of 300 images in a sequence for 10 minutes during a single inspection session.

Some examples of the data processed by means of the ThermoFit software (Tomsk Polytechnic University) are shown in Fig. 2b-f. The source IR thermogram obtained in 1 minute after switching the betatron on (no electron acceleration mode) clearly shows the power supply section with the overheated thyristors (Fig. 2b).

The statistical technique of PCA applied to all 300 images in the recorded sequence allows underlining the basic thermal phenomena in the first three components (Fig. 2c-e). The first component (Fig. 2c) is very similar to the source image in Fig. 3b because it shows areas of thermal phenomena with the highest amplitudes. The second component stresses less significant areas while the third component exhibits weak temperature anomalies in the power supply unit. We remind that PCA reduces a source sequence of N images to the corresponding sequence of N components of which only 3-5 first components reflect thermal peculiarities (anomalies) of an object under test. All other component images contain only noise. Figure 2f shows the result of applying a special algorithm of data processing which is based on determining a coefficient of correlation between each current pixel and a pixel chosen as a reference. The resulting image called correlogram exhibits image sections with a similar behavior of temperature evolution curves. In a particular power supply unit, a special attention should be paid to the section with powerful thyristors (blue green area in Fig. 2f) where improving heat sink conditions is recommended to betatron designers.

It is worth noting that advantages of cutting-edge data processing become visually illustrative in pseudo color images which can be obtained by means of the ThermoFit Pro software.

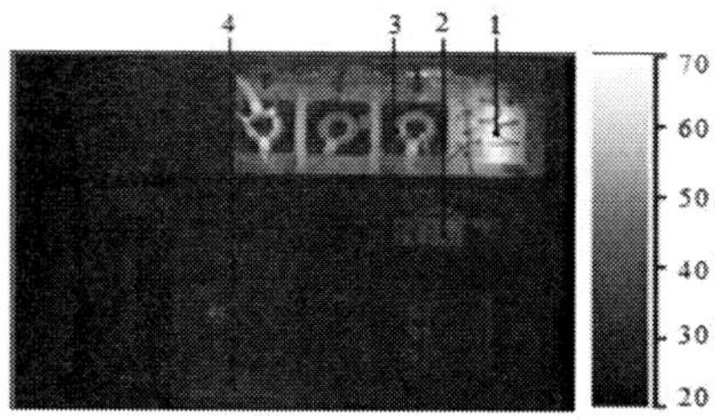

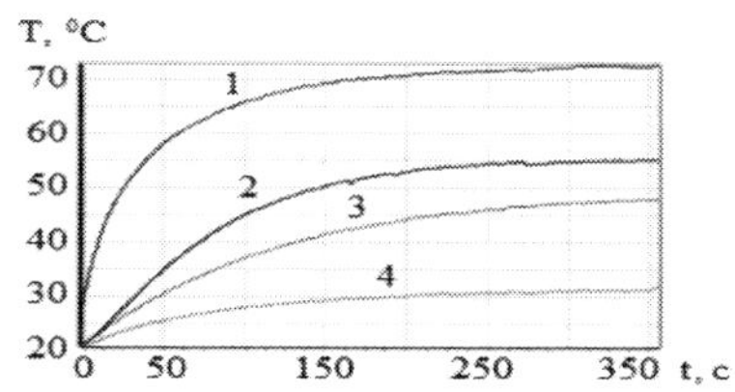

Fig. 3. Temperature evolutions in time. 1–resistors in preliminary current circuit, 2–damping throttle, 3–power contour thyristors, 4–rectifying diode module.

Figure 3 shows pixel-based temperature evolutions in some points of the IR scene. In further research, the attention will be paid to the analysis of these evolutions by applying the technique of the so-called thermal tomography [11].

4 Conclusion

A basic conclusion from the conducted research is that IR thermography is an efficient means of evaluating operation thermal conditions in betatron power supply units. This allows elaborating a well-founded approach to the designing of power supply units thus potentially improving betatron operation reliability.

References

1. M. Stein, S. Kasyanov, V. Chakhlov, J.Macleod, P. Marjoribanks, S. Hubbard, *Proc. 16th World Conference on nondestructive testing* (2004)
2. A.A. Filimonov, S.V. Kasjyanov, V.A. Kasjyanov, J. Phys. Conf. Ser. **671**, 1 (2016)
3. L. Ananyev, V. Chakhlov, M. Stein, Defectoscopy **6** (1968)
4. N. Natalinova, D. Avdeeva, V. Kazakov, V. Baranov, O. Galtseva, D. Ivashkov, MATEC Web of Conf. **79**, 01005 (2016)
5. M. Vollmer, K.-P. Möllmann, *Infrared Thermal Imaging: Fundamentals, research and applications* (Wiley-VCH, Germany, 2010)
6. X. Maldague, *Theory and practice of infrared technology for nondestructive testing* (Wiley Series in Microwave and Optical Engineering, USA, 2001)
7. A. Daniels, *Field guide to infrared systems* (Society of Photo Optical, USA, 2006)
8. H. Kaplan, *Practical Applications of Infrared Thermal Sensing and Imaging Equipment* (SPIE Press, USA, 1999)
9. X. Maldague, *Infrared methodology and technology. Nondestructive testing monographs and tracts* (Gordon and Breach Science Publishers, USA, 1994)
10. W. Minkina, S. Dudzik, *Infrared Thermography: Errors and uncertainties* (Wiley, USA, 2009)
11. V. Vavilov, *Thermal/Infrared Testing, in: Nondestructive Testing Handbook* (Publishing House Spektr, Russia, 2009)

Investigation of heat transfer at flow of a gas in the quasitriangular channel with a constant temperature wall

Ekaterina Slesareva[1,*], *Sergey* Elistratov[2], and *Mihail* Harisov[2]

[1]SB RAS, Kutateladze Institute of Thermophysics, 630090 Novosibirsk, Russia
[2]Novosibirsk State Technical University, 630092 Novosibirsk, Russia

Abstract. The experimental research of heat exchange for gas flow in the quasitriangular channel has been carried out. Experiments have been executed for a case of a constant temperature of a wall. A thermogram of the temperature field on outlet of a channel by the thermal imaging camera was registered. The visualization of the temperatures field with the use of the mesh with small cells has been performed. It was shown, that the dependence between the changes of a friction coefficient from Reynolds number for the quasitriangular channel is similar to the known dependences for the smooth pipe; the average Nusselt number have monotone correlation from unitless parameter Ψ, and this correlation lies higher of the dependence for the round channel.

1 Introduction

In the presented work the heat exchange for air flow in the quasitriangular channel with a constant temperature of the wall was investigated. The panoramic thermographic method can be used for measurement of the temperature field at the channel outlet, as complementary to traditional methods [1-3]. In this method at visualization of a temperature field in a gas stream, the wire in a fine-mesh carries out the role of the indicator of temperature (fig. 1). The mesh parameters choose so that its influence on the field of velocity and temperatures in a gas flow was minimally. The thermograms has given possibility to see the panoramic picture of heating of gas in quasitriangular channel, and a full details of the evolution of the gas heating on the channel outlet.

2 Measurement procedure

In fig. 2 the scheme of a working section is shown. Walls of the quasitriangular channel 1 are made from the three strips of a copper foil in the thickness 50 micron and length $L = 250$ mm. The channel form was set by means of inserts 4 in the form of circle segments. On an mounting disc 3 were fixed three the shaping circular segments so that internal curvature of walls of channel R_w was equaled 6.4 mm. Hydraulic diameter of such

* Corresponding author: styuardessa@yandex.ru

channel is $D_h = 1.31$ mm. The sealing of the channel cavity by the joints 8 was provided. For stabilization of a stream of air on the entrance of the channel into the input flange 5 the porous plate-membrane 6 has been installed. The constant temperature of wall T_w equal to 0°C by a thermal stabilization infill 7 (a mix of fine-grained particles of ice with water) was supported.

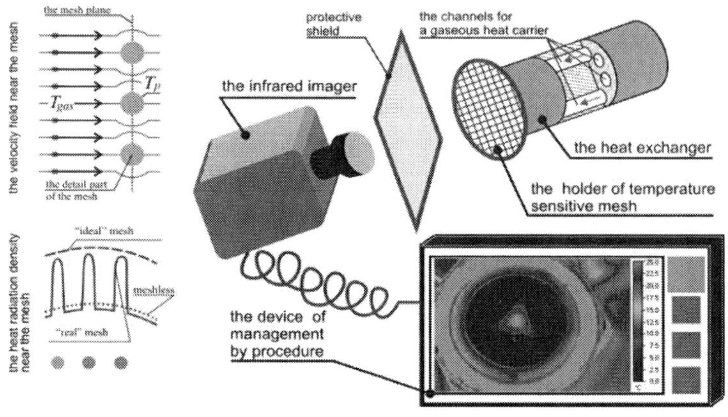

Fig. 1. Schematic picture of the panoramic thermographic method.

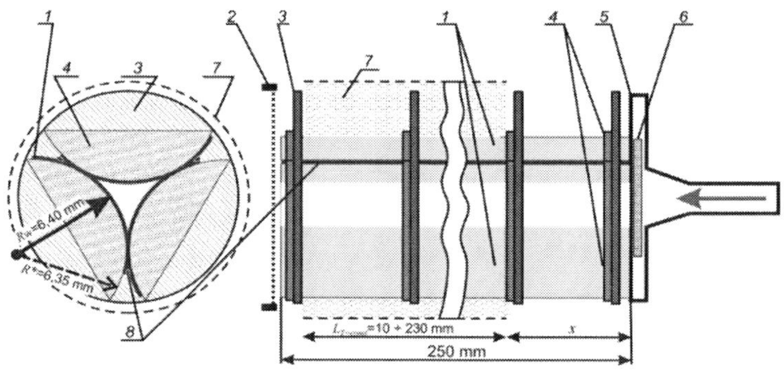

Fig. 2. Scheme picture of the test section: 1 - the wall of the quasitriangular channel, 2 - the holder of the mesh-thermal detector, 3 - mounting disc, 4 - the circular segment for a shaping, 5 - the input flange, 6 - the porous plate membrane, 7 - the infill for thermal stabilization , 8 - the sealing joint.

A during 200÷800 seconds, depending on value of flow rate and temperature air, the property of a thermal stabilization for the infill was maintained. The length of a section with constant temperature $L_{T=const}$ changed from 10 to 230 mm, respectively the area of heat exchange surface S of this section changed in range 200÷4600 мм^2. Other part of the channel the kaolinite wadding thickness 10 mm was heat-insulated. The mesh-thermode, from a brass wire 80 μm with a size of cell mesh 300×300 mm, was pressed to the outlet channel surface. Thermogram of the mesh-thermode by infrared imager NEC TH7100 (temperature accuracy 0.3 K, spatial resolution 0.1 mm, the exposure time 30 ms) have been obtained. An air in the channel by the compressor ABN R Meko720 was injected. Before an air arrived into the compressor, it's through the dehumidifier with silica gel was passed. Flow rate by the needle valve on bypass line has regulated. The value of the flow rate air in test section by the flow meter Triton-gas-SGM 4 was measured. The average velocity air was calculated by formula:

$$U = \frac{G}{\rho \cdot R_w^2 \cdot \left(\sqrt{3} - 0.5 \cdot \pi\right)}. \quad (1)$$

Static pressure on the channel entrance P_{inp} and on the channel outlet P_{out} by inspection manometers (a class of accuracy 0.15) has been measured. Measurements of a temperature of an air and a channel wall by the thermocouples accurate within 0.25 K have been measured. The average air temperature on the channel outlet by data from the thermograms has been calculated. The thermophysical properties of an air: ρ – density, ν – viscosity, C_p – thermal capacity, λ – heat conductivity, Pr – Prandtl number in depending from pressure P and temperature T were calculated based on the tabular data from the handbook [4].

3 Discussion of results

The experimental research of heat transfer at gas flow in the quasitriangular channel with a constant temperature of the wall has been carried out. The experiment has been executed for the range of a air-mass flow G from 0.06 to 0.6 g/s. Operation condition for experiments were transformed to an unitless parameters (Re, Pe, Y) by formula:

$$\mathrm{Re} = \frac{U \cdot D_h}{\nu}; \quad \mathrm{Pe} = \mathrm{Re} \cdot \mathrm{Pr}; \quad \Psi = \frac{1}{\mathrm{Pe}} \cdot \frac{L_{T=const}}{D_h} . \tag{2}$$

Used of experimental data of the static pressure on the entrance and on the outlet channel, the friction factor ζ for the case the air flow in the quasitriangular channel with a constant temperature wall has been obtained. The value of a friction factor was calculated by formula:

$$\zeta = \left(P_{inp} - P_{out}\right) \cdot \frac{D_h}{L} \frac{2}{\rho \cdot U^2} . \tag{3}$$

Figure 3 shows a change of the friction coefficient in dependence from the Reynolds number Re. As it is possible to see from graph the character of the friction coefficient change from a Reynolds number for the quasitriangular channel is similar to known dependences for a smooth pipe.

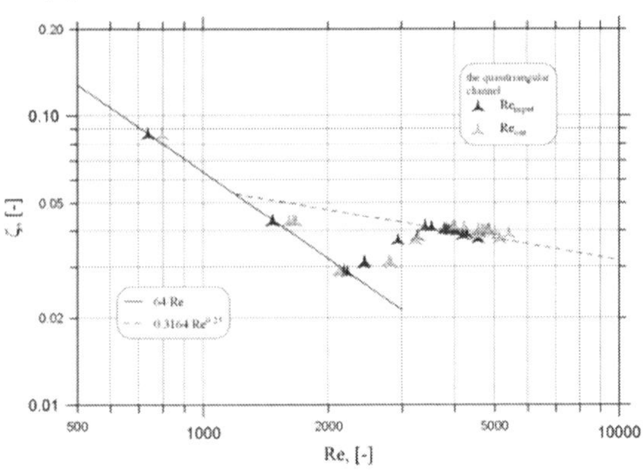

Fig. 3. The dependence between a change in friction coefficient ζ and Reynolds number Re for flow of the air in the quasitriangular channel with a constant temperature wall.

Using the data from the thermogram, the average Nusselt number $\overline{\mathrm{Nu}}$ for the area of heat exchange surface has been obtained. Calculation was performed by formula:

$$\overline{\mathrm{Nu}} = \frac{dQ}{dt} \cdot \frac{1}{S \cdot \left(T_w - \overline{T}_{air}\right)} \cdot \frac{D_h}{\lambda}; \quad \overline{T}_{air} = \frac{\left[T_{T_w=const} + T_{inp}(x)\right]}{2}; \quad \frac{dQ}{dt} = C_p \cdot G \cdot \left(T_{T_w=const} - T_{ad}\right),$$

where $T(x)$ - the temperature air on the beginning heat exchange section, x - the distance from the channel entrance to the beginning heat exchange section, T_{ad} - the temperature air on the channel outlet with adiabatic wall on all length, $T_{Tw=const}$ - the temperature air on the channel outlet with constant temperature of the wall T_w on the section in length H. Figure 4 shows the dependence between the change the average Nusselt number and dimensionless parameter Ψ for the quasitriangular channel with the constant temperature wall. You can see that Nusselt number have monotone correlation with the parameter Ψ. The dependence between Nusselt number and the parameter Ψ has been interpolated by the fitted function: $\overline{\mathrm{Nu}} = 3.51 + \frac{0.15}{(\Psi + 0.0001)^{0.85}}$. The dependence Petuhov [5] for the round channel is also shows in fig. 4. It is visible that the data for the quasitriangular channel lies higher of the dependence for the round channel.

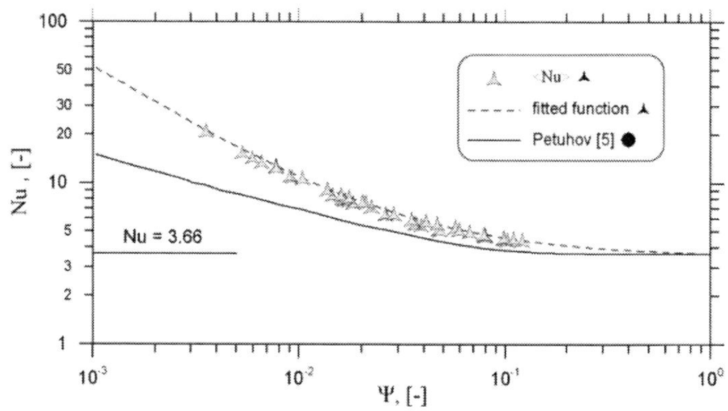

Fig. 4. The dependence between the change in average Nusselt number $\overline{\mathrm{Nu}}$ and dimensionless parameter Ψ for the quasitriangular and circular channel with constant temperature wall.

4 Conclusion

The experimental research for flow of the air in the quasitriangular channel with constant temperature wall shown, that the dependence between the change in friction coefficient ζ and Reynolds number Re was consistent with known dependences in smooth pipes. It is shown, that the average Nusselt number have monotone correlation with unitless parameter Ψ, and data for the quasitriangular channel lies higher of the dependence for the round channel. It was shown, that the panoramic thermographic method can be used for the research of the heat exchange in the channel with any form of channels for the flow section at constant temperature of the wall.

This work was supported by the Russian Foundation for Basic Research, No. 16-38-00502-mol-a.

References

1. A.V. Efimova, A.V. Zaitsev, B.P. Zhilkin, D.N. Tokarev, K.V. Zaitsev, and Kh. Dashpuntsag, Bulletin of USTU: Thermal Engineering **33**, 139 (2004)
2. S.L. Elistratov, E.Yu. Slesareva, EPJ Web of Conf. **82**, 01060 (2015)
3. V.E. Nakoryakov, S.Y. Misyura, Energy and Fuels **27**, 7089 (2013)
4. G.F. Hewitt *Heat Exchanger Design Handbook. Part 5. Physical Properties* (Hemisphere Publishing Corporation, Carlsbad, CA, 2002)

5. B.S. Petuhov *Heat exchange and resistance under a laminar flow fluid in tubes.* (Energy, Moscow, 1967)

Simulation of heat transfer in intricately-configured polymer composite structures of instrumented container type

Mikhail N. Slitkov[1], *Alexander* A. Kulkov[1], and *Anatoly* N. Garashchenko[1,*]

[1]Central Research Institute for Special Machinery, Moscow, Russia

Abstract. Method of mathematical simulation of heat transfer processes in polymer composite (PC) products with intricate configuration, being an alternative of using up-to-date commercial software complexes has been developed. On the example of PC container with instrumentation and fiberglass electric heaters located in it, a mathematical model describing unsteady temperature field (a system of nonlinear differential heat balance equations for each element) has been formulated. Features of heat transfer between elements (heaters, instrumentation, enclosing structures) were taken into account. The verification of the method was conducted by comparing of theoretical temperature distributions with results of measurements in experiments with simplified variant of the structure. The developed method is effective, in particular, for such PC products as containers, modules, bunkers and vessels. It allows us to specify optimum operation modes for heating elements, operational parameters for conditioners and funs, heat insulation characteristics for providing a given level of air temperature inside objects in winter and summer service periods.

1 Introduction

Containers, modules, bunkers and vessels that should be provided with a given heat mode, are the advanced kind of products of polymer composite materials. PC containers (assemblies) for locating instrumentation which should function under different climatic conditions are such typical products. For maintaining a given temperature mode in these assemblies, it is efficiently to use thin-layered fiberglass electric heaters (NEST) that have been widely practiced (from household application to maintaining a given temperature mode in elements of the rocket and space engineering) [1, 2].

To design such assemblies, the information about time history of temperature of all comprised elements is necessary. It is also necessary to control given heating modes. The listed PC products have, as a rule, an intricate configuration that substantially hampers the simulation of temperature fields in structures. The use of up-to-date commercial software complexes, in particular [3, 4], is impossible for a whole number of objective reasons. In such complexes, for example, not all possible variants of heat transfer between modeled

* Corresponding author: a.n.gar@mail.ru

objects and environment are provided for, or it is taken into account in the simplified form. The correction of commercial software packages is not possible.

2 Statement of the problem

It is necessary to use alternative methods of computer simulation. As such method, this paper proposes a method providing the realization of any variants of complex physical models of heat transfer processes in products. A mathematical model describing unsteady temperature field of an assembly, in particular, a PC container is based on a system of usual differential nonlinear heat balance equations for each separate assembly element. The increase of the number of such elements causes only appropriate increase of the number of equations. Let us consider the mathematical model, a problem solution method and obtaining of calculation results with its assistance on the example of a container for electronic equipment. Its schematic is presented in Fig. 1.

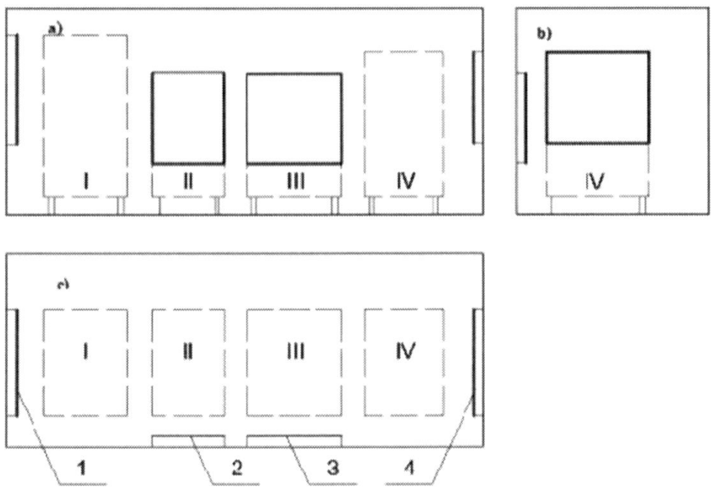

Fig. 1. Problem solution domain (inner volume of container: a) back side; b) right side; c) top view).

A box-shaped container 2000 mm in length, 700 mm in width and 1000 mm in height was adopted for the simulation. The front wall has two doors for access to the container. Four blocks of electronic equipment (I, II, III, IV) and four NESTs (1, 2, 3, 4) are located inside the container. NESTs are installed in that way that each of them radiates a heat only on a device standing opposite it and on a surface section of a wall located behind NEST. Outer and inner layers of vertical walls, a cover and a bottom are made of glass fiber reinforced plastic. The intermediate layer of walls, a bottom and covers is made of polyurethane foam (heat-insulating layer). Blocks are set on supports with a gap from floor equal to 20 mm. Distances from blocks I and IV to lateral walls are 150 mm and 100 mm respectively. The distance between blocks is 80 mm, and between blocks and longitudinal walls is 170 mm (see Fig. 1).

It is adopted that the heat in the container in winter period of time is evolved by all four NESTs as well as two instrumentation blocks (I and II). The mode of maintaining the air temperature over the range of +10 °C...+25 °C inside the container when operating in winter and summer periods at the ambient temperature from -50 °C to +40 °C was given. In summer period, solar radiation fluxes were taken into account, and an air-conditioner and a fun were used to maintain the given level of the air temperature inside the container. In winter period, heaters were switched out when reaching the upper boundary of the indicated range and switched on when reaching the bottom boundary.

To determine the temperature field of an assembly "container-blocks-NEST", including the mutual heat interaction, it was broken into 22 elements. For each element, a differential heat balance equation was formed. Since heat bonds between elements include temperatures to be determined, the equations were nonlinear and a system of equations was solved by one of numerical integration methods.

The admissibility of such approach has been validated by comparing data of model experiment with calculation results. A model container with relatively simple structure (as compared with an assembly in Fig. 1) was used. Temperature of basic elements of the container was measured. The experiments were conducted in two stages (the first – without thermal control, the second – with thermal control). Dimensions of the model container are 460 mm x 420 mm, height is 390 mm. Dimensions of the instrumentation block are 140 mm x 220 mm, height and weight are 100 mm and 4.5 kg respectively. The walls, cover and bottom are multilayered and made of packing board (inner layer), acrylic plastic (intermediate layer) and molding material (outer layer). The thickness of walls is 6 mm, of a cover is 12 mm, of a bottom is 42.5 mm. The gap between NEST and a wall is 20 mm, between a block and a wall is 35 mm, between a block and floor is 10 mm. Positioning elements in the container are shown in Fig. 2. Thermocouples inside the container were installed in the following way: three on NEST; two on a block (one on the surface, other inside); two in air at different heights; one behind NEST; one by one on bottom and vertical walls. Results of the experiments are presented in Fig. 3.

Fig. 2. General view of model container.

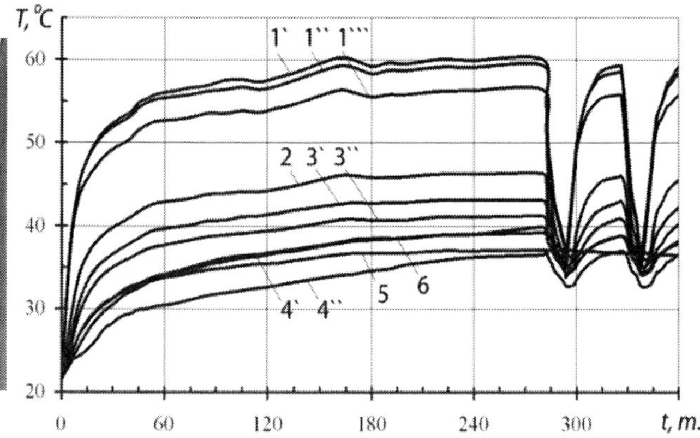

Fig. 3. Changing of element temperature in experiments with thermal control: 1', 1'', 1''' - NEST; 2 - inner wall surface in NEST zone; 3', 3'' - air; 4' - surface of block and its inner part (4''); 5 - surface of bottom and wall (6) outside NEST.

For the model container under consideration, a system of 8 differential heat balance equations was formed: 1 – NEST, 2 – wall in NEST zone, 3 – air, 4 – instrumentation block, 5 – floor in block zone, 6 – floor outside block zone, 7 – vertical walls outside NEST zone, 8 – cover). The equation stated for NEST is presented as an example:

$$C_1 \frac{dT_1}{dt} = P_1 - K_{1-2}(T_1 - T_2) - K_{1-3}(T_1 - T_3) - K_{1-4}(T_1 - T_4) - K_{1-4}(T_1 - T_4) - K_{1-4}(T_1 - T_4),$$

where P_1 is NEST power;

K_{1-2} is parameter of heat transfer by effective heat conduction from heater to wall, taking into account re-emission of surfaces and natural convection in plane slit [3];

K_{1-3} is parameter of heat transfer by natural convection from heater to air;

$K_{1\text{-}4}$, $K_{1\text{-}5}$, $K_{1\text{-}8}$ are parameters of heat transfer by radiation from heater to elements whose angle coefficients were determined with due account of positioning mutually irradiating surfaces [3].

In this statement of a problem, the temperature of NEST was supposed equal for the whole its surface. In reality, the distinction in temperatures of its upper and bottom parts can reach tens degrees at the expense of non-uniform heat removal because of natural convection (see Fig. 3). For calculating the temperature distribution in vertical direction, NEST was broken into n sections and the heat balance equation was formed for each of them. This problem should be solved separately when using already calculated temperature-time characteristics of all elements and including a dependence of natural-convective heat transfer coefficient on vertical coordinate of a heater.

3. Results and discussion

The system of differential nonlinear heat balance equations was solved by the 4th-order Runge-Kutt method. When calculating it was given that NEST power to be evolved under thermal control is 40 W (without thermal control is 30 W); specific heat capacity of instrumentation block is 850 J/kg·K; thermal control limits are +35 °C...+43 °C. The results of calculating temperatures of model container elements are presented in Fig. 4. Their comparison with the results of experiments (see, in particular, Fig. 3) showed a good correspondence that witnesses the validity of the stated calculation method.

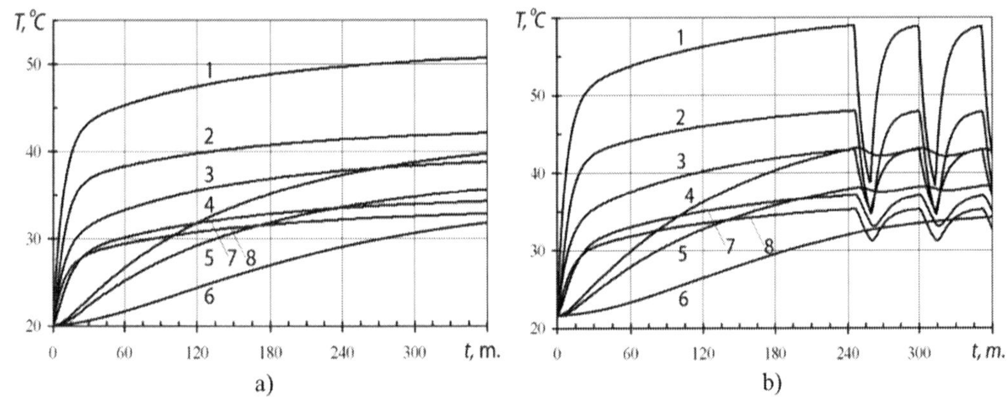

Fig. 4. Changing of element temperature in the absence (a) and in the presence (b) of thermal control: 1 – NEST; 2 – inner wall surface in NEST zone; 3 – air; 4 – block; 5 – wall in block zone; 6 – bottom outside block; 7 – walls outside NEST; 8 – cover.

After the verification of the model, the simulation was performed for more complex variant of the container (see the schematic in Fig. 1), for which a system of 22 differential heat balance equations was formed. Initial data in the form of characteristics of NESTs, instrumentation blocks and materials to be applied when making the container were used (symbols of electric heaters and blocks are indicated in Fig. 1 and in the text). For NESTs, the following heat release power was adopted: 230 W (NEST 1), 135 W (NEST 2, 3), 180 W (NEST 4). For instrumentation blocks, weight as well as heat release power were taken into account (respectively): 52 kg, 90 W (block I); 15 kg, 115 W (block II); 30 kg, 0 W (block III); 25 kg, 0 W (block IV). The simulation of temperature fields of all elements of the container at ambient temperatures corresponding to winter and summer period was performed.

For winter period conditions, six variants of ambient temperature (from -50 °C to 0 °C) were considered. In this period, the temperature inside the container was maintained over the range of +10 °C...+25 °C that were provided with the NEST operation. Fig. 5 shows, as an example, temperature values for several elements.

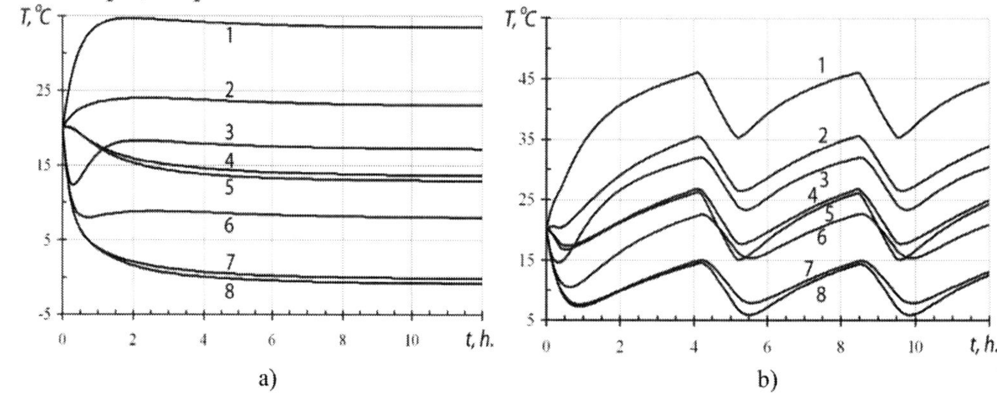

Fig. 5. Changing of block temperatures (curves 1, 2, 4, 5) and wall surfaces (3, 6, 7, 8) in block zone (see Fig. 1): II (1, 3); I (2, 6); III (4, 7); IV (5, 8) at T_c = -50 °C (a) and -30°C (b).

It is ascertained that at ambient temperatures of -50 °C and -40 °C NESTs should operate permanently to maintain the scheduled temperature mode inside the container and at ambient temperature of -30 °C the heaters can be cyclically switched out.

While analyzing the container work in summer period, the contribution of solar radiation to heating of vertical walls and a cover were taken into account. It was assumed that the air temperature inside the container is given over the range of +25 °C...+35 °C. The use of an air-conditioner (power of 400 W) and a fun (volume air consumption of 0.07 m^3/s) at night was provided for. It was proposed that turn-on and turn-out of the air-conditioner occurs under thermal control, and the fun operates only during the hours of darkness. The following variants of operating the container were modeled: without use of air-conditioner and fun; without use of air-conditioner but with operating fun; with use of air-conditioner but with switched-off fun (an example in Fig. 6).

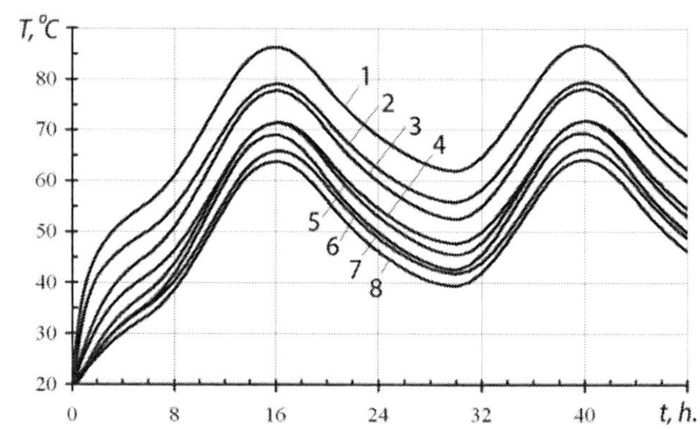

Fig. 6. Changing of temperatures of blocks and wall surfaces in block installation zone with switched-off air-conditioner and switched-on fun (symbols of curves correspond to Fig. 5).

The obtained results, for example, show that the fun power at night is not sufficient for the stable operation of instrumentation blocks. It is shown that in this case it is expedient to use the air-conditioner that operates continuously. The simulation using developed software programs allows large volume of information important for practice to be obtained. It is possible, in particular, to perform the optimization of NEST number and power that provide a scheduled temperature level depending on the number and arrangement of

DOI: 10.1051/matecconf/20179201019

instrumentation blocks. One may also choice rational materials and thicknesses of heat insulation of enclosing structures of containers and other PC products.

4 Conclusion

Thus, the developed method is an efficient instrument of simulating heat transfer processes in various PC products and can be considered as an alternative for commercial software complexes.

References

1. V.P. Gorelov, *Low-temperature composite heaters in industry and private life* (2016)
2. S.V. Baranov, O.N. Budadin, M.N. Slitkov, Russ. J. Nondestruct. **44**, 2 (2008)
3. E.M. Sparrow, R.D. Cess. *Radiation Heat Transfer* (*Hemispere Publishing Corporation, USA,* 1978)

Coal char oxidation kinetics in air medium

Alexander G. Korotkikh[1], *Konstantin* V. Slyusarskiy[1,*], and *Ivan* V. Sorokin[1]

[1]National ResearchTomsk polytechnic university, 634050 Tomsk, Russia

Abstract. Study of oxidation kinetics for three types of coal char with different carbon content in air is presented. The coal char powders of anthracite, bituminous T-grade coal and 2B-grade lignite with particle size less than 80 μm were tested. The coal char oxidation was researched by isothermal method via simultaneous TG-DSC analyzer Netzsch STA 449 Jupiter F3 in the temperature range of 1000–1200 °C. Measurements were carried out at ambient pressure. Volumetric flow rate of oxidizing medium into analyser chamber was 250 ml/min. Flow consisted of air and argon with volumetric ratio 24/1. Carbon average rate of oxidation reaction at each temperature were defined based on experimental results. Kinetic constants (the frequency factor and activation energy) were defined for Arrhenius equation modified with three submodels: volumetric model, shrinking core model and random pore model. The activation energy values for anthracite are 1,6-1,7 times higher than for chars of bituminous coal and lignite.

1 Introduction

Recently, the increased attention is paid to solid fuel conversion, in particular, for bituminous coals and lignite [1]. This data is used to improve combustion characteristics on thermal power plants and to produce syngas and hydrogen for chemical and metallurgical industry via gasification [1, 2]. Solid fuel gasification on thermal power plant would allow to resolve problem of environment pollution and utilization of flue gases which are formed in traditional combustion equipment. Thermochemical process of solid fuel gasification is carried out to obtain combustible gases (H_2, CO, C_nH_m) and realized during interaction of fixed carbon with different oxidizers. The air, oxygen, steam, carbon dioxide or their mixtures are used as oxidizing medium [3]. Depending on feed content, process temperature, fuel residence time and other factors the different syngas with needed characteristics could be obtained.

Solid fuel gasification is realized in gasifyers with gaseous oxidizer feed there the main heterogeneous reactions takes place, mainly, carbon oxidizing by steam, carbon dioxide and oxygen resulted in hydrogen, carbon monoxide and dioxide formation [4]. Modern gasification technologies apply two-staged approach [5] there processes of oxidizing and reduction are takes place in different chambers of gasifyer. This defines increased interest for carbon oxidation reaction kinetics:

$$C + O_2 = CO_2.$$

[*] Corresponding author: konstantinsv@tpu.ru

Experimental research on solid fuel gasification process as well as heterogeneous carbon reactions with air is further important as long as this reaction is closely connected to heterogeneous reactions of carbon with carbon dioxide and steam:

$$C + CO_2 = 2CO,$$

$$C + H_2O = CO + H_2,$$

$$C + 2H_2O = CO_2 + 2H_2.$$

The thermal analysis results of three coal chars of Kuznetsk and Kansk-Achinsk deposits in air medium in temperature range 1000-1200 °C. Based on experimental results the mass loss rate values were defined as well as gasification times. The kinetic constants – activation energy and frequency factor – for process were defined using Arrhenius equation and three kinetic submodels: volumetric model, shrinking-core model and random pore model. The activation energy dependence on fixed-carbon content in fuel is presented as well.

2 Experimental section

2.1 Solid Fuel Characteristics

The powders with particle size less than 80 μm of three different coal chars were used in current research. Fixed-carbon content in initial samples was defined by total mass loss rate during heating of the sample in oxidizing medium. Density of studied solid fuel samples was measured by picnometric method. Density and fixed carbon content of studied samples presented in table 1.

Table 1. Density and fixed carbon content of solid fuel samples.

Solid fuel sample	Specific density, kg/m^3	Poured density, kg/m^3	Fixed carbon content, wt. %
Anthracite	1753±56	339±2	89
Bituminous coal	1927±15	340±6	74
Lignite	1425±10	347±13	66

2.2 Thermal Analysis

The studied solid fuel samples were undergo pyrolysis by heating in SNOL 30/1100 in nitrogen atmosphere with 50 °C/min rate to 1000 °C to obtain coal char, remove humid and volatile matter from initial fuel. Solid fuel with 20 g mass were put into ceramic crucible and held in furnace for one hour and then cooled to 400 °C. Thermal analysis of obtained coal char samples with 20 mg mass was carried out via simultaneous TG-DSC analyzer Netzsch STA 449 F3 Jupiter. Studied coal char were heated in analyzer chamber in argon medium with 50 °C/min heating rate to temperatures 1000, 1100 and 1200 °C. Argon volumetric flow rate was 10 ml/min. After reaching the preset temperature the inert medium in furnace was changed on argon-air mixture with volumetric ratio 24/1. The oxidizing medium flow rate was 250 ml/min. to eliminate diffusional effects on process. The isothermal soaking time for coal char samples for studying carbon oxidation varied from 50 to 80 min.

3 Results and discussion

TG-data on studied samples in air medium are presented in form of thermographs in fig.1.

During coal char heating to preset temperature in argon the sample mass loss varied from 4 to 7 wt. % depending on initial mass and fuel grade. After hot air inlet into heated furnace the exothermal carbon oxidation reactions were registered and coupled by sample mass loss for 73 wt. % for anthracite and bituminous coal char and 85 wt. % for lignite char.

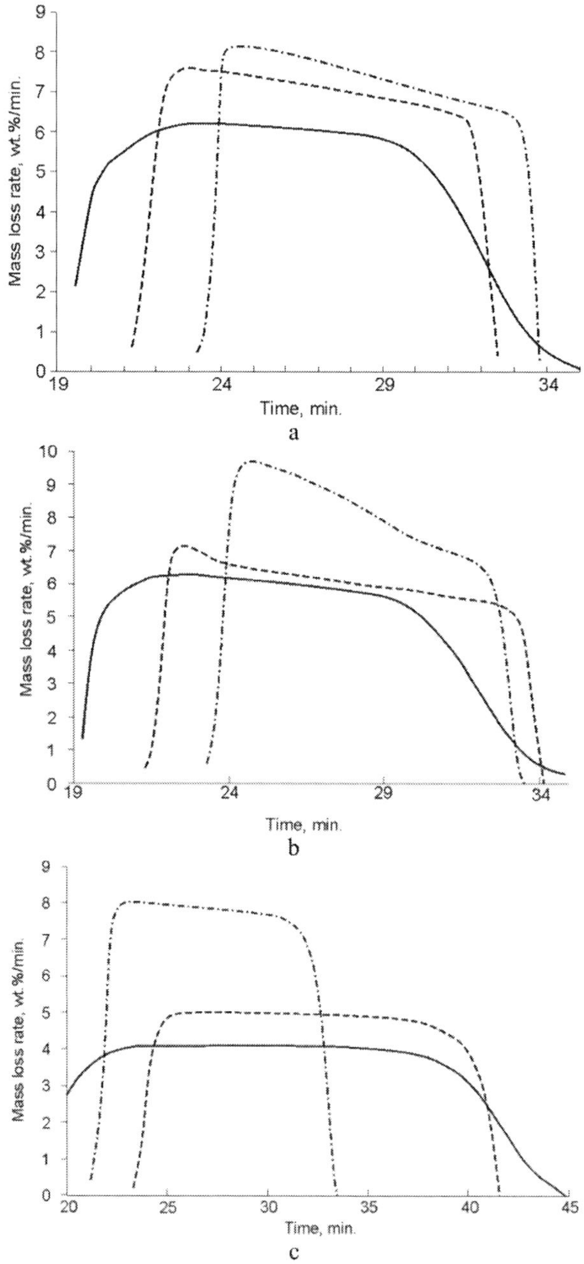

Figure 1. TG-curves for chars of anthracite (a), bituminous coal (b) and lignite (c).

The increasing of air temperature in furnace the sample combustion time values decreases while mass loss rate and carbon oxidation reaction rate increase. According to Arrhenius equation [6] the fuel mass loss rate during heterogeneous reactions are described by following equation:

$$v=A_0 \exp(-E_a/RT) f(X), \quad (1)$$

here v is the mass loss rate, kg/s; A_0 – frequency factor, kg/s; E_a – activation energy, J/mole; R – gas constant, J/(kg K); T – temperature, K; $f(X)$ – kinetic function.

Three kinetic models were used in current work to define activation energy and frequency factor. It's volumetric model, shrinking-core model and random pore model which are represented by equations 2, 3 and 4, respectively:

$$f(X)=(1-X), \quad (2)$$

$$f(X)=(1-X)^{2/3}, \quad (3)$$

$$f(X)=(1-X)(1-\psi \ln(1-X))^{0.5}. \quad (4)$$

Based on the equation 1 with application of equations 2–4 in turn the activation energy and frequency factor values were defined for carbon oxidation reaction of coal char in air medium in temperature range 1000–1200 °C. Reaction time was defined as difference between time moment with maximal mass loss rate and minimal time with mass loss rate less or equal than 0,01 from maximal value. The mass loss rates values are given in table 2 based on experimental data.

Table 2. Average mass loss rate.

Coal char sample	Mass loss rate, 1/s, at temperature		
	1000 °C	1100 °C	1200 °C
Anthracite	1.393	1.900	1.940
Bituminous coal	1.366	1.711	1.975
Lignite	0.946	1.538	1.958

Comparison of mass loss rates at different temperatures shows that they are close to each other at 1200 °C while at 1000 and 1100 °C they are significantly different. This may be the caused by influence of diffusion on process kinetics at high temperatures.

Obtained activation energy and frequency factor values (table 3) for studied solid fuel samples and are in good correspondence with literature data for high and low-grade coals [1, 7]. Activation energy dependence on fixed carbon content in initial fuel for all studied models are given in fig. 2.

Table 3. Kinetic constants for coal char combustion in air medium.

Model / Sample	Activation energy, kJ/mole			Frequency factor, 1/s		
	VM	SCM	RPM	VM	SCM	RPM
Anthracite	12.93	40.43	53.49	4.01E+06	2.72E+17	3.14E+20
Bituminous	21.63	36.68	50.06	1.16E+11	7.68E+15	1.46E+19
Lignite	23.76	28.60	26.38	1.15E+12	5.32E+11	2.27E+10

Kinetic constant value analysis shows that activation energy and frequency factor values decreases with fixed carbon content in fuel for volumetric and shrinking core models. For random pore model the kinetic constant values are close to each other for all samples. Activation energy values varies in the range of 23-28 kJ/mole. This indicates that mechanism of reaction inside pores better describes processes in coal char. Bond energy between atoms in carbonaceous matrix for coals with different level of metamorphism. In anthracite and bituminous coal the inertinite phase prevails while for lignite the vitrinite with low bond energy prevails. As long as activation energy is integral parameter it may be lower for anthracite because nearly all bonds were saved because of volatile matter low content while for bituminous coal the weakest bonds are removed together with devolatiliztion products.

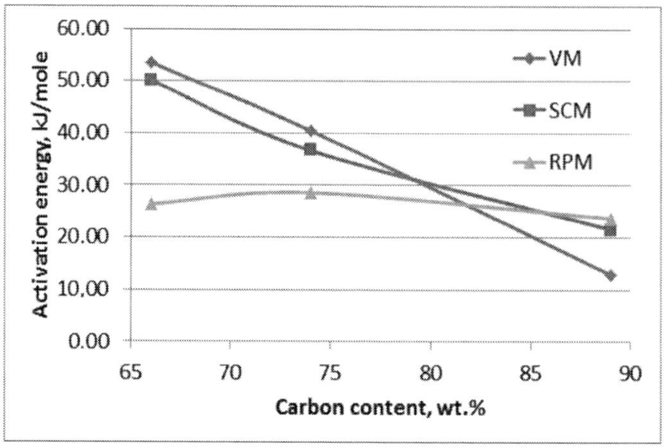

Figure 2. Activation energy dependence on fixed carbon content in initial fuel.

4 Conclusion

The oxidation process of anthracite, bituminous coal and lignite of Kuznetsk and Kansk-Achinsk deposits was studied via STA DSC Netzsch Jupiter 449 F3. The measurements were carried out in air at temperatures of 1000, 1100 and 1200 °C. The oxidation rate values tend to increase with increasing of temperature. Based on experimental results the activation energy and frequency factor were defined using 1st order Arrhenius equation with three submodels. The highest preciseness was obtained for random pore submodel. The activation energy tend to decrease with fixed carbon content in initial coal char sample.

References

1. G.L. Manjunath and R.R. Nair, Int. J. Coal Geol. **141,** 13 (2015)
2. D.G. Roberts, E.M. Hodge and D.J. Harris, Energy Fuels **24,** 5300 (2010)
3. A.G. Korotkikh, K.V. Slyusarskiy and A.A. Ditts, Russ. J. Chem. Phys. **10,** 576 (2016)
4. K. Larionov, I. Korjashov, S. Jankovskij, V. Gubin, A. Zenkov, MATEC Web of Conf. **37,** 01030 (2015)
5. N.M. Laurendeau, Prog. Energ. Combust. **4,** 221 (1978)
6. M.F Ifran, M.H. Chakrabarti and K. Kusakabe, Can. J. Chem. Eng. **91,** 1936 (2013)
7. T. Maffei, R. Khatami, S. Pierucci, T. Faravelli, E. Ranzi and Y.A. Levendis, Combust. Flame **160,** 2559 (2013)

DOI: 10.1051/matecconf/20179201021

Technique of the research of characteristics of ignition of the condensed systems to variables heat flux

Nikolay N. Zolotorev[1,*], *Vladimir* A. Arkhipov[1], and *Eugene* A. Maslov[2]

[1] National Research Tomsk State University, Research Institute of Applied Mathematics and Mechanics, 634050 Tomsk, Russia
[2] National Research Tomsk Polytechnic University, 634050 Tomsk, Russia

Abstract. This paper presents a method for determining the ignition delay time of condensed systems when exposed to radiant heat flux on the sample surface. The heat flow through the focusing lens with the ability to move along the axis was applied to the sample. Changing the geometric characteristics of the optical system can be used for measurements of the ignition of samples condensed systems in a wide range of values.

1 Introduction

One of the main characteristics in the study of the ignition of condensed systems (CS) is ignition delay time. The ignition delay time depends on the level of heat flux, chemical composition of components, the size of samples, the beam diameter of the emitter, etc.

The ignition delay time play an important role in assessment the explosion of high-energy materials, and also by development of systems of their initiation (igniters, detonators, etc.).

In work [1] the method of measurement of ignition delay time of CS is considered by the given constant radiant heat flux of xenon lamp at which the ignition delay determined by the time interval between the opening of the optical shutter and the moment of emergence of flame on sample surface registered by the ionization sensor and the photo diode. A known method of measuring the ignition delay time of the CS by the radiation of a CO_2-laser focused on the sample surface by a lens made of sodium chloride [2]. These techniques allow to take measurements of ignition delay time only at influence of heat flux with the given constant intensity.

As the results of experimental and theoretical studies [3, 4], the ignition characteristics CS essentially depend on the dynamics regime of changes in the heat flux density in the process of initiating. Regularities of ignition of the CS of a heat flux with decreasing or increasing intensity are of interest not only in respect of further development of the thermal theory of ignition, but also are important in practical applications. In actual practice heat exchange at ignition is non-stationary that is carried out in the dynamic regime.

* Corresponding author: nikzolotorev@mail.ru

In the work [5] presented the technique of research of process of ignition of CS is considered under dynamic conditions of supply of radiant energy. As radiation source in this technique the tubular gas-discharge xenon lamp is used. When opening shutter density of radiant heat flux is changed due to change of current intensity in lamp feed circuit with the special thyristor block.

2 Method of measurement

In the work the technique of measurement of ignition delay time of CS which consists in irradiation of surface of sample radiant heat flux through the collecting lens. The collecting lens is moved at a predetermined speed relative to the sample in the measurement process [6]. Before sample collecting lens is installed with the possibility of longitudinal displacement. On the irradiated surface of sample placed a diaphragm with a hole whose center is located on the optical axis of lens (Fig. 1).

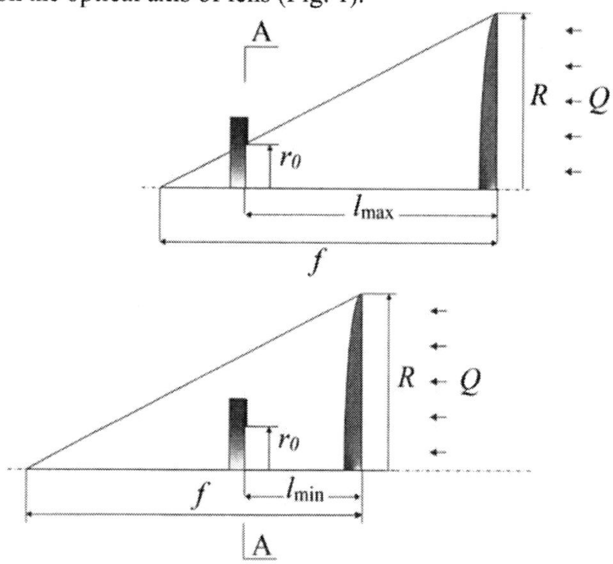

Fig. 1. Geometrical parameters of optical system.

At the time of irradiation heat flow at the opening of the shutter, the lens is moved relative to the sample. Radius of opening of diaphragm is determined by formula

$$r_0 = \frac{R}{f}(f - l_{\text{Max}}), \tag{1}$$

Variable heat flow density acting on the sample over time t=0÷t_k, calculated by the formula

$$q(t) = \frac{Q}{\pi}\left\{\frac{f}{R\left[f - (l_{\text{Min}} + ut)\right]}\right\}^2 \tag{2}$$

when moving lens from sample, or on formula

$$q(t) = \frac{Q}{\pi}\left\{\frac{f}{R\left[f - (l_{\text{Max}} - ut)\right]}\right\}^2 \tag{3}$$

when moving lens towards sample.

Values l_{max}, l_{min}, f choose according to inequality

$$l_{Min} < l_{Max} < f. \tag{4}$$

here R, f – the radius and focal distance of lens; l_{max}, l_{min} – the maximum and minimum distances between lens and surface of sample; Q – the radiant heat flux arriving on lens; u, t_k – the speed and end time of movement of lens.

Figure 2 shows the results of measurement of density of radiant heat flux depending on distance from lens l to sample surface CS. Three duplicate experiments were carried out to improve the accuracy of experimental data. The design values of $q(l)$ are given on fig. 2. Calculated values is in good agreement with the experimental data (discrepancy of results doesn't exceed 5÷7 %)

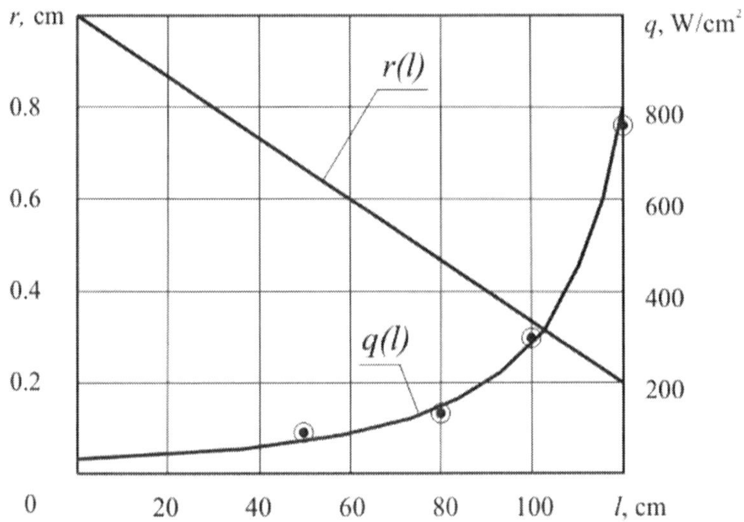

Fig. 2. Dependence of heat flux density on distance.

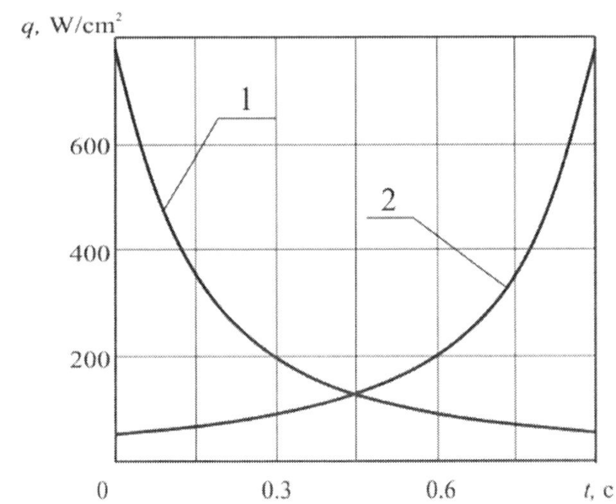

Fig. 3. Dependences of heat flux density on time.

Dependences of heat flux density on time for these conditions, are given in Fig. 3 for two options: reduction of distance from a lens to sample (1) and increase of distance from a lens to sample (2).

3 Conclusions

Thus the offered technique of research of characteristics of ignition of CS allows to expand conditions of radiation of sample:

- the technique allows to define of ignition delay time of samples of CS both at decreasing and at increasing density of radiant heat flux on sample surface.
- changing the speed of movement of the lens along the axis. This allows to vary the dynamic characteristics of the heat flow.
- at variation of geometrical characteristics of optical system (radius and focal distance of lens, diaphragm opening radius) it is possible to take measurements of characteristics of ignition of samples of CS with the broad range of values of the defining parameters.

This research carried out in 2015 was supported by "The Tomsk State University Academic D.I. Mendeleev Fund Program" grant № 8.2.46.2015.

References

1. V.A. Arkhipov, A.G. Korotkikh, V.T. Kuznetsov, E.S. Sinogina, J. Chem phys. **58**, 92 (2007)
2. L.T. De Luca, L. Galfetti, F. Severini, P. Taiariol, V.A. Babuk, B.N. Kondrikov, A.B. Vorozhtsov, G. Klyakin, Progress Combust. Deton. **151**, 384 (2004)
3. A.G. Merzhanov, V.V. Barzykin, *Some problems of propellant ignition* (Preprint of the Inst. of Chem. Phys., Moscow, 1970)
4. L.K. Gusachenko, V.E. Zarko, A.D. Richkov, J. Combust. Explos. Shock **48**, 73 (2012)
5. R.Sh. Enaleev, V.A. Mateosov, K.I. Sinaev, B.D. Dinovetskiy, R.Sh. Gainutdinov, Cheboks., Chuvash. State University Publ. **80** (1973)
6. V.A. Arkhipov, V.E. Zarko, A.G. Korotkikh, V.T. Kuznetsov, A.A. Razdobreev, Patent Rus. № 2569641 (2015)

Primary oil processing optimization

Michael Livshits[1,*], and *Alexander* Sizikov[1]

[1]Samara State Technical University, 443100 Samara, Russia

Abstract. The paper discusses oil distillation into target products with regard to the unclear separation effect. As the input data, the table of physical and chemical properties of narrow fractions is used. The problem is solved using the dynamic programming method. The methodology is realized in the authors' software product.

Primary oil processing takes place mostly in CDU (crude distillation units). There oil is separated in a number of target products: gasoline, kerosene, diesel, vacuum tower bottoms. Oil is characterized by a true boiling temperature (TBT) curve reflecting the relation between the temperature and distillation cuts' fractions. In practice, TBT is a compilation of narrow fractions with end points $[t_i, t_i + \Delta t_i]$, $i = 1, 2, ... L$, where L is the number of narrow fractions with $\Delta t = 5 \div 10$ °C.

The target product quality parameters depend on the target product temperature limits, i.e. the number of narrow fractions based on which separation is carried out [1, 2]. The problem is reduced to establishing an option of separation at which the requirements to the target products' quality are met and the maximum of some criterion is achieved. It is also necessary to take into account that the current industrial units do not provide for an absolute clear separation of oil into fractions due to which temperature overlapping occurs (Fig.1).

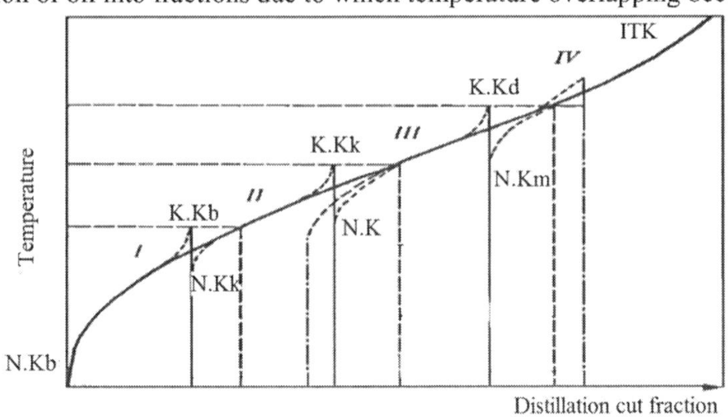

Fig. 1. Combined diagram of TBT curves of oil and product fractions.

Let's define the mathematical programming problem:

---

[*] Corresponding author: entcom@samgtu.ru

$$\max_x \left\{ \sum_{n=0}^{N} f_n(x_n, x_{n+1}) \middle| 0 = x_0 \le x_1 \le \ldots \le x_N = L \right\}. \tag{1}$$

Here N is the number of target products; x_n is the initial fraction of the n-th target product; $f_n(x_n, x_{n+1})$ is vector criterion convolution characterizing the target product. The vector criterion convolution

$$f_n(x_n, x_{n+1}) = \pi_n \hat{g}_n(x_n, x_{n+1}) - \beta_n \psi_n(x_n, x_{n+1}) \tag{2}$$

consists of two components. Component $\pi_n \hat{g}_n(x_n, x_{n+1})$ reflects the value of a respective target product; $\beta_n \psi_n(x_n, x_{n+1})$ reflects the level of its compliance with the specification.

Here π_n is the price or evaluation of the n-th product; β_n is the convolution coefficient, with the help of which the correlation between criterion components is set; $\hat{g}_n(x_n, x_{n+1}) = \sum_{i=x_n}^{x_{n+1}} g_i$ is the relative mass of the product in the oil mass; g_i is the mass fraction of the i-th narrow fraction in oil.

To assure product quality parameters, a penalty component is introduced into (2)

$$\psi_n(x_n, x_{n+1}) = \sum_{k=1}^{K} \xi_{nk} \delta_{nk}^2(x_n, x_{n+1}). \tag{3}$$

As a weighted sum of squares of relative deviations:

$$\delta_{nk}(x_n, x_{n+1}) = \begin{cases} (p_{nk} - p_{nk}^+)/p_{nk}^+, & p_{nk} > p_{nk}^+, \\ (p_{nk} - p_{nk}^-)/p_{nk}^-, & p_{nk} < p_{nk}^-, \\ 0, & p_{nk} \in [p_{nk}^-, \ p_{nk}^+], \end{cases} \tag{4}$$

Here: $p_{nk} = p_{nk}(x_n, x_{n+1}) = I_k^{-1}(p_{nk}^I(x_n, x_{n+1}))$ is the computed value of the parameter; K is the number of quality parameters; ξ_{nk} is the expert appraisal of the parameter priority; p_{nk}^-, p_{nk}^+ are the lower and higher tolerable limits of the parameter; $p_{nk}^I(x_n, x_{n+1})$ is the parameter index additively determined through indices of an appropriate parameter of the narrow fractions integrated by the product:

$$p_{nk}^I(x_n, x_{n+1}) = \sum_{i=x_n}^{x_{n+1}} \rho_{ik}^I g_i \hat{g}_n^{-1}, \tag{5}$$

where $\rho_{ik}^I = I_k(\rho_{ik})$ is the index of parameter ρ_{ik} of a narrow fraction; I, I^{-1} are the direct and inverse functions of index transformations [2]

To account for the overlapping effect due to which fractions from neighboring cuts get into the product, it is suggested to proceed from the assumption that each product is formed from the whole range of narrow fractions taken with some weight coefficients μ_{ij} depending on the distance between fractions, $|i-j|$:

$$\hat{g}_n = \sum_{i=x_n}^{x_{n+1}} \left(\sum_{j=1}^{L} \mu_{ij} g_j \right), \tag{6}$$

$$p_{nk}^{I} = \sum_{i=x_n}^{x_{n+1}} \left(\sum_{j=1}^{L} \mu_{ij} \rho_{jk}^{I} g_j \right) \cdot \hat{g}_n^{-1}, \quad k = 1, 2, \dots K. \tag{7}$$

In this instance, as a working hypothesis, it is naturally to assume that the weight coefficient decreases exponentially from $|i - j|$. Then, using normalization so that the sum of progression terms would be equal to one, we will get:

$$\mu_{ij} = \begin{cases} 0, & |i-j| > m, \\ \dfrac{1-q}{2\left(1-q^m\right) q^{|i-j|}}, & |i-j| \le m, \end{cases} \tag{8}$$

Here q is the geometric ratio; m is the number of progression terms equal to d for all fractions except for extremes (the beginning and the end), for which

$$m = \begin{cases} i, & i < d,\ j < i, \\ L - i, & i > L - d,\ j > i. \end{cases} \tag{9}$$

The values of parameters q and d are determined based on condition

$$\min_{q,d} \sum_{n=1}^{N} \sum_{k=1}^{K} \left(\frac{p_{nk} - \hat{p}_{nk}}{P_{nk}} \right)^2. \tag{10}$$

Problem (1) is solved by the dynamic programming method [3]. The process state before the n-th step means the initial fraction of the current target product, $S_{n-1} = x_{n-1}$; control is the number of narrow fractions included in the current product: $u_n = x_n - x_{n-1}$. The operator of transition into a new state is $S_n = x_{n-1} + u_n = x_n$. The step result functions should be understood as $f_n(S_{n-1}, u_n) = f_n(x_{n-1}, x_n)$. In this case, the recurrent Bellman equation will look as follows:

$$F_{n-1}(x_{n-1}) = \max\left\{ f_n(x_{n-1}, x_n) + F_n(x_n) \right\}, \quad n = 1, 2, \dots N, \tag{11}$$

where F_n is state function, $F_n(x_n) = 0$.

The method of solving the postulated problem is as follows.

1. Within the range of tolerable variations of temperature limits, a net oriented graph so that the set of its paths would correspond to the set of tolerable solutions of the problem. Nodes are grouped by levels corresponding to process steps and form sets X_0, X_1, ..., X_N, wherein sets X_0 and X_N contain one node. The distances between neighboring nodes are calculated by formula (2).

2. The computation procedure is executed from the end to the beginning: $n = N-1, N-2, \dots, 0$. At each step, for all $x_n \in X_n$ problem (11) is solved, control $\hat{u}_n(x_n)$ and potential $F_n(x_n)$ are established.

3. By determining the every next optimal point through the previous one in chain: $\hat{x}_1 = \hat{u}_0(0) \to \hat{x}_2 = \hat{u}_1(\hat{x}_1) \to \ldots \to \hat{x}_N = \hat{u}_{N-1}(\hat{x}_{N-1})$, we receive solution: $\hat{x}_1, \hat{x}_2, \ldots \hat{x}_N$.

4. We determine values of separation temperature points and coefficients of selections of target products:

$$T_n = t_{\hat{x}_n}, \quad a_n = \hat{g}_n = \sum_{i=\hat{x}_n}^{\hat{x}_{n+1}} \left(\sum_{j=1}^{L} \mu_{ij} g_j \right), \quad n = 1, 2, \ldots N. \tag{12}$$

The methodology described is realized in the authors' software product [4].The latter can be used both on its own and within a higher level system, as part of a software for technical and economic modeling of a refinery [5].

References

1. M.Yu. Livshits, A.P. Sizikov, EPJ Web Conf. **110**, 01035 (2016)
2. E.E. Dudnikov, Yu.M. Tsodikov, *Typical Problems of Continuous Production Management* (Energy, Moscow, 1979)
3. R. Bellman. *Applied Dynamic Programming* (Nauka, Moscow, 1965)
4. A.P. Sizikov. *The System of Modelling and Optimization of Primary Oil Processing (SMOPOR).* A.P. Sizikov; applicant and right holder of Samaraneftekhimproekt JSC RU).State Registration Certificate for Computer Program No. 2011618611 (15.09.2011)
5. A.P. Sizikov, Bulletin of the Samara State Technical University **7**, 38 (2010)

MATEC Web of Conferences **92**, 01023 (2017)
Thermophysical Basis of Energy Technologies - 2016
DOI: 10.1051/matecconf/20179201023

Cross-section of subsea umbilical and stress prediction of its structural components

V. V. Solovev[1,*], *P.* N. Zyatikov[1], and *I.* N. Kozyrev[1]

[1] National Research Tomsk Polytechnic University, 634050 Tomsk, Russia

Abstract. It is presented in this article analysis of two models of subsea umbilicals under axial stress. Subsea umbilicals play a crucial role in offshore petroleum production and are designed to withstand all the environmental dynamic loads during the production period as long as 25 years. Apart from dynamic loads subsea umbilicals are under thermal loads as well having impact on the structural components and changing the material properties in some cases. The cross-section arrangement is a major part of the entire process of subsea umbilicals design. That is why this research is focused on the design of cross-section and has an aim to calculate main impacts subsea umbilicals facing during exploitation.

1 Introduction

Global energy market has been significantly grown for the last 30 years which means ordinary onshore methods of production, exploration and development will face with a challenge to meet the demand. One of the potential ways to meet this constantly growing demand is to develop offshore oil and gas fields, where it requires to use reliable equipment being able to withstand all the loads for as long as 20 or 25 years. One of the most important components in offshore production system is a subsea umbilical, that is why it is necessary to perform a quality structural analysis before manufacturing to insure it suits the project conditions.

This research is focused on two different configurations of subsea umbilical with five bounded hoses same diameter made of stainless steel. These hoses are used to transport hydraulic liquids, inhibitors and some other chemicals from offshore platform towards subsea production facilities to provide the continuity of the production process and improve its integrity.

2 Cross section design

The inner diameter was chosen based on the ones provided by one of the manufactures presented on the market (see Table 1).

* Corresponding author: solowyev@gmail.com

Table 1. Stainless steel hoses sizes range.

Inner Diameter		Wall thickness (mm)	
Inches	mm	minimum	maximum
3/4	19.05	1.1	4

Designed and researched cross section designs are shown in Figure 1. The analysis was made to find the behavior of subsea umbilical under two main loads: axial and bending. The results for bending stress is not represented in this article, but will be discussed later.

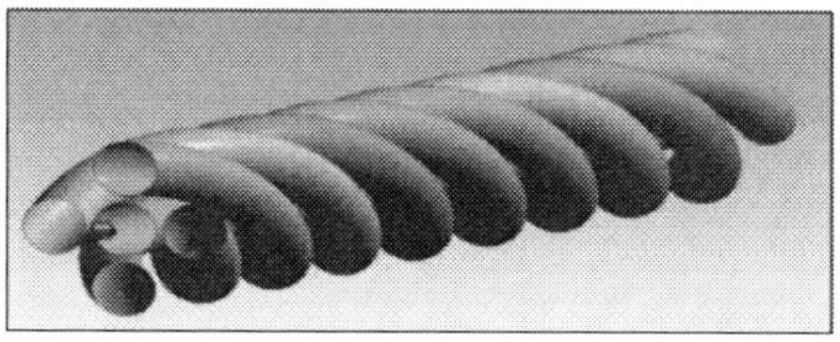

a)

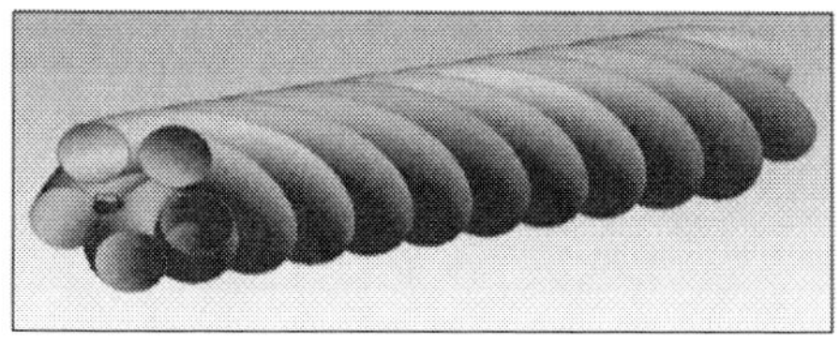

b)

Fig. 1. Two models in axial loads simulation, where red arrows show the active axial force applied to one of the end of each umbilical: a) First Model; b) Second Model.

3 Simulation results discussion

As it is shown in the Figure 2, the second model experienced smaller both the minimum and maximum loads under all five stages of stresses. However, under axial force of 15kN both model show almost equal results of maximum stress (221MPa and 209.77MPa, for the first and second models, respectively). The most obvious reason for this might be the yield stress of stain steel, which is 210MPa according to the ANSYS materials database. When the applied force was between 10kN and 15kN the first model started to experience plastic deformation. However, the second model started to deform plastically under a much higher force of about 20kN. Considering that both models consist of the same size hoses made of the same material, the only reason for such results can be the difference in structure of cross section. The second model structure can withstand a higher axial stress. The difference between starts of the plastic deformation of two models is around 5kN. Graphically the results are presented in Figure 3 below.

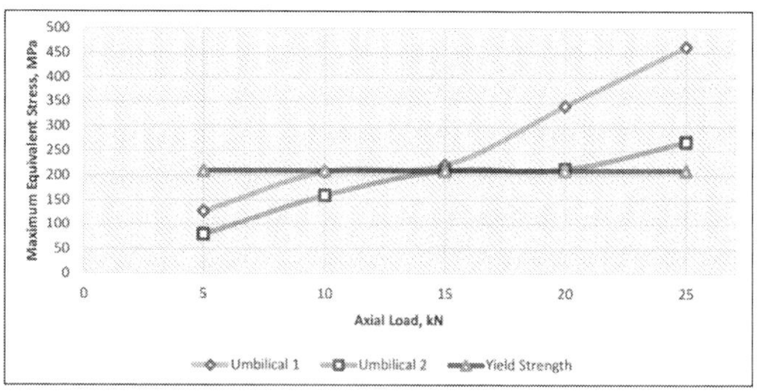

Fig. 2. Maximum Equivalent Stress (MPa) to Axial Load (kN).

Fig. 3. Two Umbilicals Axial Stress (MPa) Pattern.

4 Analytical calculation

The next step of this research is the verification of the results obtained from computer simulation which can be done by analytical calculations or by experiments. It is not possible to make the experiments at the moments due to lack of the materials and required equipment. However, there will be made analytical calculations, divided into three main stages: calculation of simplified model, where each of the metallic hoses is represented by a bounded beam, then there will be added empty cylindrical space inside along the length of each hoses, and lastly there will be added hydraulic fluid inside, which is the most similar to the real subsea umbilical working stage. Besides, there will be done some calculations to understand the effect of heat transfer in the system on subsea umbilical structure considering the fact that ambient temperature varies significantly from around +300 C near the surface to about +30 C near the sea bottom. Temperature might affect both the umbilical and hydraulic liquids transporting through it. The mathematical part is based on the variational principle of displacement, which states that virtual change of inner energy of deformation must be compensate by the same amount of external work applied to the object (Equation 1).

$$\int_V \sigma_{ij}\delta\varepsilon_{ij}\mathrm{d}V = \int_S f_i^S \delta u_i \mathrm{d}S + \int_V f_i^B \delta u_i \mathrm{dV},$$
$$\varepsilon_{ij} = \frac{1}{2}\left(\frac{\partial u_i}{\partial x_j} + \frac{\partial u_j}{\partial x_i}\right) \tag{1}$$

where σij – stress components; εij – deformation components; ui – displacement vector components; xi – present-position coordinates; fiS – surface forces vector components; fiB – mass forces vector components; V – volume of deformed object; S – surface of deformed object, which the surface forces were applied to.

System of original functions in matrix form includes components of displacement vector (Equation 2):

$$\{u\} = \{\mathrm{u}_1(\mathrm{x,y,z}), \mathrm{u}_2(\mathrm{x,y,z}), \mathrm{u}_3(\mathrm{x,y,z})\}, \tag{2}$$

also stress vector (3):

$$\{\sigma\} = \{\sigma_{11}(x,y,z), \sigma_{22}(x,y,z), \sigma_{33}(x,y,z), \tau_{12}(x,y,z), \tau_{23}(x,y,z), \tau_{13}(x,y,z)\}, \quad (3)$$

and deformation vector (4):

$$\{\varepsilon\} = \{\varepsilon_{11}(x,y,z), \varepsilon_{22}(x,y,z), \varepsilon_{33}(x,y,z), \gamma_{12}(x,y,z), \gamma_{23}(x,y,z), \gamma_{13}(x,y,z)\}. \quad (4)$$

Temperature effect will be analyzed based on the principle of heat transfer through a single layer (Figure 4a) and multi-layer (Figure 4b) cylindrical bodies.

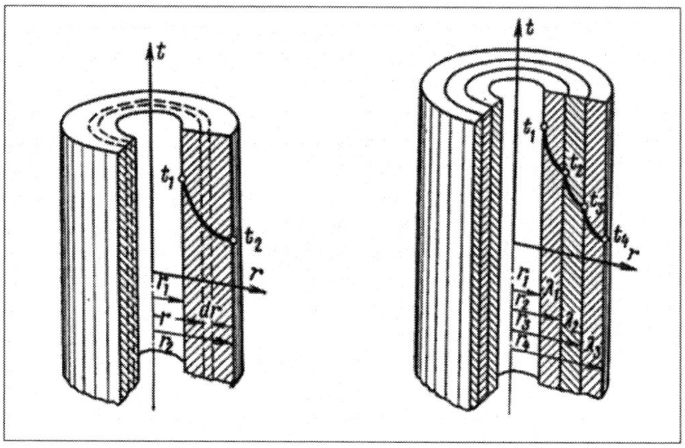

Fig. 4. Single layer (a) and multilayer (b) cylindrical structures.

In order to calculate steady-state heat transfer when through the wall of cylindrical structures it is transferred a constant amount of heat, this amount of heat going through the wall of a body with a length 'l' the following formula will be used:

$$q_l = \frac{2\pi(t_n - t_{n+1})}{\frac{1}{\lambda_n} ln \frac{d_{n+1}}{d_n}} \quad (5)$$

where t_n, t_{n+1} – temperature in between the layers, λ_n – heat transfer coefficient of a layer material, d – layer diameter.

5 Conclusion and future research

In conclusion, the results obtained after the computer simulation show that the second model is better than the first one in terms of axial force resistance. Moreover, the second model cross section design decreases the required outer diameter of subsea umbilical (67 mm and 75.8 mm for the second and first umbilicals, respectively), which makes this model better in terms of economic because it requires smaller amount of materials to produce the same length of umbilical than for the first model. It also significantly decreases the transportation and installation cost. Further studies will be focused on analytical verification of obtained data and analysis of heat transfer on the structural components and exploitation of subsea umbilical.

References

1. ANSYS Worbench Features. Accessed on 20.09.2016: http://www.ansys.com/products/Structures/Impact
2. X. Chen, S. Fu, L. Song, Q. Zhong and X. Huang, Ocean Systems Eng. **3**, 097 (2013)
3. I. Kozyrev, P. Zyatikov, V. Deeva, MATEC Web of Conferences **72**, 01049 (2016)
4. E. Kramshonkov, A. Krainov, P. Shorohov, EPJ Web Conf. **110**, 01031 (2016)

Evolution of the deformation profile of a horizontal thin ethanol layer when heated locally

Serafim Spesivtsev [1,2,*], and *Yuriy* Lyulin[1,3]

[1]Institute of Thermophysics, 630090, Novosibirsk, Russia
[2]Novosibirsk State University, 630090, Novosibirsk, Russia
[3]Tomsk Polytechnic University, 634050, Tomsk, Russia

Abstract. Thermocapillary breakdown of thin horizontal layer of ethanol when heated from a localized heat source was studied experimentally. The influence of layer depth on the breakdown process was investigated. Evolution of the layer thickness in the heating point and deformation profile were being monitored and the critical thickness of the layer was evaluated using confocal technique. Pulsations of layer thickness over the heating area before the breakdown have been found.

1 Introduction

Study of heat elimination from a pointed heat source becomes one of the most significant problems in thermal physics because of its straight association with the cooling of microelectronics [1]. Permanent development of microelectronic equipment leads to complication of the microchip structure and formation of nonuniform heat flux distribution on the chip surface. Heat flux in some small zones is much greater than the chip regular [2], of the order of 1 kW/cm². This effect could significantly deteriorate reliability and performance of the device.

Nowadays, there are a number of successful methods for cooling of local hot spots, for example spray cooling, thermoelectric coolers, boiling in microchannels. One of the most effective techniques for eliminating such high heat fluxes from a local heat source is technology based on processes with phase change such as evaporation of a thin liquid layer [3]. In this case the main mechanism of heat transfer from the heat source is an intensive evaporation due to latent heat of vaporization. Dynamics of evaporation and, therefore, the heat elimination from the heat source significantly depend on the conditions in a layer [4]. Especially, the breakdown of liquid layer leads to powerful decreasing of heat transfer from a localized hot spot [5]. Processes of liquid layer breakdown are actively investigated experimentally [5-7] and theoretically [8-10].

The aim of the present work is to study the breakdown dynamics and measure the deformation profile of an evaporating horizontal liquid layer when heated locally.

* Corresponding author: simafir94@gmail.com

2 Experimental facility

Experiments were carried out on the setup shown in Fig. 1. The working fluid is supplied to the test cell with the assistance of the syringe pump and a horizontal liquid layer is created. The thickness of liquid layer is varied from 300 to 700 μm. The layer of liquid is opened to the atmosphere and maintained on the surface of the working area using sharp edge on the substrate perimeter. Spot heating of the horizontal liquid layer takes place in the center of the substrate. The test cell consists of caprolon base, metallic substrate and the heating element. The caprolon base has a special cut on the upper side for installation of the substrate and a central through hole with a diameter of 1.6 mm. The substrate is made of stainless steel and has a diameter of 50 mm and a thickness of 1 mm. In the center of the substrate is a closed hole with a diameter of 1.6 mm and a height of 0.8 mm. The heating element is made of brass and has a round tip with a diameter of 1.6 mm and a height of 3 mm. It is tightly inserted into the closed hole of the substrate through the caprolon base. Thermal paste is used for better thermal contact between the heater tip and the substrate. The distance between the tip and the upper side of substrate is 0.2 mm. The power of heating element is controlled by the power supply. Insulating material is located on the underside of the heater to minimize heat losses. Temperature in the test cell is measured by thermocouples (type K) connected to the measuring system with an accuracy of 0.1°C. Location of the thermocouples is shown in Fig. 1. Relative humidity and atmosphere temperature are measured using the thermohygrometer Testo 645 with an accuracy of 2 % and 0.1 °C, respectively.

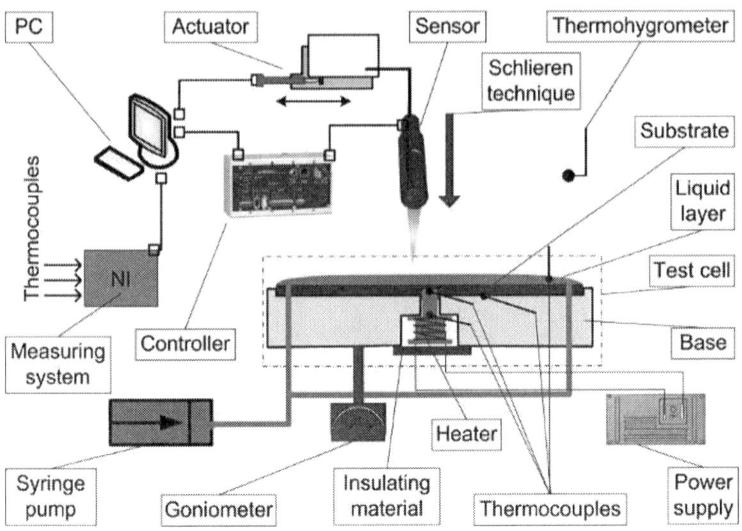

Fig. 1. Scheme of the experimental setup.

The heat flux density is determined by measuring the temperature difference between two different sections along the heater tip. The height of the horizontal liquid layer is maintained in constant position during the all experiment. Confocal system Micro-epsilon is used for measuring the layer thickness over the heating element. The system consists of the controller and the sensor. Sensors have the spatial resolution from 10 to 60 nm, the accuracy 0.35-0.7 μm, the spot diameter 6-16 μm and the measuring range 0.3-10 mm. The maximum temporal resolution is 100 μs. The sensor is fixed on the three-dimensional positioning system with high-speed linear actuator on one of the horizontal axes. Linear actuator is controlled from personal computer and by special software. The maximum speed the axis moves at is 104 mm/s. The sensor moves in range of 50 mm by steps of 1 μm. Also it moves in two other axes with the help of two hand-operated linear stages in range of 50 mm. For visualizing the deformation of the surface and recording the layer breakdown dynamics schlieren technique with Photron FASTCAM 675K-M3 high-speed camera was used (at 5000 fps, 640x640 pixels, 25 μm/pix). The test cell was installed in a horizontal position with the help of two-axis goniometer. Equilibrium contact angle of the substrate surface is defined by Young-Laplace equation [11] at room temperature of 25±2 °C and is equal to $\theta = 8\pm1$ °. Profilometer "Micro Measure 3D station" was used for measuring the roughness of substrates. Average value of the substrate roughness is equal to $R_a = 0.037$ μm.

3 Experimental results and discussion

Experiments were conducted at atmospheric pressure, temperature and relative humidity of 28±2 °C and 25±3 %, respectively. Layer thickness was ranged from 300 to 700 μm. Ethanol (95 %) was used as a working fluid. Heater temperature range was 20-105°C. The temperature values were measured at the moment of layer breakdown. Heat flux density was varied from 0 to 95.3 W/cm². Injection liquid flow rate was up to 200 μl/min. During the experiment heat flux is increased up to a critical value at which the liquid layer ruptures. At this moment heating process is stopped to prevent the failure of heating element.

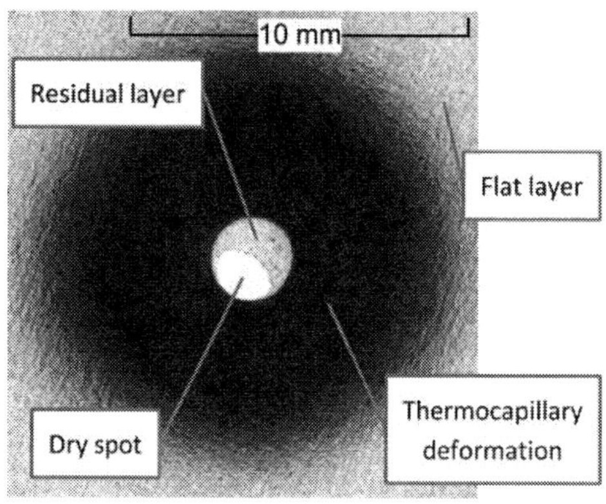

Fig. 2. Visualization of breakdown dynamics and formation of the dry spots.

It was found that breakdown process consists of several steps. At the beginning, thinning of the liquid layer over the heating area due to the effect of thermocapillary forces [12, 13] and evaporation is observed. Further thinning leads to the formation of residual liquid layer in the area of the local heating. Then the residual layer evaporates until its thickness reaches the critical and breakdown of the liquid layer occurs, Fig. 2 [5, 6]. After the breakdown the whole area of the local heating rapidly dries and quite symmetrical circular dry spot is formed.

Evolution of the layer thickness in the center of substrate is observed with the help of the confocal system. All detected steps of the breakdown process observed by the schlieren technique were also confirmed using the confocal method. Dependence of the layer thickness over the heating area on time is shown in Fig. 3.

DOI: 10.1051/matecconf/20179201024

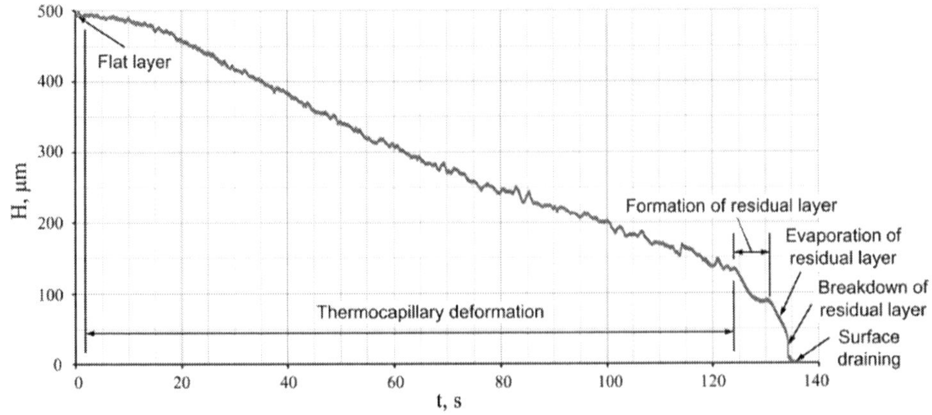

Fig. 3. Dependence of the layer thickness over the heating area on time (initial layer depth 500 μm).

Evolution of the layer thickness before the breakdown was measured for different initial layer depth in range from 300 to 700 μm (Fig. 4). Critical thickness of residual is less than 30 μm. It is assumed that critical thickness of the residual layer for present substrate mostly depends on the properties of working liquid. The time of dry spot formation is decreased when the layer thickness and, accordingly, the local heating intensity is increased. The time varies from 6 to 0.67 seconds. Temperature in the center of substrate increases and more intense evaporation of residual layer takes place. The main mechanism of the outflow of the liquid from the heating zone before the residual layer formation is the thermocapillary shear stress, and after it this is the evaporation. At small liquid thickness the viscosity suppresses thermocapillary flows.

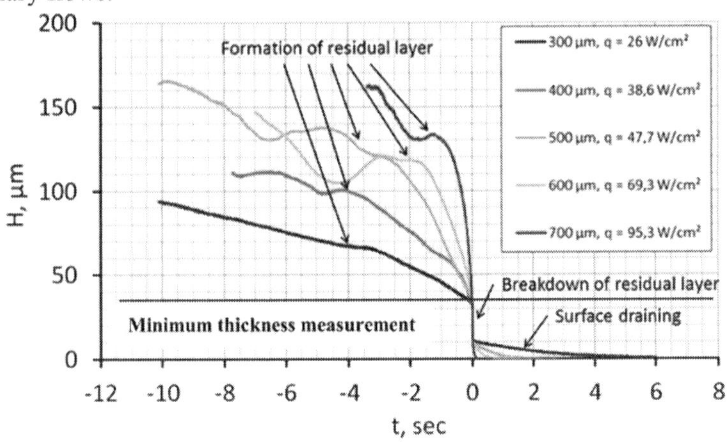

Fig. 4. Dependence of the layer thickness over the heating area on time for different initial layer depth.

Evolution of the layer deformation profile is shown in Fig. 5. Pulsations of the layer thickness before formation of the residual layer have been observed. The presence of the thickness pulsations is presumably connected with convective instability, caused by the intense heat and mass transfer in the heating area.

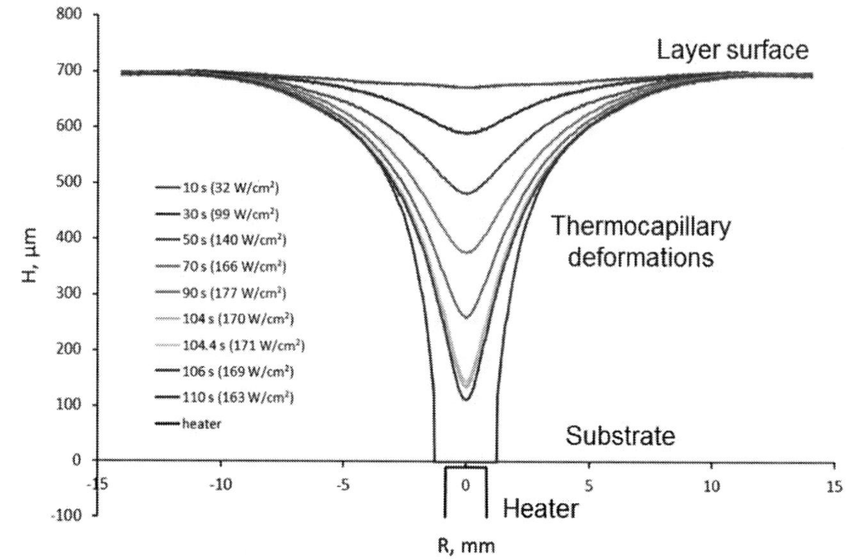

Fig. 5. Measurements of the layer surface profile in case of local heating.

4 Conclusion

Influence of the layer depth on the breakdown dynamics of liquid layer was studied. Evolution of the layer thickness in the heating point and deformation profile were being monitored using confocal technique. The existence of residual liquid layer over the heating area before the breakdown has been proved. It is found the critical thickness of residual layer is less than 30 μm. Pulsations of layer thickness over the heating area before the formation of residual layer have been observed. The main drainage mechanism for the residual layer is evaporation.

The study was financially supported by the Russian Science Foundation (Project 14-19-01755).

References

1. A. Bar-Cohen, P. Wang, J. Heat Transf. **134**, 051017 (2012)
2. R. Mahajan, C. Chiu, G. Chrysler, Proceedings of the IEEE **94**, 1476 (2006)
3. E.Ya. Gatapova, O.A. Kabov, I.V. Marchuk, Tech. Phys Lett **30**, 418 (2004)
4. A.A. Fedorets, I.V. Marchuk, O.A. Kabov, Interfacial Phenomena and Heat Transfer **1**, 1 (2013)
5. Yu.V. Lyulin, S.E. Spesivtsev, I.V. Marchuk, O.A. Kabov, Techn. Phys. Lett **41**, 1034 (2015)
6. D.V. Zaitsev, D.A. Rodionov, O.A. Kabov, Microgravity Sci. Tec. **19**,100 (2007)
7. D.V. Zaitsev, O.A. Kabov, Microgravity Sci. Tec **35**, 825 (2011)
8. V. S. Ajaev, Interfacial Phenomena and Heat Transfer **1**, 1 (2013)
9. M. B. Williams and S. H. Davis, J. Colloid Interf. Sci. **90,** 220 (1982)
10. J. P. Burelbach, S. G. Bankoff and S. H. Davis, Phys. Fluid A Fluid Dyn. **2**, 313 (1990)
11. I.V. Marchuk, V.V. Cheverda, P.A. Strizhak, and O.A. Kabov, Thermophys. Aeromech+ **22**, 297 (2015)
12. I. V. Marchuk, J. Eng Thermophys **18,** 227 (2009)
13. I.V. Marchuk, J. Eng Thermophys **24** (2015)

Assessment of the cogeneration biogas plant possibilities in the autonomous power supply system

Liudmila Sumarokova[1], *Mikhail* Surkov[1,*], *Ekaterina* Tarasova[1], and *Evgeniya* Tretyakova[1]

[1] National Research Tomsk Polytechnic University, 634050 Tomsk, Russia

Abstract. The use of biomass and wood waste for heat and power production is increasing from year to year. Waste wood is low carbon footprint, has low sulfur content and relates to renewable energy sources. The paper demonstrates the possibility of increasing the energy efficiency of power supply system of the Stepanovka settlement (Tomsk region) by means of replacing diesel power plant (DPP) by the biofuel gas piston CHP. The assessment was based on the possibility of the technical and economic comparison of power supply options in the settlement.

1 Introduction

Currently the energy efficiency of decentralized power supply systems is one of the urgent tasks required to accelerate social and economic development of the country. In Tomsk region the decentralized power plants (DPP) are the main sources of electrical energy for autonomous power supply systems. Low technical and economic indicators of most diesel power stations, high prices for diesel fuel and higher transportation rates lead to excessive cost of DPP-based electricity generation. According to experts [1, 2], the fuel component in the electricity tariff for diesel power plants is by more than 7 times higher than the cost of 1 kWh of electricity from the centralized system of power supply.

Reduction of the energy resources consumption and increase in the efficiency of their use can be implemented on the basis of high-performance poly-generation systems [3, 4, 5] In these systems, the energy of various kinds (heat, cold, and electricity) is produced in a combined way using only one primary energy source. The excess heat generated by the engine can be used to produce heat or cold for household consumers and technological needs of production. This is possible when the thermal circuit contains the compressor or absorption air conditioners, consequently it will increase the working efficiency of the installation in summer period [6, 7, 8].

Studies performed on the possibility to use different types of fossil fuels [9] have shown that the best kind of energy resource for the specified type of generation is wood fuel – wood chips. Its reserves, accessibility and technological transformation process allow efficient use of this resource for the gasification. Depending on the method of obtaining the raw materials (sawmill waste, waste in the cutting area, unmerchantable timber), the cost of

* Corresponding author: masur@tpu.ru

transportation and processing into chips - the raw material cost can vary from 2000 rubles/t. ref.fuel (waste on the upper timber landing, wood waste) to more than 7000 rub./t.ref.fuel (waste in the cutting area). The potential of wood biomass available within a radius of 30-50 km from the object completely covers the needs for this type of fuel. When planning the reconstruction of the electricity and heat supply system of Stepanovka settlement it is recommended to consider the project where the electrical energy is produced by the diesel power plant and electric and thermal energy – from biomass gas piston installation.

The article demonstrates the review of the following results in the specified topic:

- Analysis of the settlement power supply system and assessment of the heat and electricity consumption volumes.
- Technical and economic comparison of power supply system's options.

2 Analysis of the settlement power supply system and assessment of the heat and electricity consumption volumes

The current state of Stepanovka (Tomsk region) settlement utilities is characterized by a high degree of wear and tear, for most objects the depreciation rate ranges from 48 % to 82 %. On average, 68 % of utilities have served their rated resource.

The main technical and economic characteristics of DPP operation for 2015 are summarized in Tables 1 and 2. DPP analysis revealed the following:

- The working efficiency of DPP installed capacity is equal to 8-13% in real terms for 2010-2015;
- The high rate depreciation of DPP equipment which outlived its service life by more than 100%;
- The excess of the actual fuel costs over standard values.

Table 1. Diesel generators specifications.

Type of diesel generators	Kilowatt nominal	Year of commissioning	Action time for the operating period eng. hr. for 2013
DG72M №1	800	DG72M №1	43618
DG72M№2	800	DG72M №2	39235
11D-100	1000	11D-100	32746

The electricity consumption volumes were assessed according to [10]. The main technical and economic performance indicators of the DPP for 2015, and data on the volume of electricity consumption are summarized in Table 2.

Table 2. DPP technical and economic performance indicators for Stepanovka settlement in 2015 [10].

The electricity tariff for consumers, RUB / kWh	16.65
Electricity generation, thous, kWh	2768.29
DPP auxiliaries, thous. kWh	110.73
Busbar output, thous. kWh	2657.55
Process losses, thous. kWh	478.62
Net electricity supply, thous. kWh.ч.	2178.93
Specific consumption of diesel fuel, thous. kWh.	0.29
The average price of diesel fuel, RUB/ton	33286

Evaluation of heat consumption was performed based on [10] There are three boiler houses operating on wood, which supply heat to socially significant facilities – school, hospital and community centre. Actual total heat production for 2015 was amounted to 1290 Gcal. Design load is 0.46 Gcal./hour.

3 Technical and economic comparison of power supply system's options

To select the optimal power supply system we carried out technical and economic calculations of the following options:

Option 1: the production of electricity using gas generating plants, excluding by-products (heat), taking into account the capital costs in the tariff and based on budgetary funds.

Option 2: the operation of mini CHP in the cogeneration mode, taking into account the by-products and the distribution of costs in proportion to the electricity and thermal energy generation, taking into account the capital costs in the tariff and based on budgetary funds.

The indicators were calculated by the method described in the paper [9]

All the major technical and economic calculations of the options are summarized in Table 3.

Table 3. Technical and economic calculations of power supply system's options.

Rated capacity, kW	960
Specific capital expenditures, rub/kW	18412.5
Depreciation charges, rub/year	1767000
Generated electric energy, MWh /year	3 087.187
Generated heat energy, MWh /year	1 867.6
Annual electrical energy production costs, rub/year	930600
Annual heat production costs, rub/year	756000
Payback period, years	9.6
Electrical energy prime cost, rub./kW	8.69
Heat energy prime cost, rub./Gcal	5762

According to a feasibility studies for **the option 1** the average tariffs (including VAT) for electricity generated using gas generator installation excluding by-products (heat) in 2015 values are determined by the following:

1.1 with the capital costs included in the tariff – 17.65 RUB/kWh which is higher than the average tariff for the existing DPP's by 6%.

1.2 if the construction is financed from budgetary funds – 13,25 RUB/kWh which is lower than the average tariff for the existing DPP's by 20,4 %.

Given the significant reduction in the efficiency of the production of electricity only in the absence of co-generation, this **option is not feasible**.

According to the **option 2,** gas generator works as a mini-CHP with the production of electricity and heat (concomitant) energy for consumers. Option requires the allocation of costs in proportion to the generation of electricity and heat in cogeneration.

Average tariffs (including VAT) for electricity and thermal energy for 2015 amounted to:

2.1 with CAPEX included in the tariff:

- electricity – **8.69** RUB/kWh, which is lower than the average tariff for the existing DPP's by 56.4 %;

- heat - 9839.13 RUB/Gcal, which is higher than the current average tariff for boilers located in the same localities as the DPP (5762.52) by 41.4 %.

2.2 if the construction is financed from budgetary funds:

- electricity – 6.73 RUB/kWh, which is lower than the average tariff for the existing DPP's by 66.2 %;
- heat – 7 607.55 RUB/Gcal, which is lower than the average tariff for boilers (5762.52) by 24.3 %.

In the **second option** there is a significant reduction in electricity tariffs, compared with actual prices of 24.3-41.4 %. Factor increasing/reducing the tariffs for electricity and heat makes this option interesting for further planning.

4 Conclusion

One of the most promising ways of increasing the efficiency of local energy supply systems is the use of the renewable energy as a part of the regional energy mix and operating modes optimization of the main power equipment. Involvement of wood fuel in the energy mix of electricity generation will significantly reduce the need for imported liquid fuels and solve environmental and social problems.

References

1. Y. Sarsikeev, B. Lukutin, D. Lyapunov, M. Surkov, S.Obuhov, Adv. Mat. Res. **953,** 529 (2014)
2. Federal Law of 23.11.2009 No 261 - FZ "*On energy saving and energy efficiency improvements and on Amendments to the Russian Federation some legislative acts*" [in Russian].
3. K. Chua, W. Yang, T., Renew. Energ. **41**, 358 (2012)
4. J. Clerk Maxwell, *Advanced District Heating and Cooling (DHC) Systems*, (2016)
5. J. Ahrenfeldt, T. Thomsen, U. Henriksen, L. Clausen, Appl. Therm. Eng. **50**, 1407 (2013)
6. D. Chines, A. Meneghetti, Appl. Energ. **82**, 228 (2005)
7. M. Rasul, C. Ault, M. Sajjad, Energy Procedia **75**, 1065 (2015)
8. E. Beierlein, Modern Techniques and Technologies **2**, 74 (2005)
9. L. Konovalova, B. Lukutin, Increasing the efficiency of local energy supply systems Depositing VINITI (Russia, 2007)
10. Geoinformation system "Renewable energy sources in Tomsk region" http://green.tsu.ru/tomres/

Use of levitating liquid micro-droplets as tracers to study the evaporation in the vicinity of the contact line

Dmitry Zaitsev [1,2*], *Dmitry* Kirichenko[1,3], and *Oleg* Kabov[1,2]

[1]Institute of Thermophysics, 630090 Novosibirsk, Russia
[2]National Research Tomsk Polytechnic University, 634050 Tomsk, Russia
[3]Novosibirsk State University, 630090 Novosibirsk, Russia

Abstract. Self-organization of a cloud of liquid micro-droplets into an ordered two-dimensional array, levitating over a heated layer of liquid due to upward vapor flow, has been observed in several recent experimental works. In the present paper, the levitating micro-droplet array is studied under the condition when the liquid layer ruptures and a dry spot form on the heater. It has been found, that when the levitating micro-droplets are moving from wetted heater area to dry heater area and passing over the contact line, the micro-droplets levitation height is drastically increased, which is indicative of the intensive evaporation occurring in the region of the contact line.

1 Introduction

A spatially ordered structure of micro-droplets levitating above the free surface of a liquid layer spot-heated from below was discovered in [1]. The array consists of several of hundreds of drops packed into one layer, forming a hexagonal structure. The drop diameter is on the order of 10 μm; the distance between the drops is of several drop diameters, while the drops levitation height above the liquid surface is comparable to the drop size. The possible mechanism of micro-drops levitation is the Stokes force acting onto a drop from the vapor flow arising from the heated liquid surface [2, 3]. Figure 1 shows a typical image of levitating micro-droplet array over a heated layer of liquid, obtained in the present study.

Among the effective solutions for removal of high heat fluxes, there are systems of spray cooling [4, 5] as well as systems using stratified two-phase flows in micro-/mini-channels [6, 7]. Intensive evaporation from the contact line region can be a key factor for the efficiency of these systems. In spray cooling, the impact and subsequent spreading of droplets on the surface results in formation of multiple contact lines. Strong correlation between the heat transfer rate and the contact line length per unit area is observed [5, 8]. In stratified two-phase flows, the highest heat transfer rate is reached in the regime when local dry areas on the heater quickly appear and disappear [9], thus forming multiple contact lines.

* Corresponding author: zaitsev@itp.nsc.ru

In this paper the levitating micro-droplet array is used as natural tracers to study the evaporation in the vicinity of the contact line at the boundary of a dry spot formed in a horizontal water layer heated from below. A relatively large heater of size 10×10 mm^2 is used (in papers [1-3] the size of the heater was around 1 mm). With the help of high-speed imaging technique, we study the behaviour of the droplet array as it approaches and passes over the contact line region.

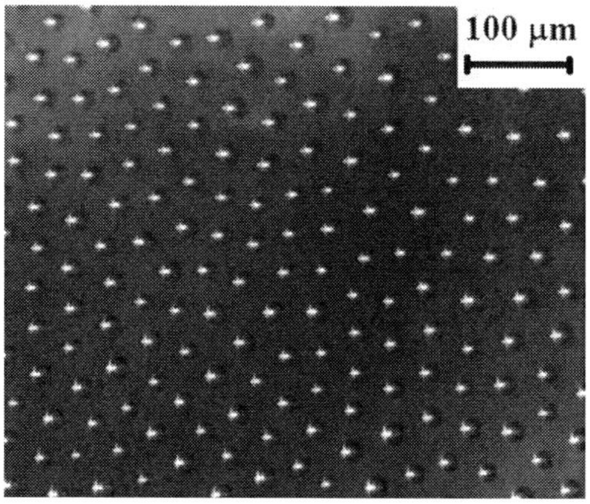

Fig. 1. Top view of an array of levitating micro-droplets, formed over a heated layer of water (heater of size 10×10 mm^2, heat flux of 27 W/cm^2, water layer thickness of 0.4 mm).

2 Experimental setup

Schematic and photograph of the experimental setup are presented in Fig. 2. The test section is a stainless steel plate 40×40×5 mm^3 in size with a flush-embedded 10×10 mm^2 copper rod, electrically heated by a nichrome wire. A layer of insulation material (mineral wool) is superimposed over the wire (nichrome wire with insulation material are not shown in Fig.2). The heater surface temperature is measured at several points by thermocouples with the accuracy of 0.1 K and frequency of 1 Hz. According to the measurements, the condition T_w=const is always met along the copper surface. In the experiments, T_w was varied from 75 to 100 °C. The heat flux is determined by the electric power released at the nichrome wire. Ultra-pure degassed distilled water (Merck Millipore) with the initial temperature of 25 °C is used as the working liquid. The test section is installed horizontally and open to the atmosphere; the temperature of ambient air is 25±2 °C.

The working surface of the test section was mechanically polished. The morphology of the surface was analysed using scanning electron microscope (JEOL JSM6700F) and atomic force microscope (Solver Pro NT MDT), Fig.3. The root mean square (RMS) surface roughness, measured on a 22 μm^2 area of the heater surface, was found to be 500 nm, with the average roughness of 0.394 nm. Thus, the roughness of the working surface is negligible compared to typical micro-droplet size (10-30 μm) and typical micro-droplet levitation height above the dry substrate (5-15 μm, see below).

The equilibrium contact angle on the copper surface was measured at different points by the method of a sessile liquid drop at the room temperature of 25±2 °C. To obtain the profile of the drop surface, the shadow method was used with a collimated light source and digital camera Nikon D800 having resolution of 6 micron/pixel. The resulting images were processed using the software by KRUSS. The advanced static contact angle measured at different points over the surface of the copper heater was 74±9°, whereas the receding contact angle was close to zero (<10°).

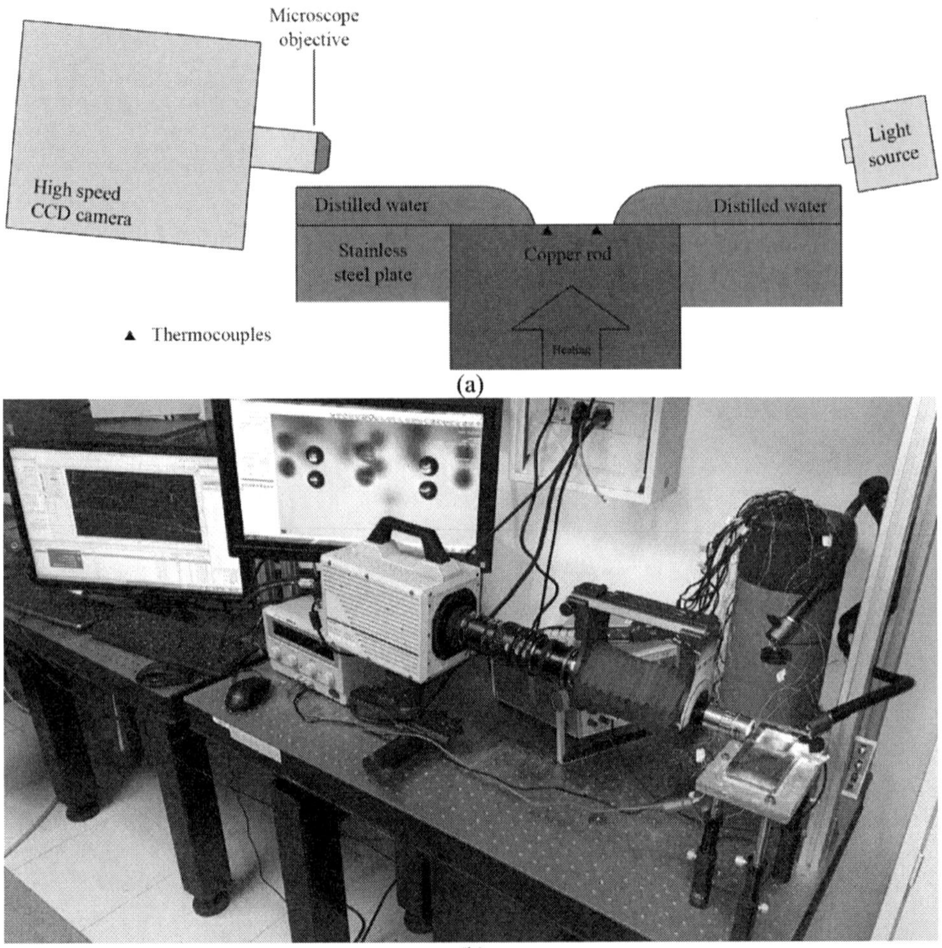

Fig. 2. Schematic (a) and photograph (b) of the experimental setup.

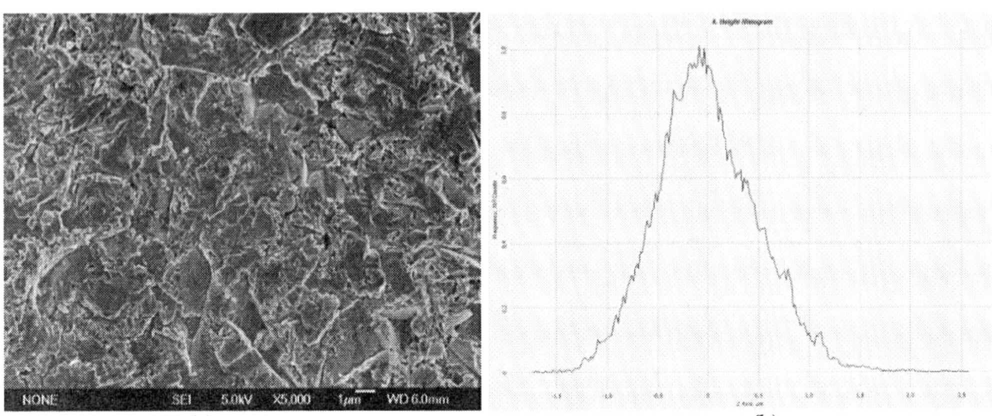

Fig. 3. Image of the heater surface obtained using scanning electron microscope (a), and height histogram obtained using atomic force microscope (b).

Optical recording was made with a high-speed camera FASTCAM SA1.1 (5600 fps with resolution of 1024x1024 pixel) coupled with a microscope objective through a set of extension tubes and bellows. Field of view of the camera was 800x800 μm^2 (resolution of 0.78 μm per pixel). The optical axis of the camera was aligned at an angle of 5° with respect to the horizon. The experimental procedure is as follows:

1) A predetermined volume of the working liquid is deposited with a syringe onto the working surface to form a liquid layer with the initial thickness of 400 μm.

2) With the help of an air jet a dry spot (about 1 mm in size) is formed on the copper surface.

3) The optical system is focused onto the contact line at the edge of the dry spot.

4) The heater is switched on, which induces formation of micro-drops levitating over the heated layer of liquid.

5) When the levitating micro-drops move to the dry spot, the recording is started.

3 Experimental results

Increase of the heat flux induces formation of the array of micro-drops levitating over the heated liquid layer (typical top view is shown in Fig. 1). Usually, the array was of an elliptical shape with the size varying approximately from 500 to 4000 μm. In most of previous studies [1-3] the micro-droplet array was localized due to the use of localized heating (the size of the heater was around 1 mm), and as such the size of the array roughly coincided with the size of the heater. In contrast, we use a heater having size much larger than typical size of the cluster, which allows us to obtain larger arrays of droplets.

Upon the formation of a dry spot in the liquid layer, the micro-droplets move down the slope of the interface under the action of gravity toward the dry area. As they get close to the contact line, the height of their levitation increases several times and eventually the micro-droplets turn out to be levitating over the heater surface, Fig. 4. After that, they disappear by evaporation within a second or so while continuing to levitate over the dry substrate.

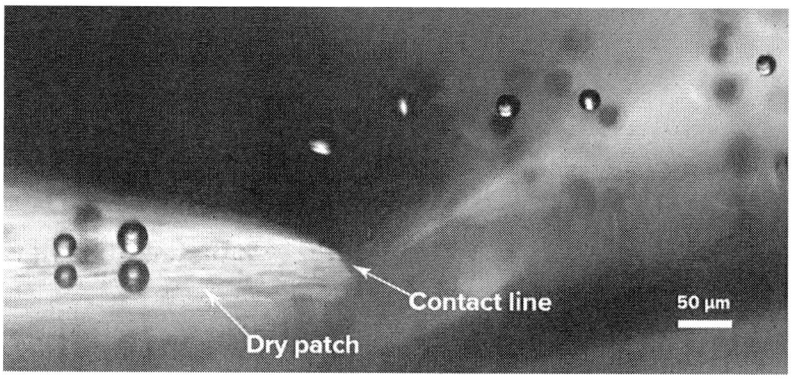

Fig. 4. Side view of micro-droplets flying over the contact line (substrate temperature of 95°C).

We revealed three common scenarios of micro-droplets transition from the wetted area to the dry area. First, a droplet can simply fly over the contact line with substantial increase of its levitation height, as depicted in Fig. 5a. The result shown in Fig. 5a was obtained for a droplet of diameter of 16 μm and substrate temperature of 79 °C. The second common scenario is when the droplet is following a circular-like trajectory in the vicinity of the contact line, with one or more full rotations, before eventually getting out of this cycle and flying to the dry area, Fig. 5b (droplet size is 20 μm, substrate temperature is 81 °C). This scenario can be connected with the vortex flow of the gas-vapor mixture formed near the contact line. Finally, some droplets can be levitating at a fixed location above the contact line, as shown in Fig. 5c (droplet diameter is 26 μm, substrate temperature is 97 °C). In this case, the dynamic force of the air-vapor flow from the contact line is balanced with the gravity force acting on the droplet. We would like to emphasize that while actual snapshots, shown in Fig. 5, were made at slightly different conditions, all three scenarios were observed under a wide variety of substrate temperatures (from 75 to 100 °C).

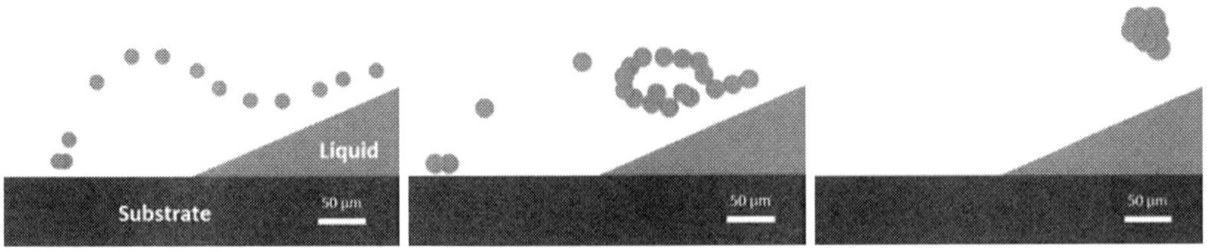

(a) (b) (c)

Fig. 5. Three common scenarios of micro-droplets transition from wetted area to dry area: simple fly-over above the contact line (a), fly-over with circular trajectory (b), and levitating at a fixed location (c). Positions of micro-droplets are shown at equal time intervals of 0.02 s. Micro-droplets are drawn in the scale.

4 Conclusions

Levitating micro-droplets, formed in the upward vapor flow above the heated liquid surface, are used as tracers to study the evaporation in the vicinity of the contact line. It has been found, that while the levitating micro-droplets are passing from the wetted substrate area over the contact line to the dry substrate area, the droplets levitation height is increased several times. From this we conclude that the specific evaporation rate in the vicinity of the contact line must be several times higher than that at a distance from the contact line. Three common micro-droplet trajectories are identified: 1) "simple" fly-over, 2) fly-over with circular motion, and 3) nearly steady-state levitation.

This work was supported by the Russian Science Foundation, project No. 1419-01755.

References

1. A.A. Fedorets, JETP Lett. **79**, 372 (2004)
2. A.A. Fedorets, I.V. Marchuk and O.A. Kabov, Tech. Phys. Lett. **37**, 116 (2011)
3. A.A. Fedorets, I.V. Marchuk and O.A. Kabov, JETP Lett. **99**, 266 (2014)
4. B. Horacek, K.T.Kiger, J. Kim, Int. J. Heat Mass Tran. **48**, 1425 (2005)
5. J. Kim, Int. Journal Heat Fluid Flow **28**, 753 (2006)
6. O.A. Kabov, Yu.V. Lyulin, I.V. Marchuk, D.V. Zaitsev, Int. J. Heat Fluid Fl. **28**, 103 (2007)
7. Zaitsev D.V., Rodionov D.A. and Kabov O.A. Tech. Phys. Lett. **35**, 680 (2009)
8. A.A. Semenov, D.V. Feoktistov, D.V. Zaitsev, G.V. Kuznetsov, O.A. Kabov Thermophys. Aeromech. **22**, 771 (2015)
9. O.A. Kabov, D.V. Zaitsev, V.V. Cheverda, A. Bar-Cohen, Exp. Therm Fluid Sci. **35**, 825 (2011)

Study of the gas driven water rivulet flow regimes in the minichannels

Olga Svetlichnaya[1, 2], *Vyacheslav* Cheverda[2, 3, *]

[1] Novosibirsk State University, 630090 Novosibirsk, Russia
[2] Kutateladze Institute of Thermophysics, 630090 Novosibirsk, Russia
[3] National Research Tomsk Polytechnic University, 634050 Tomsk, Russia

Abstract. The paper deals with the experimental study of the gas (nitrogen) driven rivulet flow regimes of water outflowing from the nozzle with a width of 10 mm into minichannel with the height of 1.6±0.02 mm and a width of 30 mm. The authors revealed a new flow regime – namely "rivulet bridge" whose occurrence is attributed to the influence of relatively large static wetting angle of water on the flow regimes. The paper presents a map of the various rivulet flow regimes characterized by different surface geometry and types of surface deformations.

1 Introduction

Against the background of the rapid development of electronics (reduction in the characteristic size of transistors) there is a need for effective cooling system that can be used in the aircraft industry (braking systems, microelectronics, etc.), automotive industry (power electronics in inverters, braking system, etc.) and in other industries.

A liquid film flow refers to a separate flow in which the fluid flows over the entire surface of the substrate. At a rivulet flow only a portion of the substrate is wetted by liquid, while the other part remains dry. Due to abnormally high heat transfer coefficient in the area of three phase contact line [1, 2], a rivulet flow has high cooling quality, since it is a particular case of the film flow with a contact line bounding the film from both sides, the line, where three phases come into contact. The fluid viscosity is significantly higher than gas viscosity. Due to this circumstance the use of rivulet flow will provide reduced resistance of the two-phase flow, as well as reduced pressure drop along the channel, and consequently, will reduce energy input for pumping of two-phase mixture through a microchannel. Despite the increased interest to this kind of problems, a number of available experimental [3, 4] and theoretical works [5] is quite small and extremely insufficient for comprehension of the physics of the processes associated with the heat transfer in gas driven rivulet flow of water in the minichannel._Following types of the rivulet flows over inclined surfaces are currently available in the literature: drip flow, straight flow, meandrine flow, flow without waves and with 2D waves, 3D and soliton-type wave flows, as well as swirling flow.

* Corresponding author: slava.cheverda@gmail.com

The aim of this work is constructing flow regime map of the gas driven isothermal rivulet flow of water in the minichannel. The results obtained will be used in the further development of the cooling system.

2 Experimental setup

Experimental studies were conducted employing setup, which is presented in Fig.1. The test bench for laboratory experiments consists of the following components: gas supply system; fluid delivery system; the system for maintaining a predetermined pressure in the test section and a regenerative system; the power supply system (not shown); a personal computer (PC-1) to control the experiment and collect experimental data; a personal computer (PC-2) for recording images from high-speed video camera; the water thermo stabilization system; the data collecting system; the optical system; the replaceable test section with the temperature-controlled substrate of the minichannel (from 10 to 100 °C).

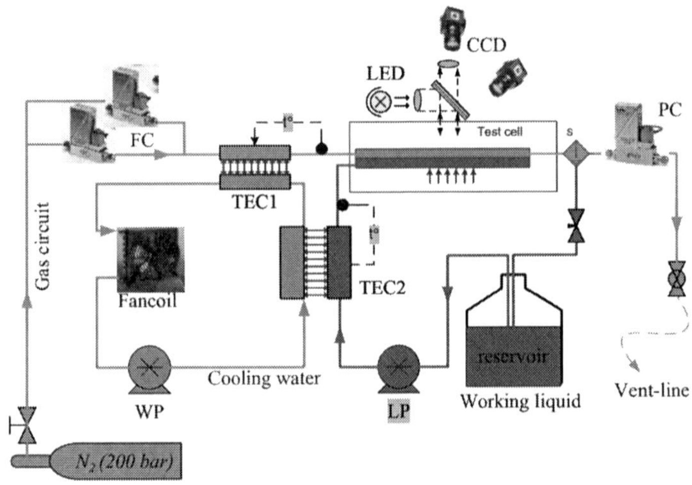

Fig. 1. The sketch of experimental setup for studying the nitrogen driven isothermal rivulet flow regimes of the working fluid (water) in a horizontal minichannel.

The frame (5) and cover (4) made of textolite serve the basis of the test section (Fig. 2a). Initial temperature of the liquid (11) and gas (12) at the inlet of minichannel, as well as temperature of the substrate (3) are varied within the range from 20 to 90 °C with an accuracy of 0.2 °C using Peltier elements (8) which are cooled by water heat exchanger (9). The inlet temperatures of the gas and liquid as well as outlet temperature of the mixture, the temperature of the substrate (2) and the ambient temperature are measured by thermal sensors.

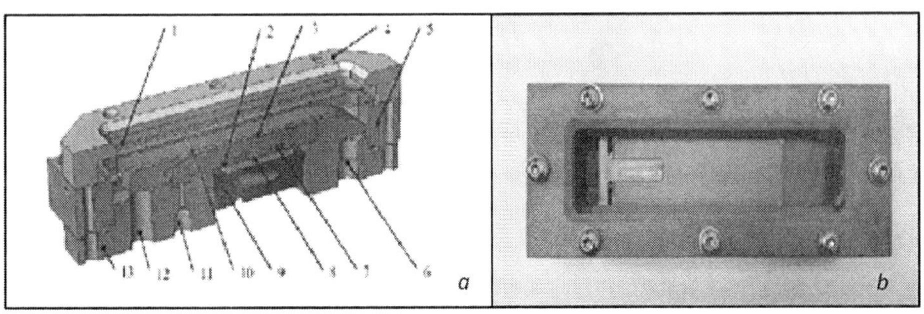

Fig. 2. Sketch (a) and photograph (2) of the test section.

The test section is closed on top by sealed cover provided with an optical glass (1), thereby forming a minichannel with a width of 30 mm and a height of 1.61±0.02 mm. Liquid is fed to the buffer chamber of the test section, passes through the nozzle between the brass plate (a blade) and the substrate and gets into the minichannel. The nozzle width for rivulet flow is equal to 10 mm. Gas (nitrogen) is fed into the test section under pressure through the nozzle (14) and drives the fluid rivulet along a rough substrate made of stainless steel (3). Further, mixture of water and nitrogen is released through the orifice (6).

Employing DSA-100 system and using sessile drop method (see Fig. 3) we measured the static wetting angle which amounted to 114.4±3°.

Also, we measured surface roughness for a better understanding of the interaction between the substrate and the liquid. The arithmetic average of the absolute values of the profile deviations within the basic length (Ra) was 0.458 μm (Fig.4.). The surface was processed with abrasive M-40 and the roughness was measured by the TR-200 instrument.

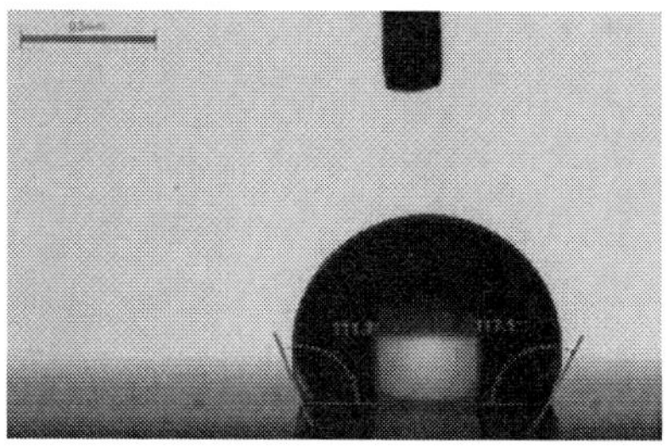

Fig. 3. Photograph of the water drop on a substrate.

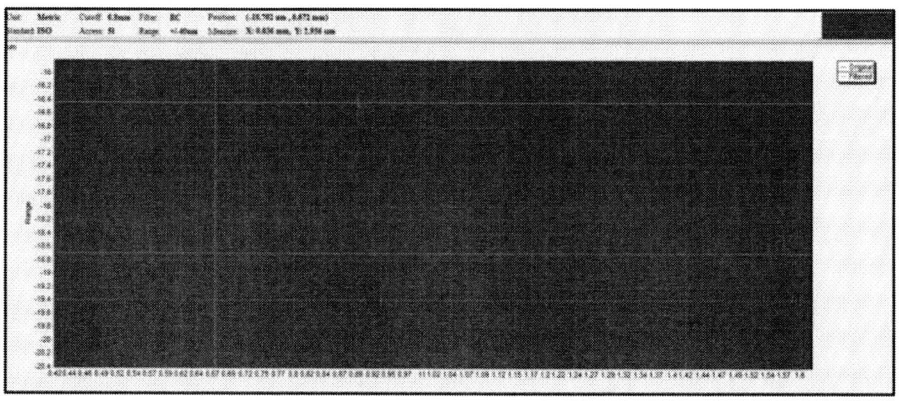

Fig. 4. The substrate surface roughness in a certain direction.

3 Results and Discussion

We have constructed regime map for variety of gas driven rivulet flows of water in the minichannel (Fig. 5). At that, we discovered a new regime, which has not previously been observed for the FC-72 fluid flow, namely "rivulet bridge". When flowing in this regime, the rivulet does not contact the side walls of the channel, though occupies its entire height. Moreover, we observed different variations in the "rivulet bridge" flow regime: pulsating flow regime, straight flow regime, divergent flow regime, and tapered flow regime. The emergence of a new flow regime is associated with the fact that the static wetting angle of water is much higher than that in the fluorinert fluid FC-72. It means that the height of the water rivulet when driven by the gas in the minichannel is much greater than that of FC-72 fluid. Hence, we can conclude that the static wetting angle significantly affects the rivulet flow regimes.

We have also observed the regimes, where the geometry of the rivulet has been previously found for the FC-72 fluid flow in the minichannel. These are straight, tapering, and pulsating flow regimes.

For better visualization we have also constructed a map representing various surface deformation regimes (Fig. 6). At that, we observed wave generation regime, already familiar to us from the experiments with FC-72 fluid, as well as a new braiding regime (Fig. 8) that is caused by a larger surface tension coefficient.

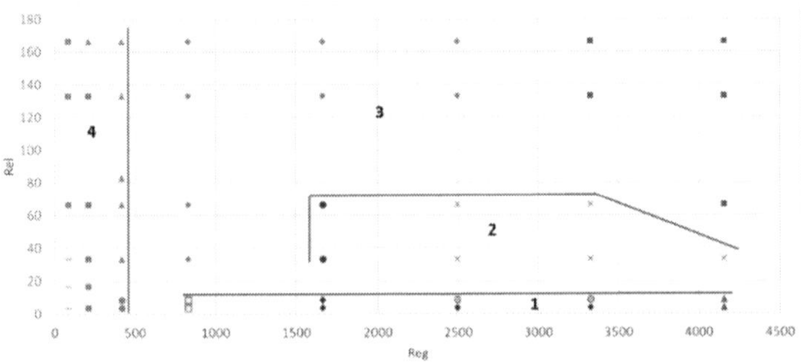

Fig. 5. Flow regime map representing varieties of the geometry of the nitrogen driven rivulet flow of water in the minichannel. 1 – pulsating flow regime, 2 – tapered flow regime, 3 – straight flow regime, 4 – rivulet bridge.

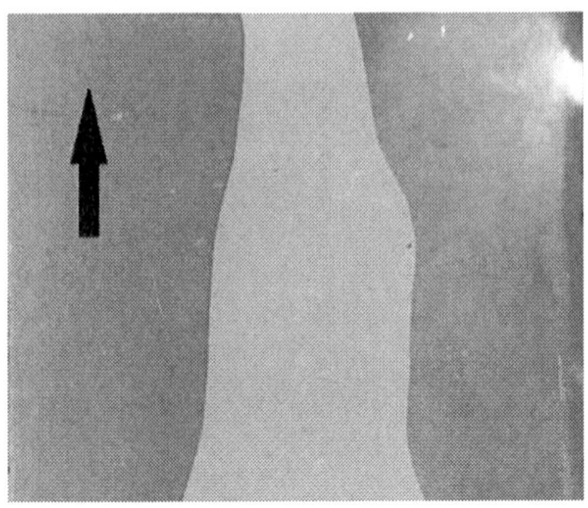

Fig. 6. The nitrogen driven water rivulet flow regime in the minichannel-rivulet bridge.

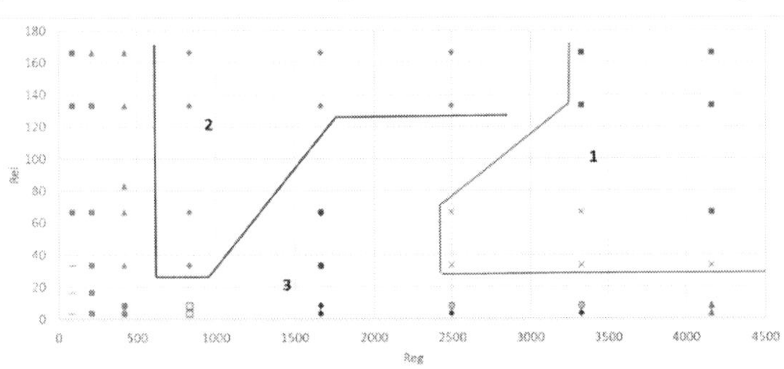

Fig. 7. Flow regime map representing varieties of deformation of the nitrogen driven water rivulet surface in the minichannel: 1 – braiding, 2 – wave, 3 –without any surface deformation.

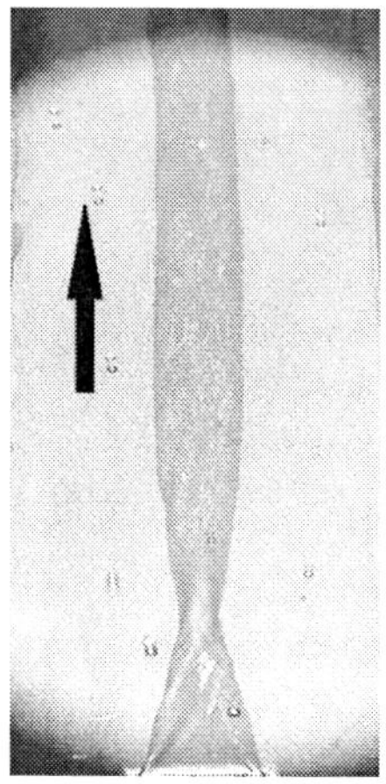

Fig. 8. Braiding surface deformation regime.

This study was performed in the framework of the Russian Science Foundation grant (Agreement No. 14-19-01755).

References

1. P. Stephan, c. Brandt, Heat Transfer Edging **25**, 78 (2004)
2. M. Potash, P.C. Wayner, Int. J. Heat Mass Transfer **15**, 1851 (1972)
3. V.V. Kuznetsov, O.V. Vitovskii, V.A. Krasovskii, High Temp **45**, 68 (2007)
4. V.V. Cheverda, A. Glushchuk, P. Queeckers, S.B. Chikov, O.A. Kabov, Microgravity sci. tec. **25**, 73 (2013)
5. M.V. Bartashevich, O.A. Kabov, V.V. Kuznetsov, Fluid dyn. **45**, 924 (2010)
6. P. Schmuki, M. Laso, J. Fluid Mec. **215**, 125 (1990)
7. S.V. Alekseenko, A.V. Bobyleb, S.M. Kharlamov, D.M. Markovich, *Experimental investigation of heat transfer in a rivulet on the inclined foil* (Lisabon, 2008)
8. K. Mertens, V. Putkaradze, P. Vorobieff, J. Fluid Mec. **531**, 49 (2005)

Mathematical modeling of the electric spark ignition of the coal-water particle fuel under conditions relevant to the internal combustion engines

Samen Syrodoy[1*], *Oleg* Penyazkov[2], *Mohamad* Assad[2], and *Nadegda* Gutareva[1].

[1]National Research Tomsk Polytechnic University, 634050 Tomsk, Russia
[2]Institute of Heat and Mass Transfer, 220000 Minsk, Belarus

Annotation. The results of the mathematical simulation of the coal-water fuel particle in the conditions of electric-spark discharge have been given. It has been found that the CWF particle can ignite even in a relatively low (Tg<500K) ambient temperature. It has been shown that the strength of the current and ambient temperature have a significant effect on the characteristics and conditions of ignition.

1 Introduction

The uncertainty in the energy markets (oil and gas) [1] has significantly increased in the recent years (10-20 years). Their heterogeneous distribution in the different regions of the world leads to an increase in the geopolitical tensions [2, 3]. Such a situation creates the conditions for the partial replacement of oil and products of its distillation (fuel oil, gasoline, alcohol, etc.) with the new ecological and effective fuel compositions [4]. Coal-water slurry can be classified to the latter. Coal in contrast to oil is more evenly distributed on the planet and can be a factor in the reduction of the "turbulent" dynamics of the prices for hydrocarbons [5]. Coal, however, is the fuel that is difficult to apply for combustion in the internal combustion engines (ICE). The attempts are known to use solid fuel (coal, wood) for internal combustion engines [6]. The latter attempts have been based on the technology of solid fuel gasification and combustion of synthetic gas obtained in the combustions chambers of the engine. However, these techniques require cumbersome and expensive reactors of the gasifiers [6]. Taking into account the constrained layout of the car, the application of technology of the coal gasification is unlikely.

One of the most promising ways to use coal as a fuel for internal combustion engines is the combustion of coal fuel in coal-water slurry [7]. This technology will allow with the minimal changes in the design of the piston or rotary engine to use coal as the main fuel for the vehicles for different purposes (cars, locomotives, river and sea vessels, etc.) [8-10]. However, so far, the processes of the particles ignition of the water-coal fuel in the conditions corresponding to the combustion chambers of the internal combustion engine while the spark discharge have not been investigated.

*Corresponding author: ssyrodoy@yandex.ru

It is known that the WCF is a substantially heterogeneous structure, consisting of liquid (water) and solid (coal, organic fuel, metals) components [11]. It should be noted that the ignition of the traditional liquid fuels (gasoline) in the combustion chambers of the internal combustion engine occurs due to a complex radiation-convective heat transfer and electric spark heating. It should be noted that the ignition of the hydrocarbon fuel occurs in the conditions of intensive physico-chemical (thermal decomposition) and phase (evaporation of water) transformations [12]. In the process of dewatering of the coal-water particle changes the electrical resistance of the latter changes, respectively, the intensity of heat release during WCF heating also changes. Taking into account the above we can say that the electric spark ignition of the water-coal fuel particle in the conditions corresponding to the combustion chambers of the internal combustion engine is a complex and still unsolved problem.

The aim of this work is the mathematical modeling of ignition of coal-water fuel particle by heating by an electric spark.

2 Mathematical model

When setting the task, it has been assumed that water-coal particle enters the combustion chamber of the internal combustion engines in the inter-electrode space (fig.1.). As a result, the spark discharge it is heated and the water evaporation process is initiated. The evaporation front moves through the surface into the particle. The resulting water vapor is filtered to the external surface of the particle. The water vapors come into thermochemical interaction with carbon coke. Further heating leads to thermal decomposition of the organic part of the fuel and volatile. The latter are filtered on the surface, near which steam-gas mixture is formed having a flammable range with the achievement of the critical values of temperatures and concentrations of fuel.

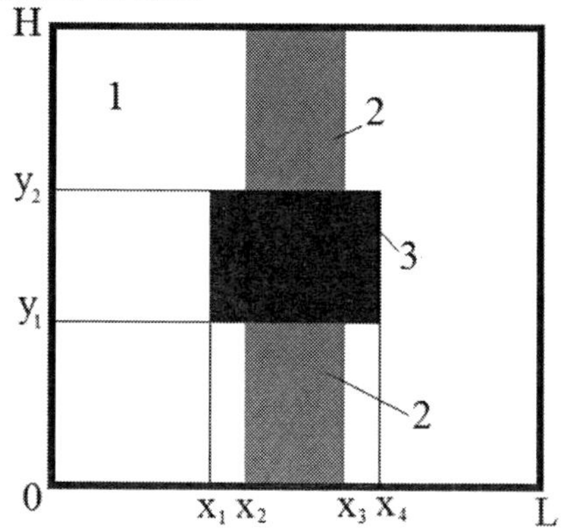

Fig. 1. The scheme of solving the problem. 1 – gas area; 2 – electrodes; 3 – WCF particle.

The mathematical formulation of the problem corresponding to the above physical model includes the system of nonstationary differential equations:

- the energy equation for the WCF particle:

$$C_1(T)\cdot\rho_1(T)\cdot\frac{\partial T_1}{\partial t}=\frac{\partial}{\partial x}\left[\lambda\cdot\frac{\partial T_1}{\partial x}\right]+\frac{\partial}{\partial y}\left[\lambda\cdot\frac{\partial T_1}{\partial y}\right]-\frac{Q_{eva}\cdot W_{eva}}{h_d}\cdot\delta(r_{eva})\pm\sum_{i=1}^{i=2}Q_i\cdot W_i-C_s\cdot\rho_s\cdot\left[U_s\cdot\frac{\partial T_1}{\partial x}+V_s\cdot\frac{\partial T_1}{\partial y}\right] \quad (1)$$

Where: $\sum_{i=1}^{n=2}Q_i\cdot W_i=Q_{w\to vol}\cdot W_{w\to vol}+Q_{C+H_2O}\cdot W_{C+H_2O}+I\cdot R^2\cdot\theta(x_1-x_2,y_1-y_2)$

$\delta(x)$ is the Dirac Function. $\theta(x_1-x_2,y_1-y_2)$ – function of Heaviside. The thermophysical characteristics (λ, C and ρ) are determined by the position of the evaporation front. The mass rate of evaporation (W_{eva} $[{}^{kg}/_{M^2 \cdot s})]$) has been calculated from the expression [13]:

$$W_{eva} = W_f \cdot \exp\left(\frac{Q_{wat} \cdot \mu \cdot \left(T_{wat} - T_f\right)}{R \cdot T_0 \cdot T_{wat}}\right) \tag{2}$$

The temperature distribution in the gas area surrounding the particle is described by the energy equation. The latter takes into account the exothermic effects of the oxidation reactions of the main combustible components by air:

$$C_g \rho_g \frac{\partial T_g}{\partial t} = \lambda_g \cdot \left[\frac{\partial^2 T_g}{\partial x^2} + \frac{\partial^2 T_g}{\partial y^2}\right] + \sum_{i=1}^{i=4} Q_i \cdot W_i \tag{3}$$

Where: $\sum_{i=1}^{n=4} Q_i \cdot W_i = Q_{CH_4+O_2} \cdot W_{CH_4+O_2} + Q_{CO+O_2} \cdot W_{CO+O_2} + Q_{H_2+O_2} \cdot W_{H_2+O_2}$

The boundary conditions of the 4 kind have been taken at the boundary of the system partition "WCF particle — gas environment":

$$\lambda_1 \frac{\partial T_1}{\partial n} = \lambda_g \frac{\partial T_g}{\partial n} \tag{4}$$

$$T_1\left(x, y, t\right) = T_g\left(x, y, t\right)$$

The temperature distribution in the electrodes is described by the equation:

$$C_S \rho_S \frac{\partial T_S}{\partial t} = \lambda_S \cdot \left[\frac{\partial^2 T_S}{\partial x^2} + \frac{\partial^2 T_S}{\partial y^2}\right] \tag{5}$$

The pressure distribution of the mixture of water vapor and products of pyrolysis of the fuel in the particle has been determined from the equation piezo conductivity:

$$\frac{\partial p}{\partial t} = \chi \cdot \left[\frac{\partial^2 p}{\partial x^2} + \frac{\partial^2 p}{\partial y^2}\right] \tag{6}$$

The filtration rate of vapor has been determined from the differential expression of Darcy's law:

$$U_s = -\frac{K_P}{\nu}\frac{\partial p}{\partial x}; \quad V_s = -\frac{K_P}{\nu}\frac{\partial p}{\partial y} \tag{7}$$

When setting the task it is accepted that water vapor enters into endothermic chemical reaction with coal. The reaction rate $C + H_2O = CO + H_2 - 118.485$ MJ/kg has been calculated from the following expression:

$$W_{C+H_2O} = k_{C+H_2O} \cdot c_{H_2O} \cdot \rho_s \cdot \exp\left(-\frac{E_i}{R \cdot T}\right) \quad (8)$$

The decomposition of the fuel with the release of volatiles begins when there is the condition on the surface of the particle $T_{sur} \geq T_{std}$ (T_{std} — the beginning temperature of thermal decomposition). The process of thermal decomposition is described by the equation of chemical kinetics:

$$\frac{\partial \eta}{\partial t} = [1-\eta] \cdot k_{w \to vol} \cdot \exp\left(-\frac{E_{w \to vol}}{R \cdot T}\right) \quad (9)$$

The rate of pyrolysis has been calculated using the following expression:

$$W_i = [1-\eta] \cdot k_{w \to vol} \cdot \rho_1 \cdot \exp\left(-\frac{E_{w \to vol}}{R \cdot T}\right) \quad (10)$$

The following reactions have been taken into account in the vicinity of the surface of the particle:

$$1. H_2 + 0.5 \cdot O_2 = H_2O + 141900 \, {}^{J}/_{kg};$$

$$2. CO + 0.5 \cdot O_2 = CO_2 + 10090 \, {}^{J}/_{kg};$$

$$3. CH_4 + O_2 = CO_2 + H_2O + 55546 \, {}^{J}/_{kg};$$

The rate of these reactions has been calculated from the mathematical expression of the law of Arrhenius:

$$W_i = k_i \cdot c_i \cdot \left(1 - \sum_{i=1}^{n-1} c_i - c_{H_2O} + c_{O_2}\right) \cdot \rho_i \cdot \exp\left(-\frac{E_i}{R \cdot T_g}\right) \quad (11)$$

The concentration of the combustible components (H_2, CO and CH_4) in the near-wall area has been evaluated by solving the diffusion equation for the corresponding component of the reaction:

$$\rho_g \frac{\partial c_i}{\partial t} = D_i \cdot \rho_g \left[\frac{\partial^2 c_i}{\partial x^2} + \frac{\partial^2 c_i}{\partial y^2}\right] - W_i \quad (12)$$

The system of equations (1–13) has been solved under the following boundary conditions, and closing ratios:

$$x_1 < x < x_2, y_1 < y < y_2, \; T(x,y,0) = T_0 = 298K, \; p(x,y,0) = p_{atm}, \; \eta(x,y,0) = \eta_0$$

$$0 < x < x_{out}, 0 < y < y_{out}, \; T(x,y,0) = T_g, \; c_i(x,y,0) = 0.$$

$$\left.\frac{\partial^2 T_g}{\partial x^2}\right|_{\substack{x=0;\\ x=L}} = \left.\frac{\partial^2 T_g}{\partial y^2}\right|_{\substack{y=0;\\ y=H.}} = 0$$

$$\left.\frac{\partial p}{\partial x}\right|_{\substack{x=x_1;\\ x=x_2.}} = \left.\frac{\partial p}{\partial y}\right|_{\substack{y=y_1;\\ y=y_2.}} = -\frac{\nu}{K_P}u_s$$

$$p\big|_{\substack{x=x_{out}\\ x=0}} = p_{atm}$$

$$\left.\frac{\partial C}{\partial x}\right|_{\substack{x=x_1;\\ x=x_2.}} = \left.\frac{\partial C}{\partial y}\right|_{\substack{y=y_1;\\ y=y_2.}} = \frac{W_i \cdot r_0}{\rho_g \cdot D_i}$$

$$\left.\frac{\partial C}{\partial x}\right|_{\substack{x=x_1;\\ x=x_2.}} = \left.\frac{\partial C}{\partial y}\right|_{\substack{y=y_1;\\ y=y_2.}} = \frac{W_{eva} \cdot r_0}{\rho_g \cdot D_1}$$

$$\left.\frac{\partial^2 C_1}{\partial x^2}\right|_{\substack{x=0\\ x=x_{out}}} = \left.\frac{\partial^2 C_i}{\partial y^2}\right|_{\substack{y=0\\ y=y_{out}}} = 0$$

The thermal conductivity, heat capacity and density of the wet and "dehydrated" parts of the fuel have been calculated similarly [14]. The formulated problem of mathematical physics has been solved by the finite difference method [15]. The approximation of the equations has been carried out with using the implicit four-point finite difference template. The discretization of the two-dimensional equations has been carried out with using a locally one-dimensional method of Samarskii A.A. [16].

3 Results and discussion

Figure 2 shows the dependence of delay time of ignition of the WCF particle on the current strength, obtained by the numerical solution of the system (1)-(19). The ambient temperature T_e=300K. It is clearly seen that even at relatively low ambient temperature (T_e<600K) the WCF particles can be ignited stably. The analysis of the results of numerical simulation shows that the current strength has a significant influence on the characteristics and conditions of ignition of the WCF particle. So, when increasing J from 60 to 220A, the time delay ignition of the fuel particles (diameter $d=1.5\cdot10^{-3}$ m) decreases 2 times (from 0.04 to 0.02 s). It may be noted, there is a substantially nonlinear character of the dependences (curves 1-3 of Figure 2). The latter indicates the significant role of the processes of heat and mass transfer occurring together in heavy phases (evaporation of water) and chemical (thermal decomposition, the interaction of water vapor and carbon coke, ignition of volatiles) in the induction period. The analysis of dependences shows, that the linear dimensions of the particle also have a significant impact on the characteristics of ignition. So, at the current of J=60 A, the diameter reduction of WCF drops from $d=1.5\cdot10^{-3}$ to $d=0.5\cdot10^{-3}$ m (in other words in 3 times) leads to decrease of the ignition delay time from 0.04 s to 0.01 s (almost in 4 times).

4 Conclusion

According to the results of mathematical modeling it has been established that hydrocarbon particles can ignite even in low temperature (T_e≈300K). It has been shown that the strength of the current, the spark discharge has a significant influence on the characteristics and conditions of ignition. The increase in J leads to a significant reduction of delay time of ignition. Taking into account the nature of the curves shown in figure 2, it is safe to say that the complex of the co-occurring processes of heat and mass transfer plays an essential role in the thermal preparation and ignition of the particles of the coal-water fuel.

According to the results of numerical solution of the system (1)-(19), it has been found that the coal-water particles are ignited stably in the conditions of the combustion chambers of the internal combustion engine. It should be added that the ignition of the fuel particles occurs even at relatively low ambient temperatures (T_e <500K). The latter confirms the possibility of using WCF as the main fuel for internal combustion engines.

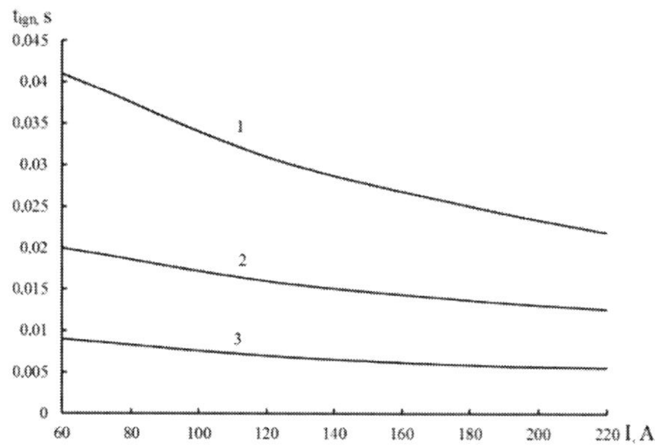

Fig.2. Delay time of ignition of the particle coal-water fuel according to the strength of the current at ambient temperature T=300K. The particle diameter of the WCF:
1 - d=1,5·10^{-3} m; 2 - 1,0·10^{-3} m; 3 - 0,5·10^{-3} m.

This work was supported by the scientific schools grant NSH-7538-2016.8.

References

1. Y. Weia, X. Guo, Energ. Econ. **56,** 88 (2016)
2. Y. Tuzova, F. Qayum, Energ. Policy **90,** 140 (2016)
3. Aad Correlje′, Coby van der Linde, Energ. Policy **34,** 532 (2006)
4. G. Lan, X. Gong, Z. Wang, L. Zhao, Y. Wang, M. Wang, Energy **96,** 69 (2016)
5. K. Johnson, B. Mohammed, Energ. Econ. (to be published)
6. A.K. Hossain, P.A. Davies P, Renew. Sust. Energ. Rev. **21,** 165 (2013)
7. B. Choongsik, K. Jaeheun, P. Combust. Inst. (to be published)
8. K. Seshadri, J.A. Caton, K.D. Kihn, J. Eng. Gas Turbines Power **114,** 515 (1992)
9. K. Wamankar, S. Murugan, Energy **91,** 1030 (2015)
10. T. Uzkan, C.E. Horton, J. Eng. for Gas Turbines Power **112,** 376 (1990)
11. G. Papachristodoulou, O. Trass, Can. J. Chem. Eng **65,** 177 (1987)
12. V.V. Salomatov, G.V. Kuznetsov, S.V. Syrodoy, N.Y. Gutareva, Appl Thermal Eng **107,** 10 (2016)
13. Ya.I.Frenkel, *Kinetic theory of liquids* (Nauka, Leningrad. 1975)
14. V.V. Salomatov, S.V. Syrodoy, N.Y. Gutareva, EPJ Web of Conferences **82,** 01037 (2015)
15. A.Zh. Samarskii, Comp. Math. Math. **2** (1962)
16. A.Zh.Samarskii, USSR Comp. Math. Math. **3,** 9 (1963)

Evaluation of scale formation in waterwall heated surfaces

Tatiana Taylasheva[1*], *Alexandra* Skirnevskaya[1], *Lyudmila* Vorobjeva[1], and *Vitali* Lebedev[2]

[1] Tomsk Polytechnic University, 634050, Tomsk, Russia
[2] Omsk State Transport University, 644046, Omsk, Russia

Abstract. This paper presents the possibility of forecasting assessments of the speed and the time of formation of depositions in the evaporator-tube elements of double-drum boilers. The values of thermal flow in the wall region of tank screens of boiler furnace are obtained, besides the velocity values of scaling metal corrosion products are obtained. Conclusions about the ability of forecasting unnominal situations and emergency risks dependent with damage to the screen surface heating pipes are made.

1 Introduction

Water is known to be the most common heat-transport medium in steam boilers. It can be explained by several factors. First of all water possess high values of heat storage capacity, density, heat conductivity and viscosity. All this allows achieving fairly high coefficients of heat exchange. Secondly, water is a preferable heat transport medium since it is cheap.

But given all its advantages water has some disadvantages. The major disadvantage is that water actively causes corrosion. It results in certain problems when it is used as a heat transport medium, for instance, corrosion of metal of various equipment of the boiler and consequent deterioration of the boiler's reliability.

Various impurities present in water including corrosion products can deposit not only in the inner surface of heated pipes but also in some other equipment that will result in the significant reduction of reliability and cost effectiveness of the boiler and even of the steam and water pipes of the whole power plant in general.

2 Description of the model and the object of research

The study is devoted to investigation of the process of scale formation and corrosion of inner heated surfaces since they require special attention.

The intensity of the above process is correlated with the quality of the boiler water, i.e. contents of impurities and their quantity. This correlation explains the strictly regulated content of each impurity both of feed water and boiler water and the lower content of impurities is, the higher pressure of the heat transport medium is.

* Corresponding author: taylasheva@tpu.ru

Thus the quantitative content of impurities is to be lower than the required one to prevent the deposit of these impurities on inner surfaces of pipes of various heating surfaces such as superheaters, condensers, various desuperheaters etc. Due to the above requirements it is necessary to clean the primary water to provide the required quantitative parameters which can be achieved by the following: filtration, natural settling, ionization, coagulation, decarbonization etc. And in order to support the sufficient quality of the boiler water, the boiler is continuously blown off.

This article studies the boiler of DKVR-20-13 GM.

The specific feature of this boiler is the availability of two drums (upper and lower) which are aimed for different tasks. The upper drum is used for steam separation and feed water supply. The lower drum is used for sludge settlement and feeding of saline parts (loop with remote cyclone separators). The feed water is supplied to the upper drum through two feeding lines and from there it is supplied to the lower drum through low heated pipes of the convection tube bank. Furnace waterwalls are fed by non-heated pipes from the lower and upper drums: the front waterwall is fed by water from down-take pipes of the upper drum and the rear waterwall is fed by water from down-take pipes of the lower drum. The steam and water mixture from the waterwalls and risers is supplied to the upper drum. All boilers of this type are provided by internal steam separators in the upper drum.

The scale formation is the most probable in the parts of the drums connected to remote cyclone separators due to the high content of salt in the boiler water.

3 Results and discussion

The study is based on the correlation of the deposit formation rates of A [1]:

$$A_i = k \cdot C_i \cdot q^n, \tag{1}$$

where k – factor of proportionality; C_i – impurity concentration, mg/dm^3; q – heat flow, W/m^2; n = 1, 2, 3.

Impurities such as hardness salts (Ca and Mg) and steel corrosion products (Fe) are the most common in the feed water. In these cases the rate of n can vary from 1 to 3.

In order to study this correlation (1) the burning process going on in this boiler was simulated with the use of FIRE 3D with the nominal capacity of [2,3]. For this purpose six waterwall pipes were selected along the total length of the lateral panel, and heat flow values were numerically calculated for them in the near-wall region for all their height. The heat flow values are presented in Table 1.

In order to visualize the obtained results the heat flow curves were built up for each pipe and they are presented on Fig.2.

Table 1. Heat flows q (kW/m^2) in near-wall part of the lateral waterwall of DKVR-20 boiler.

Pipe	**No. 1**	**No. 2**	**No. 3**	**No. 4**	**No. 5**	**No. 6**
X, m / Y, m	0.096	0.4808	1.3256	1.536	2.1704	2.3816
0.14	847.3	849.3	827.2	810.5	733.7	702.2
1.13	739.2	783.9	823.5	814.9	749.4	718.1
1.47	694.9	751.3	817.0	812.8	756.2	726.4
1.84	655.9	717.1	802.2	802.1	754.5	726.4
2.20	608.6	670.1	764.8	767.6	727.6	701.4

DOI: 10.1051/matecconf/20179201029

2.45	567.8	627.7	722.9	726.9	691.7	667.1
2.63	534.7	592.7	685.7	690.5	658.8	635.6
2.85	490.1	544.4	632.2	637.5	610.8	589.9
3.08	439.3	488.9	569.7	575.6	554.8	536.6
3.30	386.8	431.9	506.3	512.8	498.2	482.8
3.51	334.4	375.2	445.3	452.6	444.3	431.7
3.72	283.5	320.9	389.0	397.4	395.3	385.2
3.95	228.1	262.3	330.8	340.6	345.2	338.0
4.28	155.9	186.5	262.8	274.6	287.4	283.8
4.61	104.4	129.2	211.0	224.4	242.8	242.0
4.93	78.7	97.2	174.1	187.5	208.4	209.4
5.26	68.9	83.8	149.6	161.8	182.5	184.4
5.42	67.6	81.3	141.3	152.8	172.8	174.8

Values of X and Y are geometrical coordinates of the estimated area which plane coincides with the plane of the waterwall pipes location and the point of reference is the lower point of the left lateral waterwall.

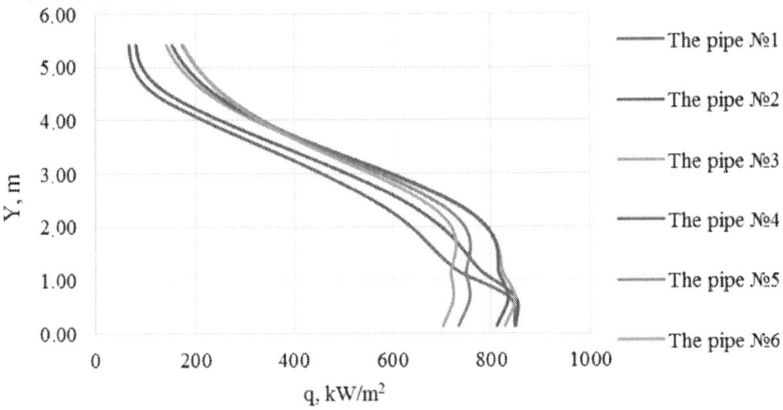

Fig. 1. Heat flow curves in the near-wall region of the lateral waterwall of DKVR-20 boiler.

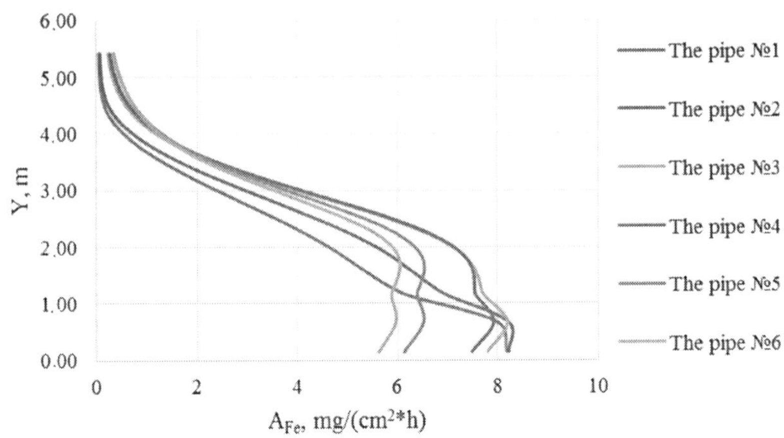

Fig. 2. The curve of deposit formation rate of ferrous corrosion products in the pipes of the lateral waterwall of DKVR-20 boiler.

Figure 2 shows the estimated deposit formation rate from which one can conclude that the maximum scale formation rate coincides with the level of burners' location, and the area of maximum deposit formation is located there.

4 Conclusions

The final purpose of this study is to predict the time of failure free operation of boilers, i.e. to define the time till formation of critical volume of deposits of various impurities. Results of this study allow predicting non-routine events and failure risks related namely to the damage of pipes within waterwall heating surfaces.

References

1. V.N. Voronov, T.I. Petrova, *Vodno-khimicheskie rezhimy TES i AES* (MEI, Moscow, 2009)
2. N.V. Vizgavljust, A.V. Starchenko, A.V. Gil, T.S. Taylasheva, EPJ Web Conf. **82**, 01040 (2015)
3. V.V. Popov, T.S. Taylasheva, V.M. Lebedev, Simulations **72**, 01089 (2016)

Heat transfer behind the backward-facing step under the influence of longitudinal pressure gradient

Tatyana Bogatko*, *Viktor* Terekhov, *Aleksey* Dyachenko, and *Yaroslav* Smulsky

Kutateladze Institute of Thermophysics SB RAS, 630090, Novosibirsk, Russia

Abstract. Results of an experimental and numerical study of the flow structure and turbulent heat transfer in a rectangular channel with longitudinal pressure gradient behind the backward-facing step. With the channel expansion the angle of rotation of the upper wall is 1.43; 2.86 and 4°, and with the channel convergence the rotation angle is 3; 5.7 and 7.6°. Experiments are carried out at Reynolds numbers, calculated on the step height and velocity in front of the backward-facing step, Re_H = 4 000; 8 000; 12 000. It was determined, with the increase of pressure gradient the maximum value of Nusselt number increases for the converging channel and decreases for the expanding channel.

1 Introduction

This study focuses on the properties of a separated flow behind the backward-facing step under conditions of imposed longitudinal pressure gradient. The most common are channels with pressure gradient in the flow paths of compressors. These are interblade channels of impellers, guide vanes and straighteners, as well as transition channels with different configuration of the flow part. Separation usually leads to harmful consequences, causing loss of energy, onset of instability, etc. To further improve the efficiency of thermal power equipment and to develop the fundamentals of separated flows it is crucial to conduct comprehensive experimental and computational studies of new opportunities of intensification of heat and mass transfer in the separation gradient flows behind the backward-facing step. Of interest are both expanding and converging channels. This report contains the results of an experimental study of the separated flow behind the backward-facing step in a flat channel with flow acceleration and deceleration. The longitudinal gradient was created due to the position of the upper channel wall behind the backward-facing step. The pressure fields and heat transfer coefficients in recirculating and relaxation areas behind a drastic channel expansion were measured. It should be noted that data on similar themes available in the literature concern the flow dynamics [1-3], and data on heat transfer are virtually absent.

The report also presents results of the numerical simulation of the effect of positive and negative longitudinal pressure gradients on hydrodynamic flow characteristics and turbulent

* Corresponding author: tbogatko@gmail.com

heat transfer in a pipe with sudden expansion. A theoretical approach to the description of turbulent separated flows is rather complicated and remains topical to the present time. Extensive research for creating the engineering methods of separated flow calculation was conducted in [4-6]. The calculations were performed within the model of incompressible fluid based on the system of Reynolds averaged stationary equations of Navier-Stokes and energy equations.

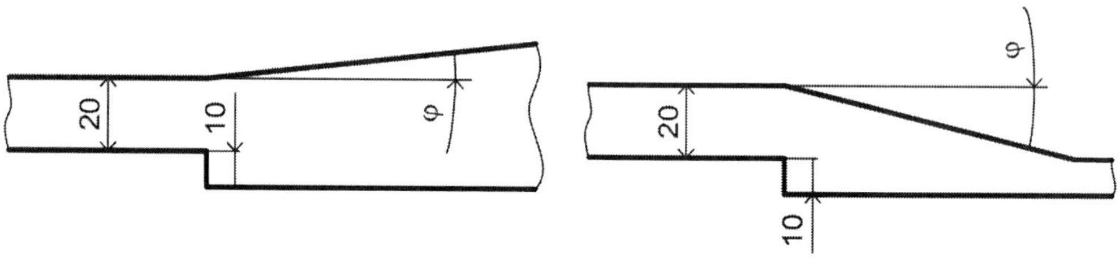

Fig. 1. Schematic diagrams of experimental section in an expanding (left) and converging (right) channels.

Fig. 1 shows schemes of experimental measurements in expanding and converging channels. With the channel expansion the angle of rotation of the upper wall is 1.43; 2.86 and 4°, and with the channel convergence the rotation angle is 3; 5.7 and 7.6°. In the latter case, the length of the inclined wall (L=380; 200 and 150 mm) decreases, and its lower end is always located at a distance of 10 mm from the bottom channel wall. Experiments are carried out at Reynolds numbers, calculated on the step height and velocity in front of the backward-facing step, Re_H=4000; 8000; 12000. The calculations required for the work are performed using the k-ω SST turbulence model [4], most adequately describing the turbulent separation flows [5]. The problem statement is two-dimensional, and the flow is stationary. The geometry and initial conditions correspond to experimental parameters.

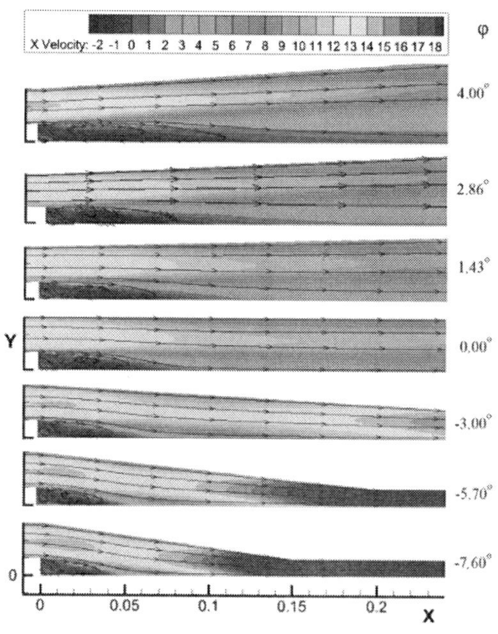

Fig. 2. Velocity field in expanding and converging channels, Reh=8000.

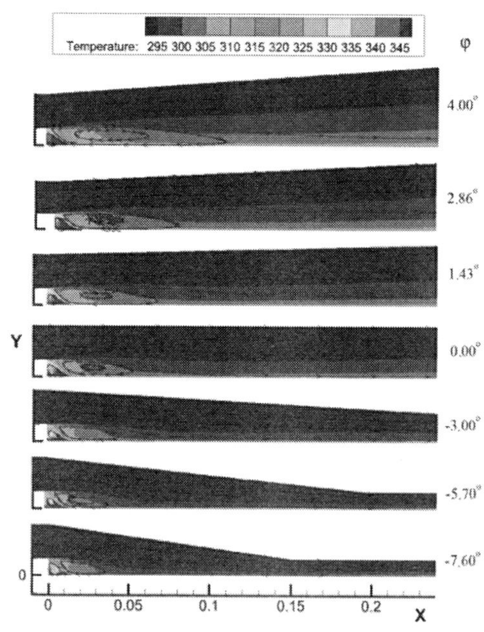

Fig. 3. Temperature field in expanding and converging channels, Reh=8000.

2 Results

For different values of the longitudinal pressure gradient, the flow character behind the backward-facing step varies considerably. The main difference is observed along the length of the recirculation zone. Fig. 2 shows the flow patterns in the region of sudden expansion and the field of the longitudinal velocity component. It is seen that the length of the recirculation zone increases with increasing adverse pressure gradient and greatly reduces at imposed favorable pressure gradient. In addition, this figure clearly shows the deceleration and acceleration of the flow at a turn of the upper cover of the channel. In this section of the channel convergence there is a significant, two-fold acceleration of the flow.

Fig. 3 with streamlines superimposed on the temperature field clearly demonstrates how the thermal boundary layer increases (at decelerating flow) and decreases (at accelerating flow). Accordingly, the heat removal from the surface in this section of the channel varies significantly.

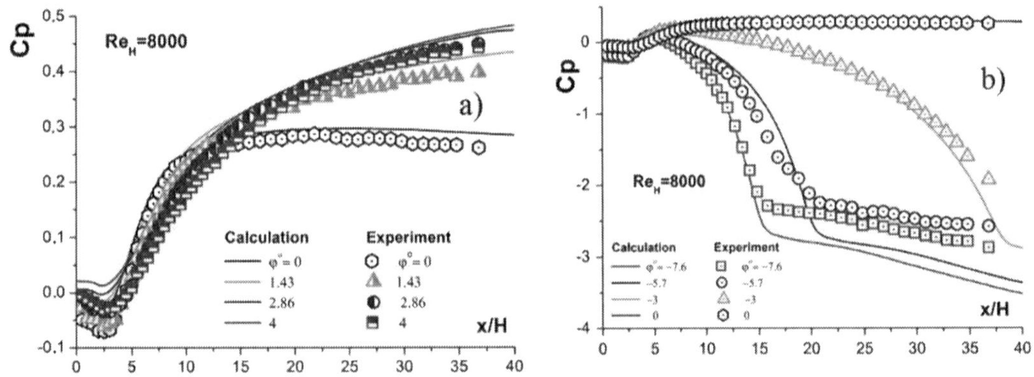

Fig. 4. Distribution of the pressure coefficient in the expanding (a) and converging (b) channels.

Distributions of pressure coefficients for both cases are shown in Fig. 4. With an increase in the angle of channel expansion the pressure ratio increases, but the increase takes place almost to the value φ =2.86°. A further increase in the angle has no influence on the value of pressure coefficients. With the growth of the channel convergence angle the pressure ratios decrease. When L=200 and 150 mm (φ=5.67 and 7.5°) the pressure distribution along the confusor length changes, and the region of flow relaminarization is clearly seen.

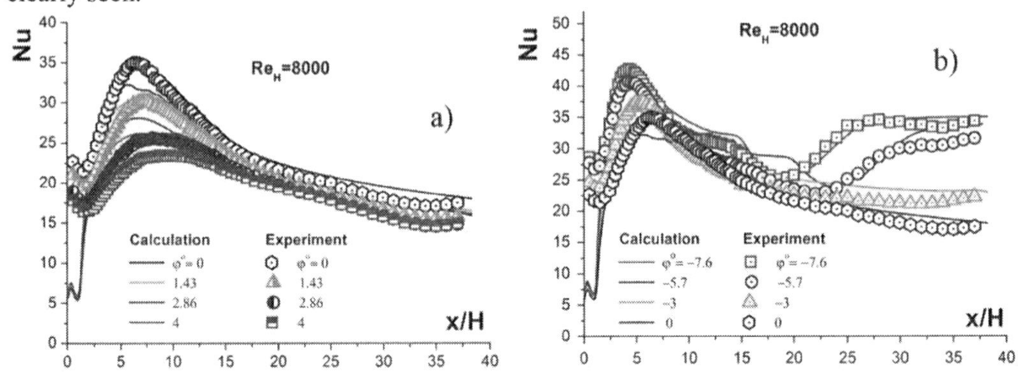

Fig. 5. Distribution of heat transfer coefficient in the expanding (a) and converging (b) channels.

Fig. 5 demonstrates changes in the Nusselt number along the longitudinal coordinate behind the backward-facing step. At adverse pressure gradient, with increasing angle the flow decelerates increasingly, and the maximum heat transfer coefficient decreases.

At favorable pressure gradient the situation is opposite. Maximum heat transfer increases, but less intensively than in case of adverse pressure gradient. At x/H = 25–30 for angles φ=5.67 and 7.5°, similar to pressure ratios, the Nusselt numbers clearly reflect the pronounced flow relaminarization.

Figure 6 shows maximum heat transfer coefficients depending on the Kays parameter $K = \frac{\nu}{U_0^2}\frac{dU_0}{dx}$. The zero parameter of Kays corresponds to zero-pressure-gradient flow, negative values correspond to diffuser flow, and positive values correspond to confluence.

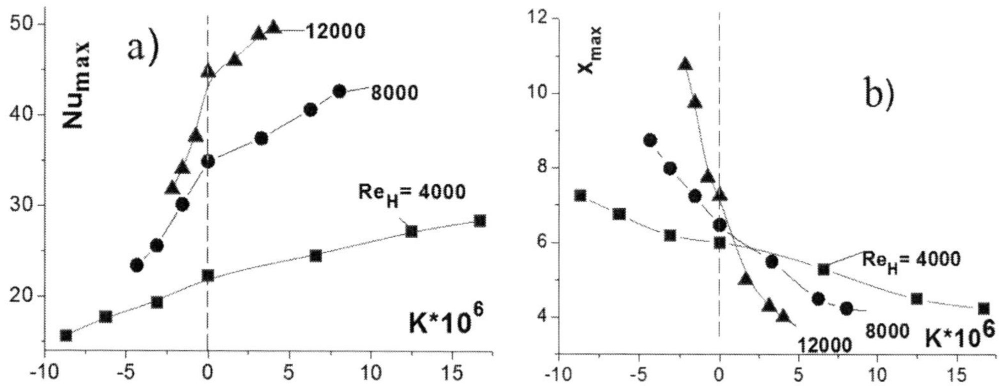

Fig. 6. Dependence of maximum heat transfer coefficient (a) and its coordinate (b) in the expanding and converging channels on Kays parameter.

With the increase of pressure gradient the maximum value of Nusselt number increases for the converging channel and decreases for the expanding channel. The dependence of the Nusselt number on the Reynolds number is apparent. At increased Reynolds number, not only the absolute value of Nu_{max}, but the angle of the lines describing the experimental data increase. This suggests that with increasing Reynolds number the effect of longitudinal pressure gradient on heat transfer is greater.

An important parameter in the separation flow is the length of the return flow area, which correlates with the length of the maximum heat transfer x_{max}. In accordance with Fig. 6, the value of x_{max} increases at growing adverse pressure gradient and decreases at an increase in favorable pressure gradient. In this case x_{max} is strongly dependent on the Reynolds number.

As can be seen from Fig. 6b, at zero-pressure-gradient flow (K=0) the Re number influences the position of the maximum heat transfer (x_{max}=6-7.5), which is confirmed by experiments of other authors [6-8].

In accordance with Fig. 6b, the gradient flows have qualitatively different flow patterns with a strong dependence on the Re_H number.

This work was supported by the Russian Science Foundation (Project RNF 14-19-00402).

References

1. Lo, Kin Pong, J.E. Christopher, K.E. John, Exp. fluids **53**, 1317 (2012)
2. D.M. Driver, H.L. Seegmiller, AIAA J. **23**,163 (1985)
3. D.A.D. Kuehn, AIAA J. **18**, 343 (1980)
4. F.R. Metnter, AIAA J. **32,** 1598 (1994)

5. V.I. Terekhov, T.V. Bogatko, J. Appl. Mec. Tec. **56**, 471 (2015)
6. A.I. Leontiev, S.I. Kaskov, Yu.F. Gortyshov, I.A.Popov, V.V. Olimpiev, ASME, Heat Transfer Division (2006)
7. V.K. Migay, *Modelling of heat exchange power equipment* (Energoatomizdat, Moscow, 1987)
8. Yu.F. Gortyshov, I.A. Popov I.A., Therm. Eng. **53**, 249 (2006)

The issue of ensuring the safe explosion of the spent orbital stages of a launch vehicle with propulsion rocket engine

Valeriy I. Trushlyakov[1,*], *Konstantin* I. Zharikov[1]

[1] Omsk State Technical University, 644050 Omsk, Russia

Abstract. A method for increasing the safe explosion of the spent orbital stages of a space launch vehicle (SLV) with a propulsion rocket engine (PRE) based on the gasification of unusable residues propellant and venting fuel tanks. For gasification and ventilation the hot gases used produced by combustion of the specially selected gas generating composition (GGC) with a set of physical and chemical properties. Excluding the freezing of the drainage system on reset gasified products (residues propellant+pressurization gas+hot gases) in the near-Earth space is achieved by selecting the physical-chemical characteristics of the GGC. Proposed steps to ensure rotation of gasified products due to dumping through the drainage system to ensure the most favorable conditions for propellant gasification residues. For example, a tank with liquid oxygen stays with the orbital spent second stage of the SLV "Zenit", which shows the effectiveness of the proposed method.

1 Introduction

According to [1], 36 % of explosions in near-Earth space are associated with the destruction of explosive fuel tanks, due to existing residual pressure in the fuel tank and the fuel component of the unused residues. Pressurized gas boost fuel tanks are subjected to thermal loading, thereby increasing the internal pressure in the fuel tanks and lines due to the expansion of the gas and the evaporation of the fuel component of the unused residues.

In the growth process of the internal pressure in the fuel tanks and lines, the failure stress of the construction material increases, reaching values exceeding the strength of the construction material, which leads to explosions. Reduction of strength of the material structure of the fuel tanks, other than increasing the temperature, act as contributory factors in space: corpuscular beams, ultraviolet radiation, solar flares, etc. [2].

The analysis of the destruction of the fuel tanks' upper stages contributed to the introduction of international organizations' (IADC, COPUOS) [3] effective measures to reduce explosions in near-Earth space: making the energy stored in the spent stage passive, the removal of unused fuel component residues [4, 5], withdrawal from orbit [6, 7], and the drainage of excess gas turbocharging. However, when resetting the vapor mixture, for example, oxygen from the fuel tank (vapors' oxygen and pressurization gas-helium)

* Corresponding author: vatrushlyakov@yandex.ru

through the drainage system into the surrounding space, there is a sharp decrease in pressure and decrease in gas temperature, which leads to condensation of vapor from the oxygen and its freezing, which results in the drain pipe freezing.

2 Formulation of the problem

In order to ensure the safe explosion of the orbital spent stage of SLV with undeveloped remnants of propellant the following steps are invited to be implemented:

a) selecting the composition of GGC combustion products which flow as a result of the fuel tank that will not interact with liquid propellant unusable residues;

b) upon reaching the predetermined pressure in the fuel tank, as determined by its strength, provide relief obtained by gasification products (evaporated remnants of propellant+pressurization gas+composition of combustion of the GGC) through the drainage system in the near-Earth space;

c) selection of physical-chemical parameters of the GGC to provide gasification conditions of the unusable propellant residues and preservation of the gaseous phase of the propellant vapor when resetting gasification products through the drainage system;

d) due to the discharge of gasified products through the drainage system to make a turn of the orbital spent stage of SLV to ensure the most favorable conditions for gasification of liquid propellant residues in feeding the hot gas tank.

3 The criteria for selecting the composition of the GGC

The source of the heat carrier (HC) supplied to the fuel tank are hot gases-products of the combustion of the GGC.

At this point in the missile industry a different GGC [8, 9] has been used for more than a decade with the desired characteristics and to select the optimal of the GGC and consequently reception of HC a number of requirements must be entered:

1) Products of combustion in the GGC must not react chemically with the propellant component, nor gas pressurization – (G_I);

2) Products of combustion in the GGC must not be composed of the gas product converted into the solid phase at the boiling temperature of oxygen (90-120K) – (G_{II});

3) The HC must have a maximum thermal capacity – (G_{III});

4) To transfer the maximum amount of heat in the unusable propellant residues - HC should have the lowest molecular weight – (G_{IV}).

Table 1 lists the possible compositions of the GGC formally satisfying the above requirements, similar to the compositions described in [10-15], and the composition temperature of the HC, calculated in the software package [16] at a pressure of 2.5 atm.

Table 1. Products of compositions of the GGC, the temperature products of combustion of the GGC, including the condensed phase (s) remaining in the gas generator.

№	Investigation of the GGC	Temperature products of combustion, K	Number of HC required for gasification of 1 kg of liquid oxygen, kg	Number of GGC required for gasification of 1 kg of liquid oxygen, kg
1	64.7 % NaN_3 + 35.3 % $KMnO_4$	1515	0.074	0.18
2	67.6 % NaN_3 + 33.4 % CrO_3	1798	0.059	0.14
3	50 % NaN_3 + 50 % Fe_2O_3	1125	0.109	0.34
4	60 % NaN_3 + 40 % CaO_2	1740	0.062	0.154

5	64 % $NaClO_3$ + 26 % BaO_2 +6 % SiO_2 + 4 % Mg	1466	0.081	0.28
6	86 % $NaClO_4$ + 10 % Na_2O_2 +4 % Si	1480	0.081	0.19
7	77.5 % $NaClO_4$ + 17 % CaO_2+ 5.5 % Mg	1457	0.082	0.20
8	78 % $NaClO_3$ + 7 % BaO_2 +4 % SiO_2 + 9.5 % Fe + 1.5 % Mg	1475	0.081	0.26

4 Technology of gasification process in the tank for the spent stage of the space launch vehicle

The basic technology of the process of gasification of the residual propellant in the fuel tank of liquid oxygen in the sample stage of the SLV is implemented in the following sequence: after turning off the main propellant rocket engine to the fuel tank from the remnants of propellant (liquid and vapor phase) the pressurization gas with total pressure of about 3 atm is served of the HC with a temperature of 1100-1700 K and speed of about 900 m/s [4].

As a result of the heating and evaporation of the liquid phase of the propellant, the pressure increases, the temperature gradient is aligned to the thermodynamic system "products of combustion the GGC+vaporized residual propellant+gas helium pressurization".

In the future, we are considering the possibility of heat transfer from the gas generator housing to the drainage system via the heat bridge to prevent freezing of the vapor on the inner surface of the drainage pipe (see. Fig. 1).

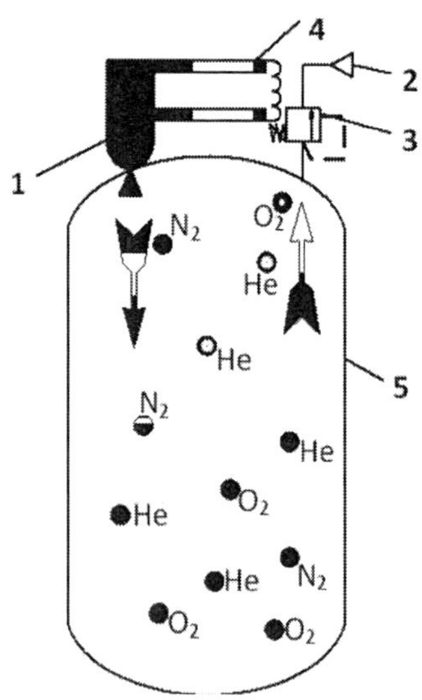

Fig. 1. Schematic diagram of the installation of the gas generator on the fuel tank and drainage system: 1 - gasifier of the GGC; 2 - nozzle reset gasification products from the tank; 3 - drain-safety valve; 4 - thermal bridge for the transfer of residual heat from the gas generator to the drainage system; 5 - fuel tank.

5 The process of releasing gasification products into the environment

Upon reaching the opening pressure of the drain-safety valve the gasification products of the fuel tank are released into the environment. For example, the fuel tank of liquid oxygen of the second stage of the SLV "Zenit" held a preliminary assessment of the thermal loading of the fuel tank [17] and input of the HC to the fuel tank [18].

The curves of change in partial oxygen pressure, shown in Fig. 2, deliver the supply efficiency of the HC to the fuel tank.

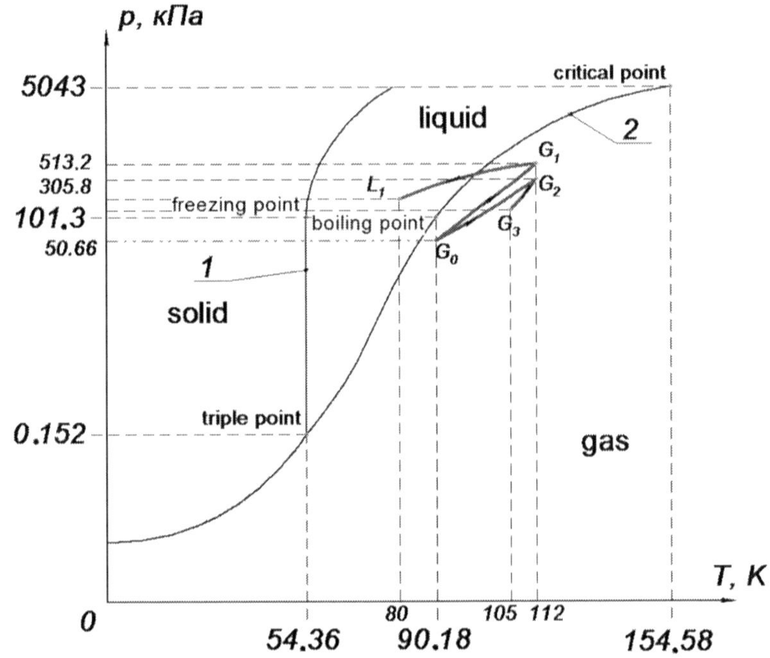

Fig. 2. Phase diagram and the phase equilibrium line: 1 – solid and liquid states; 2 – liquid and gas states; the values of the partial pressures: G_0 – at the time off the main propellant rocket engine; G_1 – the thermal loading of the fuel tank; G_2 – the entering of the HC; L_1 – release of the gas-vapor mixture; G_3 – release of gasification products.

6 Conclusions

1. Provide safe explosion of the spent orbital stage of the space launch vehicle with unusable remnants of the liquid propellant in the tanks is proposed to be carried out on the basis of ventilation by means of hot gases produced by the combustion of the GGC.
2. Prevention of the freezing of the drainage system on release of the gasification products (vaporized residual propellant + residues pressurization gas + hot gases) into the surrounding space is achieved through the selection of physical-chemical properties.
3. The most effective of GGC are oxygen or nitrogen generation HC № 2 and 4.
4. The use of reactive energy discharged gasification products allows for the most favorable boundary conditions of liquid propellant in the tanks for the process of gasification.

Research carried out with the financial support of the Ministry of Education of Russia, agreement №14.577.21.0157 from 28.11.2014, (unique identifier RFMEFI57714X0157).

References

1. Tech. Report on Space Deb., A/AC.105/720 [Internet]. [Cited 2016 Oct 21]. Available from: http://www.orbitaldebris.jsc.nasa.gov/library/references.html.
2. B. A. Campbell, S. W. McCandless jr., *Introduction to space sciences and spacecraft applications* (Gulf Professional Publishing, 1996)
3. Update of the IADC space deb. mitigation guidelines. IADC-11-02. Beijing, May (2014)
4. V. I. Trushlyakov, D. B. Lempert, M. E. Bel'kova, Combust. Explo. Shock **51**, 326 (2015)
5. Report by the Secretariat. Committee on the Peaceful Uses of Outer Space [Internet]. [Cited 2016 Sep 15]. Available from: http://www.unoosa.org/pdf/reports/ac105/AC105_620R.pdf
6. V. T. Trushlyakov, Ya. T. Shatrov, Flight **7** (2014)
7. K. Takase, M. Tsuboi, S. Mori, K. Kobayashi, Mitsubishi Heavy Industries Technical Review **48**, 11 (2011)
8. D. B. Lempert, G. N. Nechiporenko, G. B. Manelis, Combust. Explo. Shock **47**, 45 (2011)
9. S. Pourmortazavi, S. Hajimirsadeghi, I. Kohsari, M. Fathollahi, S. Hosseini, Fuel **87**, 244 (2008)
10. N. Matsuda, N. Hirata, S. Iyoshi, Patent 5780767 US (1995)
11. V. D. Sasnovskaya, A.P. Razumova, V. V. Aleshin, Patent 2052283 RF (1993)
12. I. A. Smirnov, A. T. Logunov, V. V. Shapovalov, V. I. Vanin, Patent 2149136 RF, (1998)
13. A. Ya. Malyshev, Ye. V. Levakov, A. A. Tatynov, V. K. Raevskiy, A. I. Sivertsev, Patent 2151735 RF, (1998)
14. H. Ellern, *Modern Pyrotechnics: Fundamental Journal of Applied Physics* (Chemical Publishing Company Inc., New York, 1961)
15. A. A. Shidlovskiy, *Principles Of Pyrotechnics* (Mashinostroenie, Moscow, 1973)
16. B. G. Trusov, *The XIV Int. Symp. on Chemical Thermodynamics* (MGTU, Moscow, 2002)
17. V. I. Trushlyakov, K. I. Zharikov, Fire and Explosion Saf. **25**, 7 (2016)
18. V. V. Shalai, V. I. Trushlyakov, V. Yu. Kudentsov, Thermal processes in the art **6**, 67 (2014)

Theoretical substantiation choice of "comfortable" air temperature in industrial premises

Mikhail A. Ulchiekov[1,*], and *Elena E.* Bulba[1]

[1]National Research Tomsk Polytechnic University, 634050 Tomsk, Russia

Abstract. In this paper solved the heat conduction problem for the system - the object of heat - insulation - environment (indoor air) considering natural convection. New variant substantiation of a choice the room temperature is proposed based on mathematical modelling.

1 Introduction

Energy saving is a priority problem in many sectors of the economy. The solution of global (essentially) energy saving task is only possible as a result of the decision of a large number (several hundreds) of specific targets to reduce energy costs in all industries and in the social sphere. One of these (belonging to the category of the most significant, perhaps) is the task of forming a comfortable [1] for the working climate of the production facilities. Most importantly, perhaps, the parameter that determines to a large extent the degree of "comfort" climate is air temperature TA. But now, the selection of this normative parameter [2, 3], is conducted on the basis of subjective feelings of the worker. But "comfortable" temperature in industrial premises can affect a large group of factors: the thermal insulation properties of clothing, work intensity, air circulation and others. Therefore, the definition of "comfortable" temperature of production facilities on the results of mathematical modeling "comfortable" temperature industrial premises is actual direction of research.

The purpose of this paper is mathematical modeling of the heat sink from the worker and the establishment of numerical values of T_A that correspond to the conditions of "thermal comfort".

2 Formulation of the problem

It was assumed [1] that a healthy person the main part of the body temperature T_I can fluctuate within a relatively small range of 35.5 °C to 37.4 °C. Upon reaching temperatures that are beyond the scope of this range, we can talk about overcooling or overheating of worker (which does not meet the conditions of "thermal comfort").

* Corresponding author: ulchieckovmihail@mail.ru

In formulating the problem was considered (and for the use of additional layers of clothing – seven-layer) five-layer plate (as the heat sinks on the other coordinates are negligible). Fields of solutions includes: a layer of skin, which consists in turn of the layers: subcutaneous fat (thickness Δ_1), dermis (thickness Δ_2), the epidermis (thickness Δ_3)), the air gap between human skin and clothing (Δ_4), and as an external heat insulator – clothing (Δ_5). Skin body cover adjoins as the outer layer to the main part of person, which constant temperature. On the external surface of clothing the heat exchange with the environment where the temperature $T_A=const$ by natural convection.

The solution to this problem is reduced to the calculation of skin surface temperature T_S of the worker, which will determine the "comfort" of the thermal regime.

The mathematical model of the investigated process is a system of five equations of heat conduction with boundary and initial conditions given below.

The mathematical formulation of the problem:

$$\begin{cases} \rho_1 c_1 \dfrac{\partial T_1}{\partial t} = \lambda_1 \dfrac{\partial^2 T_1}{\partial x^2}, \quad 0 < x < x_1; \\ \rho_2 c_2 \dfrac{\partial T_2}{\partial t} = \lambda_2 \dfrac{\partial^2 T_2}{\partial x^2}, \quad x_1 < x < x_2; \\ \rho_3 c_3 \dfrac{\partial T_3}{\partial t} = \lambda_3 \dfrac{\partial^2 T_3}{\partial x^2}, \quad x_2 < x < x_3; \\ \rho_4 c_4 \dfrac{\partial T_4}{\partial t} = \lambda_4 \dfrac{\partial^2 T_4}{\partial x^2}, \quad x_3 < x < x_4; \\ \rho_5 c_5 \dfrac{\partial T_5}{\partial t} = \lambda_5 \dfrac{\partial^2 T_5}{\partial x^2}, \quad x_4 < x < L; \end{cases}$$

Here ρ_i, λ_i, c_i – the density, thermal conductivity and specific heat of the material system of i-layer, Ti – current temperature value ($i = 1 \dots 5$).

On the left border of the system (at $x = 0$) is accepted the condition of constancy of the main body part of worker temperature (boundary conditions of the first kind). On right boundary is accepted boundary condition of the third kind, taking into account the heat transfer from the outside to worker environment - the room air. At separation boundaries between the layers were written the conditions of the fourth kind. There is considered the radiation from the skin.

The initial and boundary conditions:

$$\begin{aligned} & x = 0: \ T = T_1, \ T_1 = const; \\ & x = x_1: \ \lambda_1 \frac{\partial T_1}{\partial x} = \lambda_2 \frac{\partial T_2}{\partial x}, \ T_1 = T_2; \\ & x = x_2: \ \lambda_2 \frac{\partial T_2}{\partial x} = \lambda_3 \frac{\partial T_3}{\partial x}, \ T_2 = T_3; \\ & x = x_3: \ \lambda_3 \frac{\partial T_3}{\partial x} = \lambda_4 \frac{\partial T_4}{\partial x} + \varepsilon\sigma((T_S)^4 - (T_A)^4), \ T_3 = T_4; \\ & x = x_4: \ \lambda_4 \frac{\partial T_4}{\partial x} = \lambda_5 \frac{\partial T_5}{\partial x}, \ T_4 = T_5; \\ & x = L: \ \lambda_5 \frac{\partial T_5}{\partial x} = \alpha(T_L - T_A); \\ & t = 0: \ T_1 = T_1^0, \ T_2 = T_2^0, \dots\dots, \ T_{x1} = T_{x1}^0, \ 0 < x < x_3; \\ & \qquad T = T_0^a, x_3 < x < x_4; \\ & \qquad T = T_0^{cl}, x_4 < x < L; \end{aligned}$$

Here α - heat transfer coefficient, ε - reduced emissivity surface of human skin, σ - Stefan-Boltzmann constant.

The heat transfer coefficient was by criteria expressions for conditions of natural convection near a flat wall [5]:

$$Gr = \frac{\beta \cdot l^3 \cdot g \cdot (T_L - T_B)}{\nu};$$

Here Gr - the number of Grazgof, l - a characteristic size (height of a man), ν - kinematic viscosity coefficient (air), β - the temperature coefficient of volume expansion of the air.

In laminar flow the Nusselt number for air is defined by the relation ($10^{-3}<Gr*Pr <5*10^2$) [6]:

$$Nu = 0,46 \cdot Gr^{0,25};$$

If the transitional flow regime ($5*10^2 <Gr*Pr <2*10^7$) [6]:

$$Nu = 0,46 \cdot Gr^{0,25};$$

If the turbulent flow regime ($Gr*Pr >2*10^7$) [6] :

$$Nu = 0,133 \cdot Gr^{0,33};$$

Where from [6, 7]:

$$\alpha = \frac{Nu \cdot \lambda}{l};$$

The system of differential equations with appropriate initial and boundary conditions is solved by finite difference method, using an algorithm [8,9] developed for solving heat transfer problems in the areas of multi-layer discontinuous thermal conductivity.

At the solution of the problem the temperature dependence of thermal characteristics of materials and layer materials are not taken into consideration, since the latter was only a change from $1\ K$ to $10\ K$.

Also assumed, the object is stationary (no additional internal heat).

3 Results of numerical modeling

At the modelling T_A varies in real possible range (from $277\ K$ to $304\ K$) the thickness of the outer insulation (from $1*10^{-3}\ m$ to $5*10^{-3}\ m$).

$$\lambda_1 = 0,2\ Bm/м \cdot K,\ \lambda_2 = 0,45\ Bm/м \cdot K,\ \lambda_3 = 0,21\ Bm/м \cdot K,\ \lambda_4 = 0,024\ Bm/м \cdot K,$$
$$\lambda_5^{wool} = 0,03\ Bm/м \cdot K,\ \lambda_5^{cotton} = 0,05\ Bm/м \cdot K,\ \lambda_5^{dpbf} = 0,0097\ Bm/м \cdot K$$
$$\rho_1 = 850\ кг/м^3,\ \rho_2 = 1000\ кг/м^3,\ \rho_3 = 1200\ кг/м^3,\ \rho_4 = 1,2\ кг/м^3,$$
$$\rho_5^{wool} = 120\ кг/м^3,\ \rho_5^{cotton} = 100\ кг/м^3,\ \rho_5^{dpbf} = 161\ кг/м^3$$
$$c_1 = 2250\ Дж/кг \cdot K,\ c_2 = 3200\ Дж/кг \cdot K,\ c_3 = 3600\ Дж/кг \cdot K,\ c_4 = 1005\ Дж/кг \cdot K,$$
$$c_5^{wool} = 2260\ Дж/кг \cdot K,\ c_5^{cotton} = 1200\ Дж/кг \cdot K,\ c_5^{dpbf} = 1380\ Дж/кг \cdot K.$$

The thicknesses of the layers of skin accepted in accordance with [7]:

$$\Delta_1 = 0,013\ m,\ \Delta_2 = 0,0005\ m,\ \Delta_3 = 0,001\ m$$

Typical initial temperatures:

$$T_0^{cl}=301\ K,\ T_0^{a}=303\ K,\ T_0^{x1}=307.$$

The duration of the modeled process chosen within the limits: $t = 7000\ c$.
The degree of emissivity of human skin [13]:

$$\varepsilon = 0{,}7.$$

Fig. 1 shows the temperature field in the investigated system at various ambient temperatures under free convection of air surrounding the worker. As outer insulation adopted a single-layer woolen clothing $\Delta_5=3\ mm$.

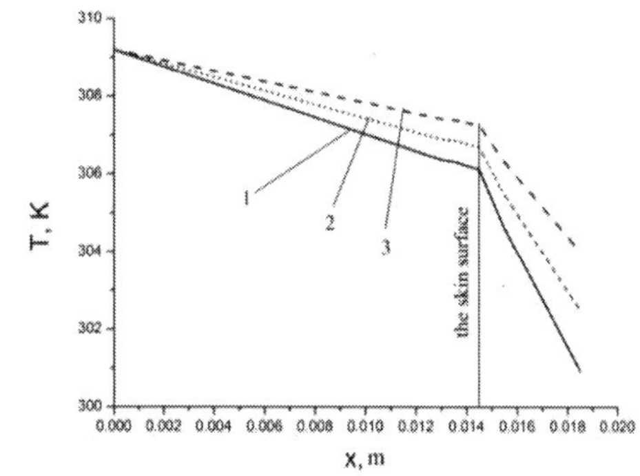

Fig. 1. The temperature distribution along the x coordinate at: 1 – $T_A=291\ K$; 2 – $T_A=294\ K$; 3 – $T_A=297\ K$.

It can be seen that as the ambient temperature falls and the temperature worker surface of the skin is reduced. From the analysis of the dependence of $T(x)$ (fig. 1) and comparing them with data on the perceptions worker at different temperatures external to the environment it is possible to draw a conclusion about the possibility of assessing the numerical value of "comfortable" for worker T_A.

It is obvious that the worker does not feel the skin temperature of the body and senses the intensity of heat removal from the surface - q. This heat flow depends on the heat-insulating properties of the clothing and the size T_A. Therefore, based on the obtained results (fig. 1) we can conclude that the "comfortable" be q values in the range of $28.2\ W/m^2 \leq q \leq 42{,}3\ W/m^2$.

Fig. 2 shows the dependence of the temperature of the skin at time with the different T_A. It is found that the T_S falls in time to a certain meanings appropriate to the conditions of heat exchange at the boundary $x = L$, and then does not change with increasing t.

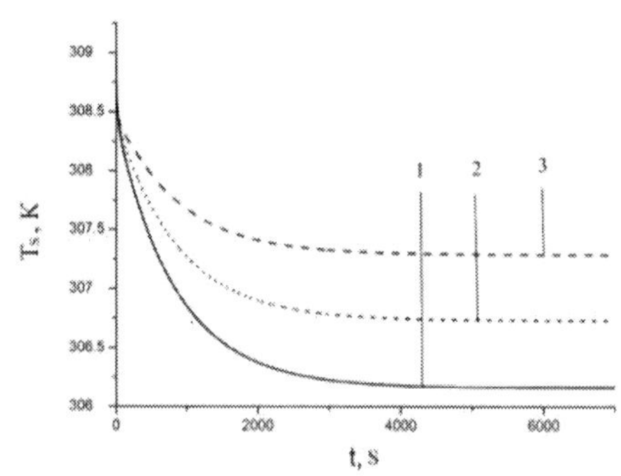

Fig. 2. T_S dependence on time at: 1 – $T_A=291\ K$; 2 – $T_A=294\ K$; 3 – $T_A=297\ K$.

Fig. 3 shows the dependence of temperature of skin surface of the worker on the magnitude of T_A. Based on the data obtained experimentally (comfortable for a person is the temperature of the skin surface from *306 K* to *307 K*) [1] it can be concluded that the range of temperatures from 290 K to 296 K can be considered comfortable for the person in general. In actual practice, most likely, this interval depends essentially on the characteristics of clothing (layer thicknesses, the thermal conductivity of them), and it is from *4 K* to *6 K*.

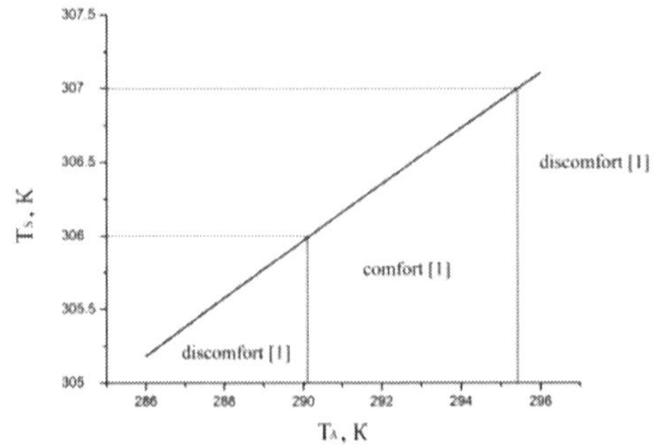

Fig. 3. The dependence of the T_S on the ambient temperature (single layer wool clothing).

Also analyzed the dependence of the T_S on the thickness of external insulation, at different ambient temperatures. It is found that the change in the T_S ranges from *0.34* to *0.85 K* Δ_5 with growth from *1 mm* to *5 mm* (Fig. 4). It is clearly seen significant as might be expected, the effect of this parameter on the value of the T_S.

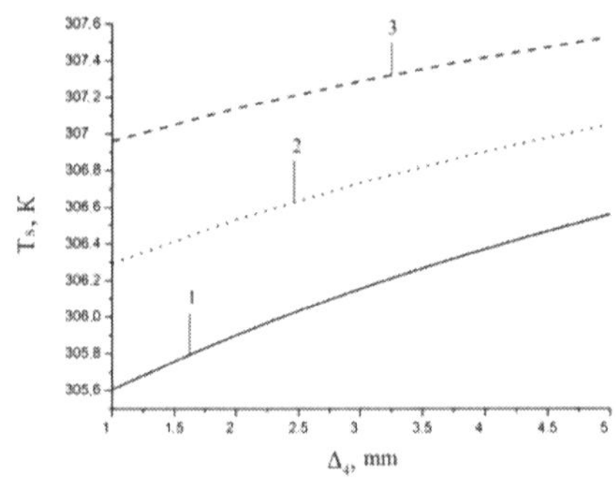

Fig. 4. The dependence of the T_S on the thicknesses of the thermal insulation: 1 – T_A=*291 K*; 2 – T_A=*294 K*; 3 – T_A=*297 K*.

Fig. 5 shows the relationship "comfortable" [1] of the room temperature and the thickness of the the thermal insulation layer. This makes it possible to evaluate the usefulness of a particular clothing at work in industrial environments.

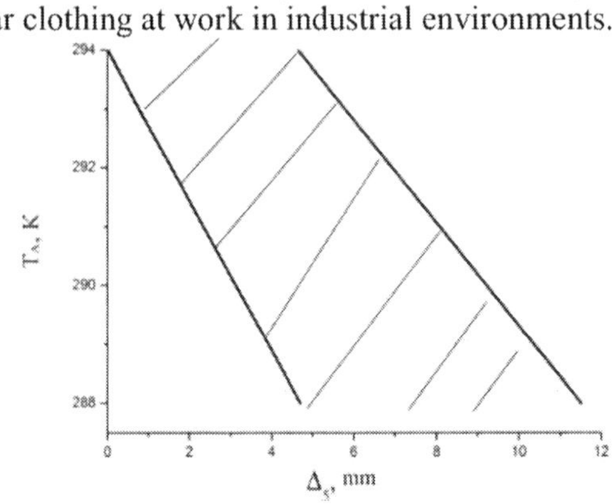

Fig. 5. The dependence of the thickness of the insulation (clothing) from the "comfortable" T_A.

In addition to the influence of the thickness of the individual layers the thermal insulation as well the impact of the number of layers of clothing and their thermal characteristics to "comfortable" temperature T_A is investigated. Thus in fig. 6 T_S dependence on the ambient temperature is considered, where the external insulation system appear the "cotton inner wear (thickness *1 mm*) - air layer (thickness *1 mm*) - woolen outer layer (thickness *3 mm*)". It can be seen that the interval comfortable ambient temperature is changed from *288 K* to *294 K*.

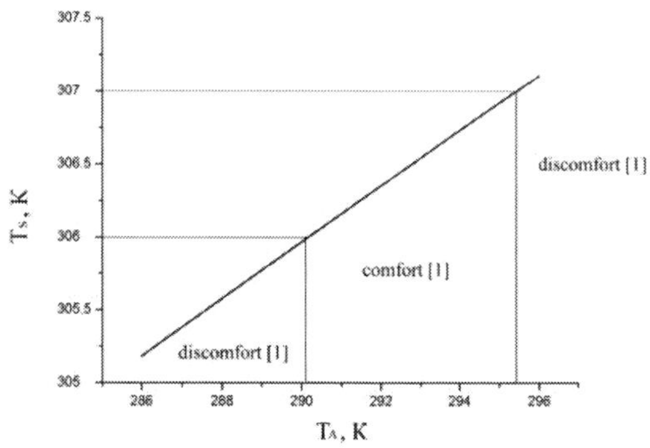

Fig. 6. The dependence of the T_S on the ambient temperature (double-layer clothing - cotton and wool).

Fig. 7 shows a similar relationship for the system – "cotton inner wear (thickness 1 mm) – air layer (thickness 1 mm) – basic-fleecy fabrics layer (thickness 3 mm)". Due to the fact that the thermal conductivity of this material is substantially less than the thermal conductivity of wool, comfortable temperature range is from 280 K to 288 K.

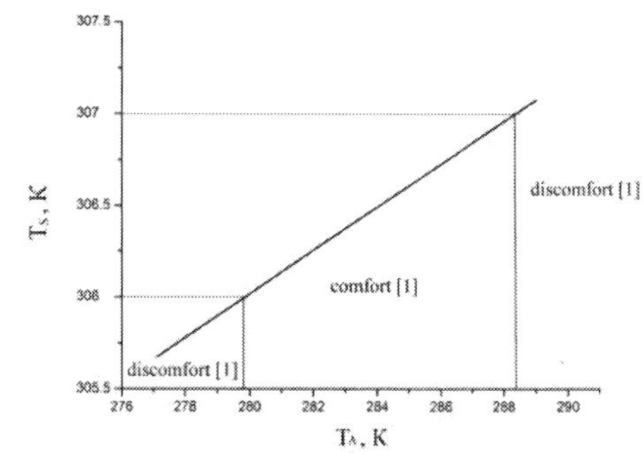

Fig. 7. The dependence of the T_S on the ambient temperature (double-layer clothing - cotton and basic-fleecy fabrics).

4 The main results and conclusions

1. Heat conduction problem for the system "heating facility - the thermal insulation layer - the external environment" is solved, taking into account the natural convection in the outer clothing the border and radiation at the human skin.
2. Depending on the temperature of working surface of the skin from the outside air temperature at different thicknesses of the layers of external insulation were obtained at different sets clothing and different thermal characteristics of materials.

3. Insulation thickness corresponding to the "comfortable" ambient temperatures were set.
4. Ambient temperature, corresponding to the conditions of "comfort" for a variety of clothing options in natural convection mode are shown.

The work was supported by the Russian Federation President's grant (NSH project - 7538.2016.8).

References

1. V. N. Platova, *Biologia 8* (High School, Moscow, 2008)
2. P. Domingues, P. Carreira, R. Vieira, W. Kastner, Matec Web of Conferences **45,** 6 (2016)
3. M Wallance, R. McBride., P. Mhaskar, J. House, T. Salsbury, Chem. Eng. Sci. **69,** 45 (2012)
4. M. Taleghani, M. Tenpierik, A. Dobbelsteen, Solar Energ. **103,** 108 (2014)
5. Robert E. Simons, Electroning cooling **6** (2001)
6. F. P. Incropera, D. P. DeWitt *Fundamentals of Heat and Mass Transfer* (John Wiley and Sons, NY, 1985)
7. F. Xu, T. J Lu, K. A., Sef. Acta Mech. S. **24** (2008)
8. O.V. Vysokomornaya, G.V. Kuznetsov, P.A. Strizhak, Rus. J. Ph. Ch. B **5** (2011)
9. V.L. Strakhov, A.N. Garashchenko, G.V. Kuznetsov, V.P. Rudzinskii, Comb., Exp. Sh. W. **37** (2001)
10. I.K. Kikoin *Tables of physical quantities* (Atomizdat, Moscow, 1976)
11. A.P. Babichev, A.P. Babushkin, A.M. Bratkovsky *Physical values*, (Eneroatomizdat, Moscow, 1991)
12. M.V. Nazarova, S. Yu. Boyko, Modern problems of s. and ed. **5** (2009)
13. U. Kukelev, A. Ustinov, Resources and Technology **3** (2005)

MATEC Web of Conferences **92**, 01033 (2017) DOI: 10.1051/matecconf/20179201033
Thermophysical Basis of Energy Technologies - 2016

Mathematical model of nonstationary hydraulic processes in gas centrifuge cascade for separation of multicomponent isotope mixtures

Alexey Orlov[1], *Anton* Ushakov[1,2,*], and *Victor* Sovach[2]

[1]National Research Tomsk Polytechnic University, 634050 Tomsk, Russia
[2]SC "PA Electrochemical Plant", 663690 Zelenogorsk, Russia

Abstract. The article presents results of development of a mathematical model of nonstationary hydraulic processes in gas centrifuge cascade for separation of multicomponent isotope mixtures. This model was used for the calculation parameters of gas centrifuge cascade for separation of silicon isotopes. Comparison of obtained values with results of other authors revealed that developed mathematical model is adequate to describe nonstationary hydraulic processes in gas centrifuge cascades for separation of multicomponent isotope mixtures.

1 Introduction

Nonstationary hydraulic processes can appear in the operation of gas centrifuge (GC) cascade for separation of multicomponent isotope mixtures (MCIM). During these processes values of pressure and flows of process gas changes in stages of cascade and it is necessary to minimize losses of cascade productivity. At present, there is mathematical model of nonstationary hydraulic processes [1], which has such disadvantage as a limited field of application: modeling is only possible in the case of long cascade and short time periods, when disturbances have not come at withdrawal stages of cascade. Using of this model is impossible for the modeling of nonstationary hydraulic processes in short cascades. For elimination of this disadvantage we have developed the mathematical model of nonstationary hydraulic processes in GC cascade for MCIM separation. This article is devoted to its description.

2 The description of the mathematical model

MCIM separation takes place into a cascade (Fig. 1), consisting of S separation stages numbered by i index ($i=\overline{1,S}$) and connected by counter-current type [2]. The cascade has three flows: input feed flow F, output light fraction flow P and heavy fraction flow W.

* Corresponding author: ushakovaa2015@sibmail.com

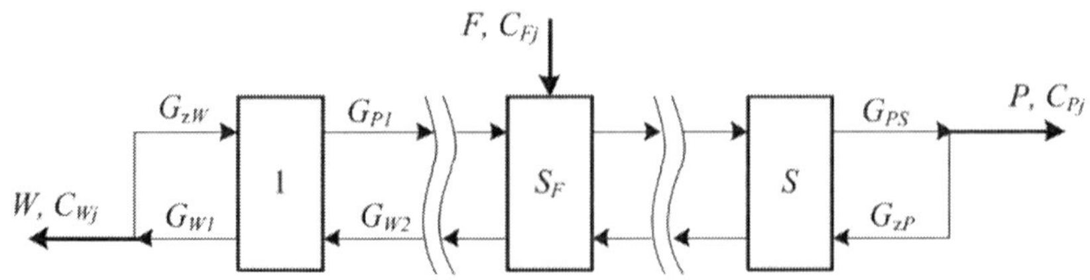

Fig. 1. Separation cascade.

During stationary hydraulic process the flows of cascade are interrelated by material balance equation:

$$F = P + W \tag{1}$$

The stage has three flows: input feed flow G_{Fi}, output light fraction flow G_{Pi} and heavy fraction flow G_{Wi}:

$$G_{Fi} = G_{Pi} + G_{Wi}\,. \tag{2}$$

For square cascade there are dimensionless parameters [3]:

$$y_P = S \ln \chi_0\,, \tag{3}$$

$$\tau = \frac{G_F t \ln^2 \chi_0}{H'}\,, \tag{4}$$

where y_P is cascade length; χ_{0i} is separation factor per unit difference between the mass numbers; τ is time; H' is cascade holdup, kg; t is time, s.

The method of calculation of the nonstationary hydraulic processes is based on the following points:

1. Each stage is composed of finite number of capacities. MCIM is distributed in these capacities.
2. At any time the value of process gas temperature and pressure are the same at any points of capacity.
3. Basic equations of the mathematical model are the balance equations of process gas in each capacity.

In the regards, the holdup of cascade is distributed in 4S capacities (Fig. 2).

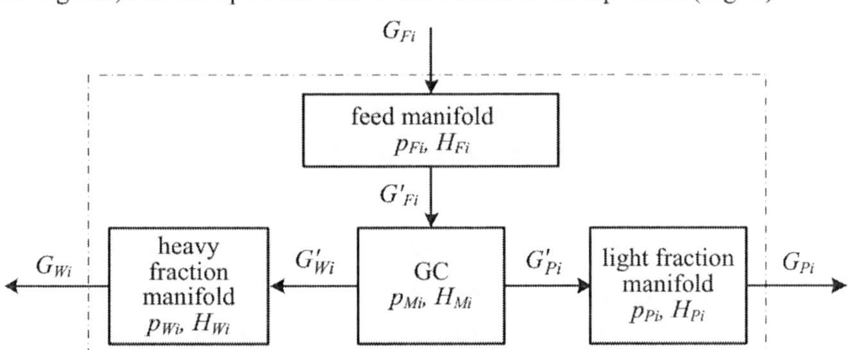

Fig. 2. Capacities of ith stage.

Differential equation describing nonstationary hydraulic processes in capacity which has holdup H and pressure of process gas p is written in the general form as:

$$\frac{dH}{dt} = G_{\text{IN}} - G_{\text{OUT}}, \tag{5}$$

$$H = a \cdot p, \tag{6}$$

$$a = \frac{\mu V}{RT}, \tag{7}$$

where H is capacity holdup, kg; a is specific capacity holdup, kg·Pa^{-1}; μ is molar mass of process gas, kg·kmol^{-1}; p is pressure of process gas, Pa; $R = 8314$ J·kmol^{-1}·K^{-1} is universal gas constant; G_{IN} is input flow, kg·s^{-1} ($G_{\text{IN}}=f(p)$); G_{OUT} is output flow, kg·s^{-1} ($G_{\text{OUT}}=f(p)$).

Replacing derivatives to difference equations by backward Euler method [4], we have received:

$$a\frac{p^{(k)} - p^{(k-1)}}{\Delta t} = G_{\text{IN}}\left(p^{(k)}\right) - G_{\text{OUT}}\left(p^{(k)}\right), \tag{8}$$

$$\Delta t = t^{(k)} - t^{(k-1)}, \tag{9}$$

where Δt is time step; k is the index of time layer.

The calculation have been realized as iterative process: on each iteration cycle calculation of pressure and flows have been done from first stage to S-th stage. The initial approximation on k-th time layer is the values of pressure and flows on previous (k-1)-th time layer. The results of calculation are values of pressure in each capacity and flows at every time layers. We have realized the algorithm described herein as software product, developed on Delphi programming language using Embarcadero Delphi XE2.

3 The results of calculation cascade for silicon isotope separation

We have calculated parameters of nonstationary hydraulic process in GC cascade for silicon isotope separation for the purpose of verification of developed mathematical model and created software product. We have seen nonstationary hydraulic process caused by short increase of feed flow of cascade during period of time τ=0…0,02. Then initial values of cascade flows have set. The values of relative changes of stage holdup on different time moments which have received by developed model and approximate analytical expression [1] are given on Fig. 3.

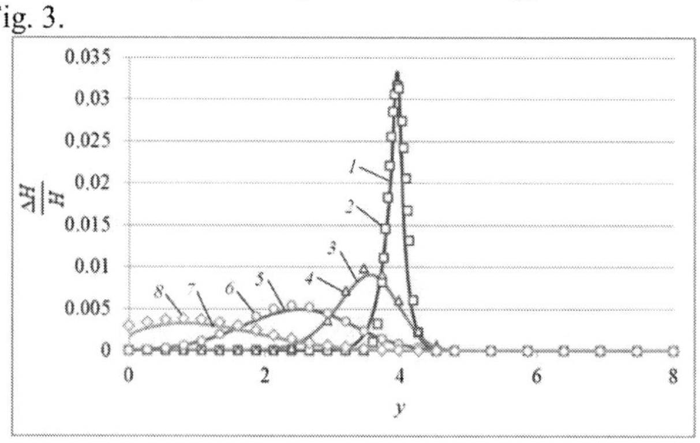

Fig. 3. The values of relative changes of stage holdup on different time moments: *1, 3, 5, 7* – developed model; *2, 4, 6, 8* – the model [1]; *1, 2* – τ=0,02; *3, 4* – τ=0,1; *5, 6* – τ=0,3; *7, 8* – τ=0,75.

As it's seen on Fig. 3, holdup of S_F-th stage (y=4) has increased. The excess of holdup moves along cascade to stage of withdrawal of heavy fraction flow. Wherein there are insignificant changing of stage holdup located between S_F-th and S-th stages (y=4…5) during period of time τ=0…0,3.

Thus the results obtained by developed model are well correlated with the results of other researchers [1], which demonstrate the adequacy of developed mathematical model.

4 Conclusion

1. The result of research is creation of mathematical model of nonstationary hydraulic processes in gas centrifuge cascade for separation of multicomponent isotope mixtures, representing the differential equation system of first order with given initial conditions.
2. The solution algorithm of equation system describing nonstationary hydraulic processes is developed and realized as a software product.
3. We have made the calculation parameters of nonstationary hydraulic process in gas centrifuge cascade for silicon isotope separation for the purpose of verification of developed mathematical model and created software product. Comparison of obtained values with results of other authors revealed that developed mathematical model is adequate to describe nonstationary hydraulic processes in gas centrifuge cascade for separation of silicon isotope separation.

References

1. D.M. Levin, Yu.P, Zabelin, *Physical and chemical processes during separation of atoms, molecules and in laser, plasma and nanotechnologies conference report collection* (CRIatominform, Moscow, 2008)
2. K. Cohen, *The theory of isotope separation as applied to the large scale production of U* (McGraw Hill, New York, 1951)
3. G.A. Sulaberidze, V.A. Palkin, V.D. Boricevich, V.D. Borman, A.V. Tichomirov, *Theory of cascades for separation of binary and multicomponent isotope mixtures* (MEPhI, Moscow, 2011)
4. U.G. Pirumov, *Numerical methods* (Drofa, Moscow, 2003)

On optimum temperature conditions for operation of portable aliminum-air batteries

Maxim Vlasenko[1,*], *Aleksey* Bobylev[1], *Petr* Galkin[2], *Sergey* Kharlamov[1], and *Dmitry* Markovich[1]

[1]Kutateladze Institute of Thermophysics, 630090 Novosibirsk, Russia
[2]Nikolaev Institute of Inorganic Chemistry, 630090 Novosibirsk, Russia

Abstract. For construction of portable aluminium-air batteries (AABs) for field applications the data on performance of such batteries at low temperatures are necessary. Such experimental data are presented in this article. Lower temperature limit for effective operation of portable AABs is estimated. Possibility of AABs self-heating to the desired temperature is also considered.

1 Introduction

Aluminum-air battery (AAB) is an electrochemical power source in which aluminum is used as fuel and air oxygen is used as oxidant. Interest in the development of AABs is connected substantially with the fact that such batteries potentially have one of the highest specific energies in comparison with other types of batteries. At present, the main investigations concerning AABs are aimed at developing electrochemical generators with power sufficient for automotive applications [1, 2]. An alkaline electrolyte is used in such generators and they are obligatory equipped with a loop of forced electrolyte circulation and regeneration to stabilize the electrolyte composition and temperature.

The effect of alkaline electrolyte composition and temperature on aluminum anode polarization and performance of aluminum-air cell with forced electrolyte circulation at heightened temperatures was investigated in detail in [2, 3]. When concerning design of portable AABs, the results obtained in [4, 5] are useful. In these articles it is shown that main limitation of specific capacity at heightened temperatures for aluminum-air cells without any loop of electrolyte circulation is connected with anode passivation that occurs at the initial stage of aluminum hydroxide precipitation from electrolyte. At the same time for construction of portable AABs intended for field applications the data on performance of such cells at low temperatures are necessary. Such experimental data are presented in this article.

2 Experimental technique

[*] Corresponding author: vlasenko.maksim.g@gmail.com

The experiments were carried out with prismatic cells, two sides of which were air-breathing gas diffusion cathodes (Fig. 1). The total area of the cathodes was 8 cm2. The electrolyte fill volume was 15 ml. The aluminum anodes in the form of plates 2 cm wide and 3 mm thick were placed between the cathodes through the center of the cells. The initial thickness of the gap between the anode and cathode was 4 mm. Water solutions of NaOH with the initial concentration from 3 to 8 M were used as electrolyte. Since aluminum anodes are prone to strong corrosion in the alkaline electrolytes, in our experiments we dope the electrolyte with 0.02 M sodium stannate as corrosion inhibitor which allowed reducing the rate of aluminum corrosion by more than one order of magnitude in a wide temperature range.

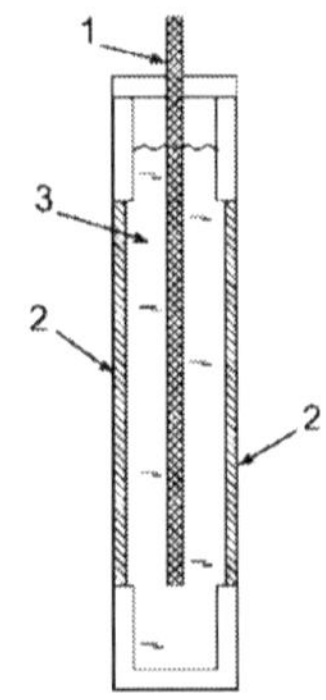

Fig. 1. Layout of the experimental cell. 1 – aluminum anode, 2 – air cathode, 3 – electrolyte.

The cells were discharged through resistive load of 2 Ohm, which provided discharge current densities at a level of 60–90 mA/cm^2 at operating voltages of 1.0-1.4 V. Simultaneously with the operating voltage of the cells, we measured the running temperature of the electrolyte in the cells. Since the discharge of the cells is accompanied by release of a great amount of heat, the electrolyte temperature in the cells with natural cooling can be several tens degrees higher than ambient temperature and can significantly vary during discharging. To stabilize the temperature of electrolyte, the cells were placed into the climate chamber with forced convection of air whose humidity was maintained at a level of 50%. Under such conditions, the electrolyte temperature in the working cells exceeded the ambient temperature by not more than 2-3 °C. The capacity of the cells was determined as time integral of discharge current until the point when voltage across the resistive load began to decrease rapidly. An example of discharge characteristics of the cells with different concentration of electrolyte are shown in Fig. 2.

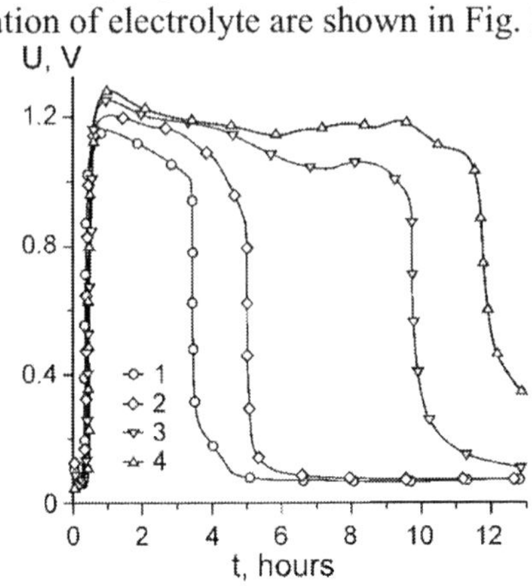

Fig. 2. Discharge characteristics of the cells with analytical grade aluminum anode at T = 32 °C. NaOH concentration: 1 – 3 M, 2 – 4 M, 3 – 6 M, 4 – 8 M.

 DOI: 10.1051/matecconf/20179201034

The working material of anodes was analytical grade aluminum (with aluminum content not less than 99.82 %) and anode alloys of composition Al – 0.4 % Mg – 0.2 % In – 0.02 % Ga (AMIG system). According to the data of [5], in contrast to pure aluminum the latter alloy is not passivated and continues to work at high working current densities after the beginning of aluminum hydroxide precipitation which results in sufficient increase of aluminum-air cells capacity.

3 Experimental results

The effect of electrolyte operating temperature on the cell specific capacity is shown in Fig. 3. As can be seen, at electrolyte temperature lower than +30 °C the cell specific capacity diminishes with decrease of the temperature so rapidly that at +15 °C it is not more than 20-25 % of the specific capacity at +40 °C independently of the anode alloy composition. In the range of operating temperatures between +30° and + 50° the cell specific capacity has a weak temperature dependence and is equal to about 400 Ah per litre of electrolyte when anode is made of pure aluminum and about 500 Ah/L for AMIG alloy anodes.

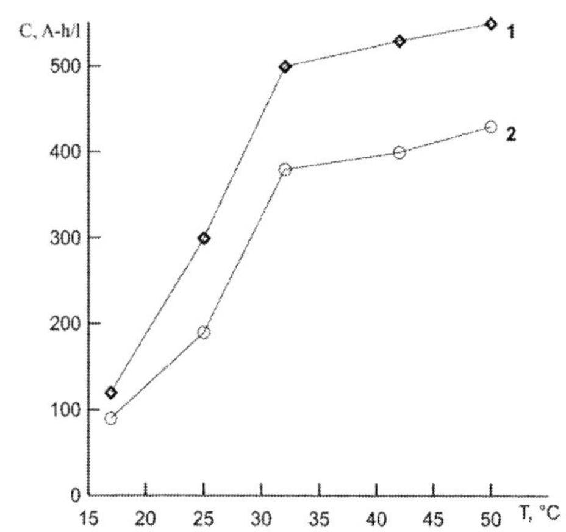

Fig. 3. Dependence of specific cell capacity on electrolyte temperature. 1 – AMIG alloy anode, 2 – analytical grade aluminum anode.

Thus, to ensure high specific capacity of AABs the electrolyte temperature in operating battery should be maintained at the level higher than +30 °C at any ambient temperature. In working AAB such temperature of electrolyte can be maintained due to the heat released during its operation [3]. However when AAB remains for a long time in standby mode at low ambient temperatures the problem of its quick self-heating to operating temperatures, for instance, by operating in short circuit mode, is arising. Our studies of dependence of short-circuit current density on the cell operating temperature showed that this dependence is exponential and for different anode alloys short-circuit current grows 2.2-2.5 times per temperature increase of 10° (Figure 4). Therefore, when electrolyte temperature falls below +20 °C quick self-heating of AABs to the operating temperatures appears impossible, and this means that at low ambient temperatures an AAB in standby mode has to use a part of its energy to maintain inner temperature at a sufficiently high level.

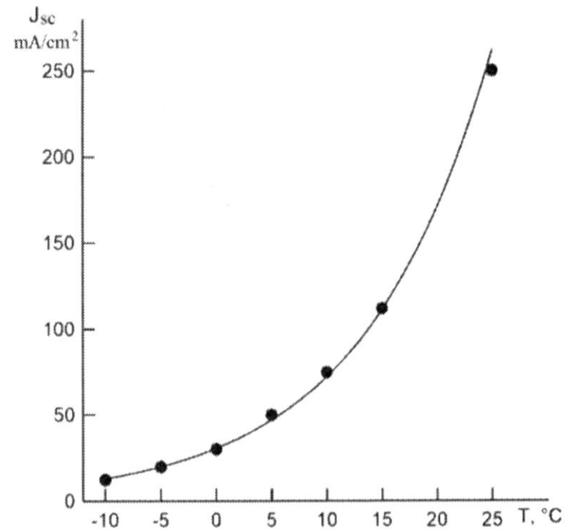

Fig. 4. Dependence of short-circuit current density on the cell operating temperature. Analytical grade aluminum anode.

In conclusion, our results show that when designing portable aluminum-air batteries for field applications it is important to consider that electrolyte temperature in AAB has to be maintained at the level higher than +30 °C. Non-fulfillment of this requirement may result in sharp deterioration of AAB performance.

This work is supported by program of RAS (project "Development of the fundamentals of aluminum-air batteries for the Arctic" (0322-2015-0001)).

References

1. S. Yang, H. Knickle, J. Power Sour. **112**, 162 (2002)
2. A. Zhuk, A. Ilyukhina, A. Ilyukhin, B. Kleimenov, M. Nagayama, Elektrokhim. Energet. **13**, 83 (2013)
3. A. Ilyukhina, A. Zhuk, B. Kleymenov, A. Ilyukhin, M. Nagayama, Fuel Cells, **16**, 384 (2016)
4. S. Alekseenko, A. Bobylev, P. Galkin, D. Markovich, S. Kharlamov, J. Eng. Thermophys. **23,** 280 (2014)
5. S. Kharlamov, A. Bobylev, P. Galkin, K. Dobroselsky, M. Vlasenko, D. Markovich, J. Eng. Thermophys. **24**, 313 (2015)

MATEC Web of Conferences **92**, 01035 (2017) DOI: 10.1051/matecconf/20179201035
Thermophysical Basis of Energy Technologies - 2016

Aspects of calculating the temperature for the graphite stack of the RBMK reactor

Alexander Vorobiev[1,*], and *Alexandra* Antonova[1]

[1]National Research Tomsk Polytechnic University, 634050 Tomsk, Russia

Abstract. The effect of thermal contact resistance in the elements of the graphite stack of the RBMK reactor on moderator temperature is studied. It is shown that estimation of the thermal contact resistance improves the accuracy of calculation of the graphite moderator temperature.

1 Introduction

One of the factors that determine the operational reliability and safety of nuclear power plants with water-graphite reactors is the reliability of the graphite moderator (stack) designed for the entire period of the reactor operation.

The graphite stack of the RBMK-1000 reactor has the shape of a cylinder with a diameter of 11.8 m and a height of 7 m made up of 2488 columns. The columns have vertical holes with pipes arranged inside as fuel channels (FC) and CPS channels. Fuel assemblies inside the channels are cooled by water with a temperature of 250...280 °C [1].

During operation of the reactor, a portion (5...6%) of its total thermal power is released in the graphite stack. This heat is also removed to the FC tubes. To ensure the thermal contact with blocks, the tubes are encircled with split graphite rings (Fig. 1), and the sweep gas is supplied through the gaps between the tubes and blocks (reference composition of 10% nitrogen and 90% helium, by volume) [1, 2].

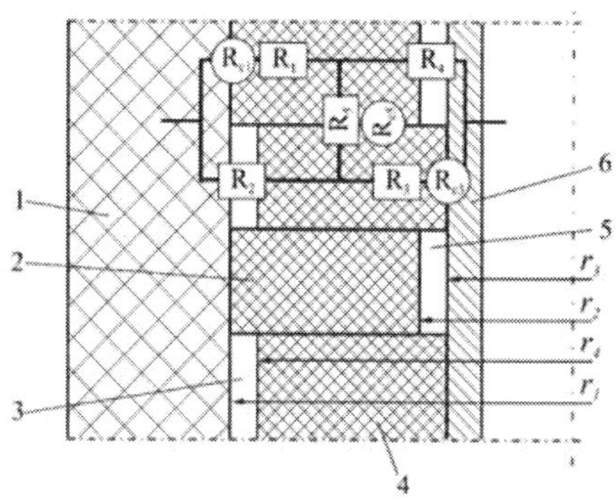

Fig. 1. Schematic diagram of FC tube arrangement in a graphite stack: graphite block (1); outer graphite ring (2); gas gaps (3, 5); inner graphite ring (4); R_i is thermal resistance; R_{ci} is contact

* Corresponding author: worob@tpu.ru

thermal resistance; r_i is radius.

The design and schematics of the RBMK reactor plant make possible to change the composition of the sweep gas [1].

The temperature field in the graphite stack primarily depends on the reactor power. The nitrogen content in the nitrogen-helium mixture (hereinafter, the sweep gas composition) affects the graphite temperature as well.

The graphite stack temperature is one of the main parameters that limit the RBMK reactor operation. In practice, it is assumed that the maximum graphite temperature should not exceed the threshold temperature of 750...800 ºC in all operational conditions of the reactor [2, 3]. In addition, variations in the average temperature of the graphite stack (moderator) cause changes in the reactor reactivity, as the RBMK units have a positive effect on reactivity relative to the moderator temperature [1, 3].

Consequently, estimation of the temperature of the RBMK graphite stack is a crucial task.

2 Calculation technique

The temperature field of the water-graphite reactor stack is three-dimensional. The programs for numerical calculations of two-dimensional and three-dimensional temperature field are very complex. Therefore, various simplifications and assumptions should be performed to estimate the maximum temperature of graphite [4, 5].

In the known method of calculating the maximum temperature of the RBMK reactor moderator [6], thermal resistances of bearings and gas gaps act like electrical resistances. The circuit of resistances being made and simplified, an equivalent electric and hence thermal resistance of the gap graphite block–fuel channel tube can be evaluated (Fig. 1).

3 Calculation results

The results of variants calculations of the temperature of an individual graphite block obtained by the considered technique are shown in Fig. 2. The thermal conductivity of graphite in longitudinal λ_p and transverse λ_t directions was assumed to be 30 W/(m·K).

The analysis of the data in Fig. 2 shows that when the nitrogen content in the sweep mixture grows from 10 to 100%, the maximum corner temperature of the graphite block t_{ang} increases by 5...15 °C depending on the value of the volumetric heat release in graphite q_v.

These figures are significantly different from the experimental data reported in [2], where the average coefficient of the effect of the mixture composition on graphite temperature is estimated in the range of 0.9...1.2 °C/%.

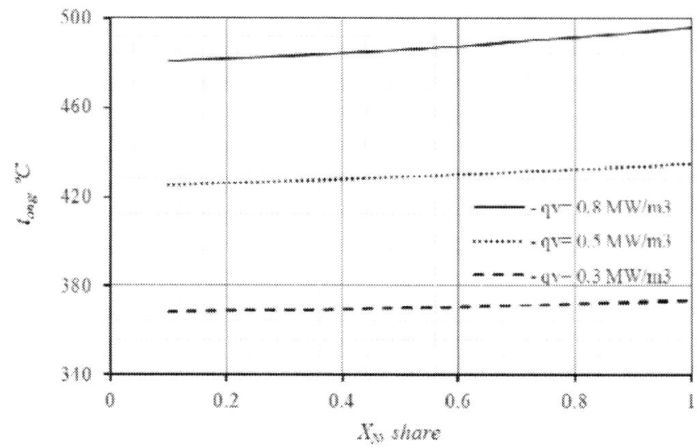

Fig. 2. Dependence of the maximum temperature of the graphite block (in the corner of graphite block) on the sweep mixture composition with no account taken of the contact thermal resistances.

It has been suggested that the reasons for this discrepancy can be unaccounted thermal resistances R_{ci} at the junctions graphite block–outer ring, outer ring–inner ring, and inner ring–FC tube (Fig. 1).

In order to determine the source of discrepancies between the calculated and experimental data related to the graphite temperature, the technique described above has been verified through inclusion of the above-mentioned additional contact thermal resistances. To calculate the latter, the criterion equation from [7] was used

$$\alpha_k = \frac{\lambda_{mix} \cdot Y}{h_0} + 8 \cdot 10^3 \cdot \lambda_{eq} \cdot \left(\frac{P_{cnt}}{3 \cdot \sigma_{str}} \cdot K \right)^{0,86}, \qquad (1)$$

where λ_{mix} is thermal conductivity of the gas mixture; Y =3.33 is relative size of the gap; λ_{eq} is equivalent thermal conductivity of the contact materials; h_0 is the total roughness of the contact surfaces; K is the coefficient that depends on value h_0; P_{cnt} is contact pressure; σ_{str} is tensile strength of the most plastic material among contact materials.

The improved technique was used to calculate the temperature of an individual graphite block by varying the composition of the sweep mixture X_N and contact pressure P_{cnt}. The accepted graphite thermal conductivity $\lambda_p = \lambda_t$ = 30 W/(m·K), and the volumetric heat release in graphite q_v = 0.55 MW/m³. The calculation results are plotted in Fig. 3.

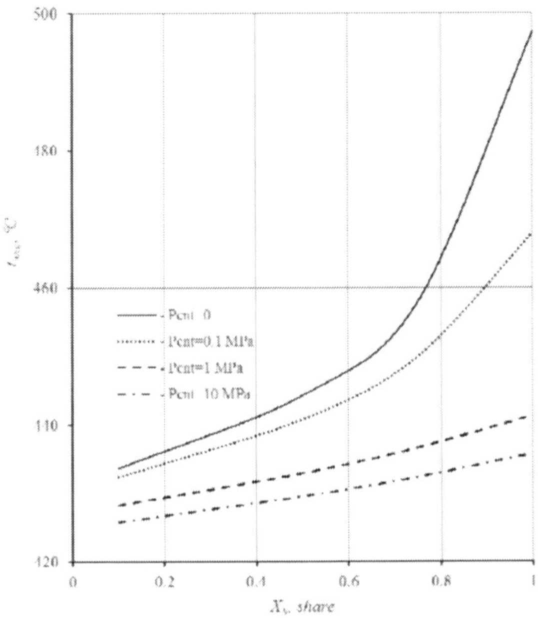

Fig. 3. Dependence of the maximum temperature of the graphite block (in the corner of the graphite block) on the sweep mixture composition at different values of contact pressure.

The analysis of the graphs in Fig. 3 and comparison with the data shown in Fig. 2 indicate that consideration of the contact thermal resistances significantly (1.07...1.15-fold) increases the specified temperature of the graphite moderator. The degree of the effect of the nitrogen content in the sweep mixture on the temperature level of the graphite moderator significantly (3...5-fold) increases as well.

4 Conclusion

An algorithm was updated and a program was developed for calculating the temperature of the graphite moderator in the RBMK reactor, which takes into account thermal resistances at the junctions of the elements of the graphite stack and FC.

This algorithm can significantly increase the precision of the engineering method for calculating the temperature of the graphite moderator of the RBMK reactor.

References

1. N.A. Dollezhal, *Channel nuclear power reactor* (Atomizdat, Moscow, 1980)
2. A.V. Vorobiev, A.M. Antonova, Proceedings of TPU **322**, 182 (2013)
3. A.V. Vorobiev, A.M. Antonova, M.P. Vinogradov, EPJ Web Conf. **110**, 01053 (2016)
4. A.M. Race, *Thermal-hydraulic calculation of the reactor RBMK-1000 reactors: textbook* (Izd. MEI, Moscow, 1982)
5. D.V. Gvozdyakov, A.O. Tanishev, S.A. Shvab, V.N. Martyshev, A.V. Kuzmin, MATEC Web Conf. **37**, 01024 (2015)
6. M.A. Skachek, *Thermal-hydraulic calculation of the RBMK-1000 reactor* (Univ. MEI, 1982)
7. Y.P. Shlykov, E.A. Ganin, S.N. Tsarevsky, *The thermal contact resistance* (Energy, Moscow, 1977)

Research of wood waste concentration influence on composite fuel energy characteristics based on low-grade 2b coal from "Borodinskoe" deposit

S. Yankovskiy[1,*], *A.* Matveev [1],*A.* Tolokolnikov[1], and *A.* Zenkov

[1] National Research Tomsk polytechnic university, 634050, Tomsk, Russia

Abstract. A pyrolysis process of different coal samples into three different types of oxidizing medium (argon, air and steam) were studied by means of thermogravimetry. Two coal types with different metamorphism degree (lignite and bituminous coal) were used. The experimental procedure was carried out in non-isothermal conditions in temperature range from 373 K to 1273 K with 20 K/min heating rate. Purge gas consisted of argon and oxidizer with volumetric ratio 1:24 and had 250 ml/min flow rate. Kinetic parameters were defined by fitting procedure using Arrhenius-type equation with the first order reaction rate on conversion distribution function. The ignition and burnout indexes were calculated to evaluate sample reactivity at different oxidizing mediums. The highest reactivity values in same atmosphere were obtained for lignite because of higher particle special surface area. For same samples the higher reactivity was obtained in steam medium. The conclusion about some kind of synergy between carbon oxidation and pyrolysis could be made.

1 Introduction

Production capacity of timber industry increases every year in Russian Federation [1]. Results of quality products production are accompanied by a parallel increase in the amount of wood waste in the form of sawdust, bark, wood chips and veneer shorts [1], conservation of which poses a threat to the environment [1].

In turn, the use of wood waste in small-scale power generation is not highly relevant because of the resulting materials low calorific value and applied rarely.

The process of concentration influence of one fuel or another in a mixed composition on calorific value, ash content, volatile substances release and combustion kinetics is poorly researched. Application of this composition in the form of briquettes leads to a decrease of the combustion temperature, ash entrainment, consumption of traditional fuels in decentralized areas and an increase of energy equipment efficiency. As a result, transportation component is reduced, and the use of local fuel will lead to synergistic effect.

* Corresponding author: jankovsky@tpu.ru

2 Experimental research

The researched samples preparation was carried out by milling followed by sieving. Particle size of investigated coal powder was less than 80 microns, timber – less than 200 microns [2].

At the preliminary stage of primary fuels gasification research, their element analysis was defined with the help of scanning electron microscope (СЭМ) JEOL JCM-6000 [3]

Fig. 1. The figure shows an image of the composite fuel with an increase in 30µm.

The elemental composition of the composite fuel is shown in Figure 2.

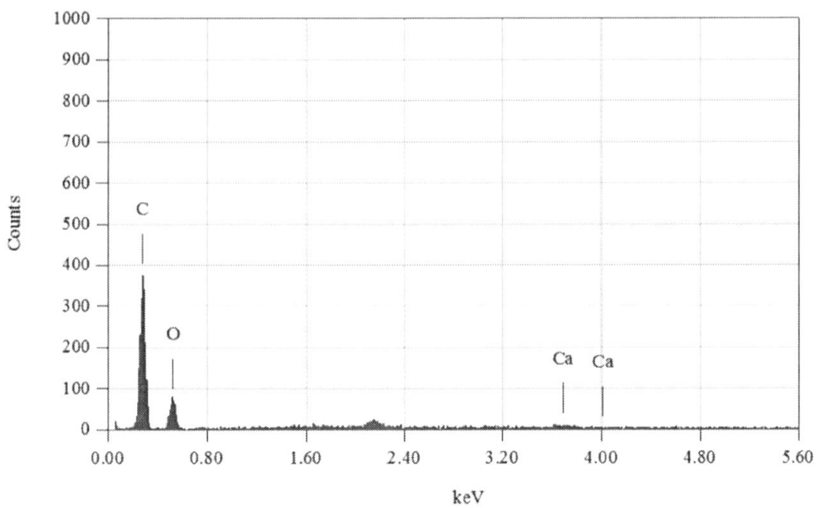

Fig. 2. The elemental composition of the composite fuel.

These milled 2B coal samples (Borodinskoe deposit) and timber production waste were mixed in Pulverisette 6 planetary mill [4-5] in the proportions of 10% of sawdust, 90 % of coal; 25 % of sawdust, 75 % of coal; 50 % of sawdust, 50 % of coal; 75 % of sawdust, 25 % of coal; 90 % of sawdust, 10 % of coal. Technical analysis of obtained mixed fuel samples was conducted.

As a result of a series of experiments, dependence of mixed fuel calorific value on its composition was obtained.

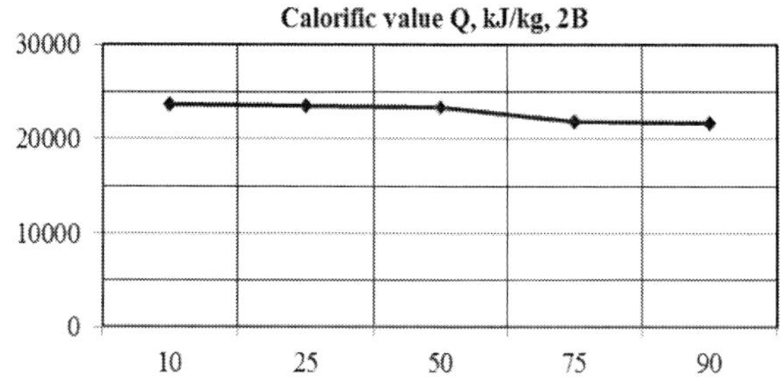

Fig. 3. Dependence of composite fuel calorific value from the percentage ratio of the primary fuel and timber production waste.

Figure 3 shows that mixing of 2B grade coal with fine timber in 50/50 percentage ratio reduces total calorific value by 2 %, and the concentration of 90/10 % leads to a reduction of calorific value by 8 %.

The dependences of ash content and volatile substances release changes of mixed fuel from its composition are obtained.

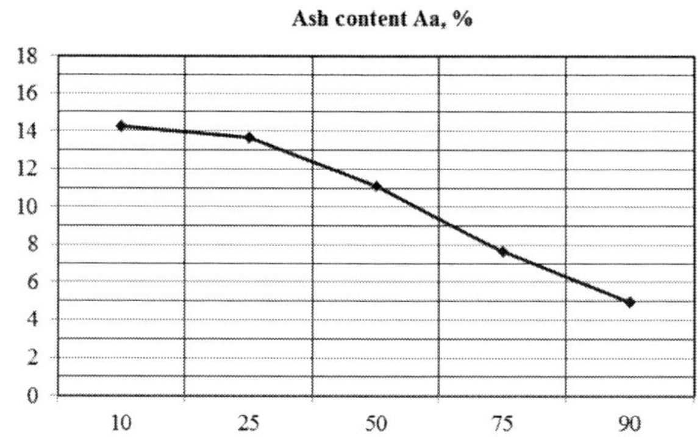

Fig. 4. Changes in ash content release of composite fuel from the percentage ratio of the main fuel and timber waste.

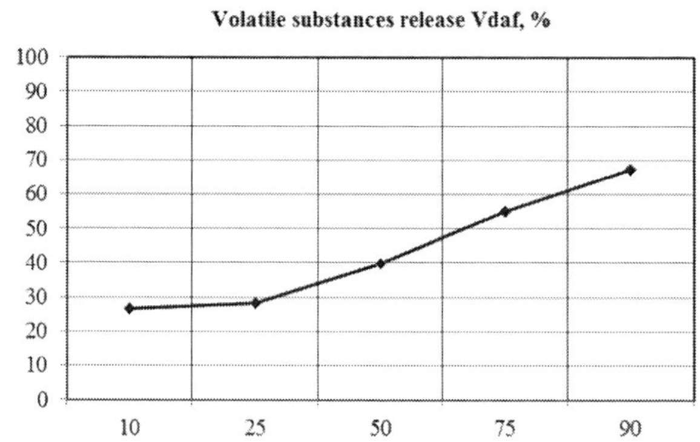

Fig. 5. Changes in volatile content substances release of composite fuel from the percentage ratio of the main fuel and timber waste.

Figure 4, 5 shows that ash content is considerably reduced with an increase of wood concentration: ash content drops to 75 % at 90 % concentration of timber within the initial fuel. An increase of wood content in composite fuel based on 2B grade coal increases the volatile substances release up to 41 %.

3 Conclusion

The obtained results allow us to determine fine timber (wood industry waste) concentration effect on the kinetic parameters of 2B grade coal. It is revealed that an increase in wood concentration in the fuel reduces the ignition temperature, activation energy and increases combustible gases intensification. Calorific value varies within 2 % when sawdust concentration in mixed fuel is changed from 10 % to 50 %.

The work has been supported by the Russian Federal focused program "Research and development on priority orientation of science and technology in Russia 2014-2020", unique identifier 'Applied Research and Experimental Developments' project RFMEFI58114X0001.

References

1. Doklad o povishenii effektivnosti lesnogo kompleksa [Report on improving the effectiveness of timber complex], 138 (2013)
2. S. Jankovskiy, D. Luzhkovoj, K. Larionov, A. Matveeva, MATEC Web Conf. **37**, 4 (2015)
3. Dzh. Goulstejn, *Rastrovaja jelektronnaja mikroskopija i rentgenovskij mikroanaliz* (Mir, Moskva, 1984)
4. A. Zenkov, S. Yankovsky, A. Matveeva, S. Lavrinenko, G. Alexander, MATEC Web Conf. **72**, 1131 (2016)
5. A.G. Korotkikh, K.V. Slyusarskiy, MATEC Web Conf. **23**, 01020 (2015)

Use of a thin liquid film moving under the action of gas flow in a mini-channel for removing high heat fluxes

Dmitry Zaitsev[1,2], *Egor* Tkachenko[1,3], *Evgeniy* Orlik[1], and *Oleg* Kabov[1,3,*]

[1]Institute of Thermophysics SB RAS, 630090, Novosibirsk, Russia
[2]National Research Tomsk Polytechnic University, 634050, Tomsk, Russia
[3]Novosibirsk State University, 630090, Novosibirsk, Russia

Abstract. Intensively evaporating liquid films shear-driven in a mini- or micro-channel under the action of cocurrent gas flow are promising for the use in modern cooling systems of semiconductor devices. In this work, we investigated the influence of liquid and gas flow rates on the critical heat flux in a locally heated film of water, moving under the action of air flow in a mini-channel. In experiments a record value of critical heat flux of 870 W/cm^2 was reached. Heat spreading into the substrate and heat losses to the atmosphere in total do not exceed 25 % at heat fluxes above 400 W/cm^2. A comparison with the critical heat flux for water flow boiling in the channel shows that, for shear-driven liquid films the critical heat flux is almost an order of magnitude higher.

1 Introduction

The development of modern microelectronic equipment requires efficient cooling systems, as it is necessary to remove high heat fluxes, up to 1 kW/cm^2, from local areas of the processor [1]. Thin and ultrathin (less than 10 μm) liquid films moving under the action of forced gas flow in mini-channels are promising for the use in thermal control systems of modern semiconductor devices [2]. Works [3-5] established the basic laws of flow and crisis phenomena in the liquid film, moving under the action of gas flow in a horizontal mini-channel under local heating. First experiments with shear-driven liquid films showed that such films are less prone to thermocapillary rupture than the ones moving under the action of gravity. This is because the film rupture and the formation of dry spots can be controlled by the gas flow. As a result, the critical heat flux for shear-driven liquid films can be several times higher than that for falling liquid films. In works [6-7] it is found, that in two phase flows in rectangular channels the stratified flow sub-regime, when a very thin liquid film is moving along the channel walls under the action of cocurrent gas flow, can exist for the channel heights as small as 100 μm and less. Authors of [8] propose an efficient way of reduction of the viscous resistance in two-phase flows in micro-channels. All these studies confirm the prospects of using shear-driven liquid films for cooling of microsystems with high heat generation. The present paper presents new experimental data

[*] Corresponding author: kabov@tpu.ru

DOI: 10.1051/matecconf/20179201037

proving that shear-driven liquid films are capable to remove heat fluxes as high as 870 W/cm^2 from a heat source of 1x1 cm^2.

2 Experimental setup

Figure 1 shows the design of the test section. Its main part is a stainless steel plate with a pressed-in copper rod. The rod surface has a shape of a square head 1x1 cm^2, imitating the surface of a computer chip. The rod is heated using a nichrome coil, wound around its lower part. This design of the heater provides constant temperature on the surface of the rod, T_w=const (as confirmed by thermocouple measurements). The work area is covered with a transparent cover of optical glass (not shown in Fig. 1), thus forming a flat channel. The channel height is 1.0-1.2 mm, and the width is 40 mm.

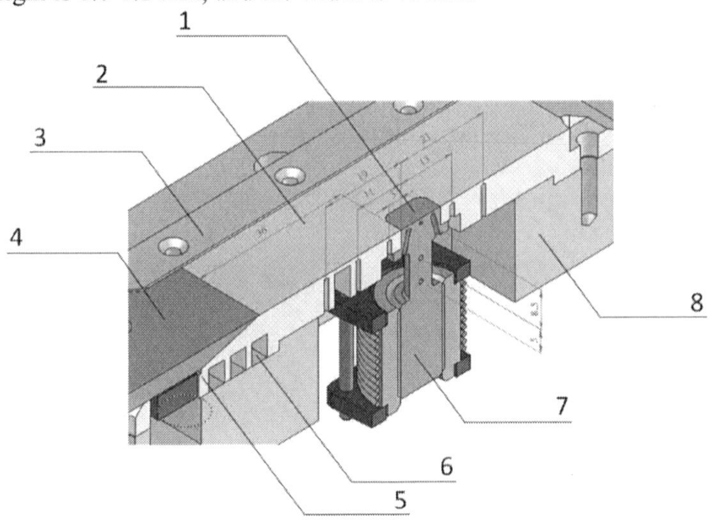

Fig. 1. The scheme of the test section: 1 — source of local heating; 2 — stainless steel plate; 3 — channel height regulator; 4 — knife; 5 — liquid nozzle; 6 — thermal stabilizer; 7 — copper rod; 8 — textolite plate. Locations and distances for thermocouples are shown.

Gas is introduced into the work area from a compressor. The fluid is supplied from the thermostat, enters the channel through the liquid nozzle and moves over the stainless steel plate as a film under the influence of gas friction. After passing the work area the gas is vented to atmosphere, and the fluid flows back to the thermostat. Distilled water with initial temperature of 24 °C is used as the working fluid. Air with temperature of 24-27°C and relative humidity of 15-30 % is used as working gas.

Thermocouples, embedded in the stainless steel plate and the copper rod, allow determining the temperature of the working surface. Heat flux is determined via the electric power, generated on the heating coil. The thermal conductivity of copper is 400 W/mK, which more than an order exceeds the one of stainless steel (15 W/mK); this provides moderate heat spreading from the heater to the steel plate. On the evaluation using measurements of thermocouples embedded in the steel plate, the heat spreading into the plate is not more than 15 % when q>200 W/cm^2. To reduce heat losses to the atmosphere, the heater was wrapped with a layer of insulating material (not shown in Fig. 1). On evaluations using measurements of thermocouples embedded along the copper rod, the heat losses to the atmosphere does not exceed 10 % at q>400 W/cm^2. Therefore, the heat spreading into the steel plate and heat losses to the atmosphere in total do not exceed 25 % at q>400 W/cm^2.

3 Results

This work studies the behavior of the liquid film, moving under the action of the gas at local heating, at different values of heat flux q = 0-870 W/cm^2, Reynolds number of the liquid Re_l = 8.5-113 and superficial gas velocity in the channel U_{Sg} = 3.8-93 m/s. Figure 2 illustrates the process of film rupture with the increase in heat flux at Re_l = 30 and U_{Sg} = 30 m/s. At a certain threshold value of the heat flux the film disrupts with the formation of dry spots on the substrate. As a rule, initially dry areas are formed below the heater, along its side edges (Fig. 2a). Here we deal with thermocapillary film rupture, since dry patches initiate at temperatures well below the saturation temperature (the heater wall temperature is indicated at the capture of photos in Fig. 2). Thermocapillary tangential stresses at the film surface, induced by the dependence of the liquid surface tension on the temperature, and tangential stresses, induced by the flowing gas, are the main competitive forces at the gas-liquid interface. The film is thinnest along the lateral edges of the heater where temperature gradient at the film surface attains its maximum values. Film rupture is most likely to occur in this area. At a slightly higher heat flux, dry areas appear on the heater. With a further increase of the heat flux the heater is covered with metastable intensively evaporating thin liquid film with rapidly appearing and disappearing small dry spots (Fig. 2b). Upon reaching the critical heat flux, the heater suddenly dries up, and its temperature begins to increase sharply (Fig. 2c).

a) q = 112 W/cm^2, T_w = 84°C b) q = 342 W/cm^2, T_w= 129°C c) q =CHF= 375 W/cm^2, crisis

Fig. 2. Rupture and crisis in a water film moving under the action of the gas flow; the heater is 1x1 cm^2, the channel height is 1.2 mm, Re_l = 30, U_{Sg} = 30 m/s, the flow is directed top-down.

Figure 3 shows data obtained on the critical heat flux depending on Reynolds number of liquid and superficial gas velocity. It is seen that with increasing Reynolds number of the liquid and superficial gas velocity, the critical heat flux increases. In experiments, a record value of critical heat flux 870 W/cm^2 was reached (Fig. 3, point Re_l=113, U_{Sg}=71 m/s). This critical heat flux for water film moving under the action of the gas flow was achieved at a mass flow rate of water 100 kg/m^{2}s. In this case, as shown in [9], the critical heat flux for subcooled water boiling in the mini-channel at the same mass flow rate of liquid (at atmospheric pressure and subcooling to the saturation temperature of 75 K) reaches a value 90 W/cm^2 (Fig. 3, line 7). According to [9], to reach a heat flux of 870 W/cm^2 at water flow boiling in the mini-channel the mass flow rate of the liquid shall be 5000 kg/m^{2}s, which 50 times exceeds the mass flow rate in our experiment. Fig. 3 also presents data on the critical heat flux for water pool boiling (formula of Kutateladze, line 5) as well as for water pool boiling at subcooling to the saturation temperature of 75 K (line 6) [10].

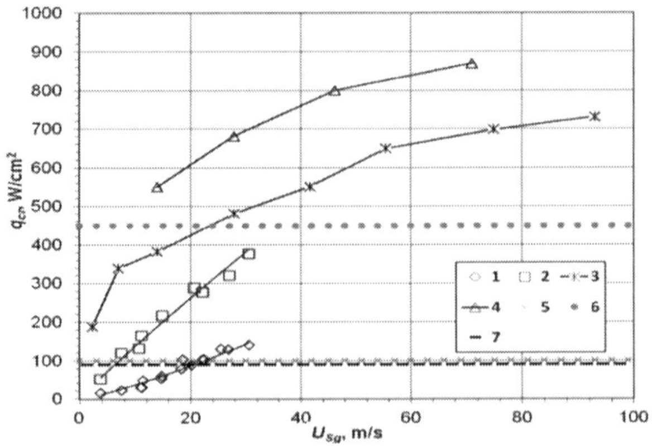

Fig. 3. Dependence of the critical heat flux on superficial gas velocity, U_{Sg}, Reynolds number of the liquid Re_l, and the channel height H. 1 – $Re_l = 8.5$, H = 1.2 mm; 2 – $Re_l = 30$, H = 1.2 mm; 3 – $Re_l = 62$, H = 1.0 mm; 4 – $Re_l = 113$, H = 1.0 mm; 5 – water pool boiling (formula of Kutateladze $q_{cr} = 0.14 r \sqrt{\rho_v} \sqrt[4]{\sigma g(\rho_l - \rho_v)}$, [10]); 6 – water pool boiling at subcooling to the saturation temperature of 75 K [10]; 7 – water flow boiling in a mini-channel for mass flow rate 100 kg/m²s, at atmospheric pressure and subcooling to the saturation temperature of 75 K [9].

4 Conclusion

The critical heat flux for thin water films, moving under the friction of gas flow, is almost an order of magnitude higher than that for pool boiling and flow boiling in the channel. In the experiments, the record value of the critical heat flux of 870 W/cm² for thin film flows was reached. This confirms the prospects of using the shear-driven liquid films in modern cooling systems of semiconductor devices with high local heat generation. Total heat losses to the atmosphere and heat spreading into the substrate in our experiments do not to exceed 25 % at heat fluxes higher than 400 W/cm².

This work was supported by the Ministry of Education and Science of Russia (Agreement No 14.604.21.0053, project identifier RFMEFI60414X0053).

References

1. A. Bar-Cohen, P. Wang, ASME. J. Heat Transf. **134**, 051017 (2012)
2. O.A. Kabov, Yu.V. Lyulin, I.V. Marchuk and D.V. Zaitsev, Int. J. Heat Fluid Flow **28**, 103 (2007)
3. D.V. Zaitsev, D.A. Rodionov and O.A. Kabov, Tech. Phys. Lett. **35**, 680 (2009)
4. O.A. Kabov and D.V. Zaitsev, Multiphase Sci. Tech. **21**, 249 (2009)
5. O.A. Kabov, D.V. Zaitsev, V.V. Cheverda, A. Bar-Cohen, Exp. Therm. Fluid Sci. **35**, 825 (2011)
6. E.A. Chinnov, F.V. Ronshin, O.A. Kabov, Int. J. Aip. Conf. Proc. **80**, 57 (2016)
7. E.A. Chinnov, F.V. Ronshin, O.A. Kabov, Interfacial Phenomena and Heat Transfer **3**, 243 (2015)
8. E.Ya. Gatapova, V.S. Ajaev, O.A. Kabov, JETP Letters **101**, 160 (2015)
9. W. Zhang, T. Hibiki, K. Mishima, Y. Mi, Int. J. Heat Mass Tran. **49**, 1058 (2006)
10. S.S. Kutateladze, *Heat Transfer in Condensation and Boiling* (Mashgiz, Moscow, 1952)

On the issue of fire hazard the fuel oil facilities of thermal power plants

Arkadiy V. Zakharevich[1, a], and *Dmitriy* N. Tsymbalov[1]

[1] National Research Tomsk Polytechnic University, 634050 Tomsk, Russia

Abstract. The methodology and results of an experimental study sets out the law ignition of typical liquid fuel – mazut, single heated up to high temperatures metal particles and porous particles of the metal oxides. The installed dependences of the numerical values of the ignition delay times of the temperature of porous particles (obtained after the welding) steel and solid particles of regular shape. The showed influence of the shape and structure of the particles (solid and porous) on the regularity of the process of mazut ignition.

1 Introduction

High requirements for reliability and safety of the equipment of thermal power stations (TPS) are traditionally defined by strict rules and regulations [1, 2]. The wear level is currently the main equipment of many thermal power stations of the Russian Federation is close to the maximum permissible operating standards. Moreover, there is no scientific basis of the main elements predictive modeling reliability of the power equipment of TPS. Under these conditions are particularly important factors affecting the safety and reliability of thermal power stations (including fire safety). All organic fuel used in thermal power plants are a fire hazard. Accordingly, the regulated norms and rules for handling them [1, 2]. But in spite of these provisions, fires in thermal power facilities still occur [3, 4]. Installed in recent years [5] regularities of mazut ignition indicate that the specific use of this type of fuel requires more attention to the rules of storage, handling and transportation. But so far there is no evidence to suggest that all of the factors that determine the fire hazard of fuel oil facilities of TPS, established and systematized.

It may be noted that the results of experimental [6-8] and theoretical [9-11] studies the conditions and characteristics of the liquid [6, 9, 10] and solid [7, 8, 11] fuels ignitions, the established the possibility of fires during the local heating (primarily single heated to a high temperature metal and nonmetallic particles (T_p) falling on the surface of the fuel.). Temperature range of local heat sources (respectively, and fire risk) covers an area significantly smaller values T_p compared to the temperature of the particles formed during welding and cutting metals

Well established [12, 13], it is possible to some ignition distillate fuels in the vapor phase under intensive heating the mixture vapor of the liquid fuel with an oxidizing agent (oxygen) for local heating of a single heated up to high (over 1000 C) temperatures particle.

[a] Corresponding author: bet@tpu.ru

The aim of this paper is to analyze the results of experimental studies of the conditions of ignition of mazut at thermal power station by heated single particles.

2 Experimental technique

In the planning and organization of the experiment were considered possible conditions of interaction heated to a high temperature of the particles formed during the welding and cutting of metal structures, with mazut film. But this is, first, unlikely. Second, for large characteristic sizes of tanks with fuel oil, the metal particle must quickly sink without initiating combustion. Ignition of flammable liquids is possible in the event of a spill on a solid surface and formation of sufficiently thin films. For these reasons all experiments were performed with films of mazut brand T-100 with a thickness of about 2 mm.

In a similar experimental [5-7] and theoretical studies [8-17] the main characteristic of the process was the dependence of delay time of ignition (τ_{ign}) from the temperature of the heat source T_p. Accordingly, in the experiments, the results of which are given below, also a function of purpose was adopted by the dependence $\tau_{ign}(T_p)$. In General, the methodology of the experiment was identical to that described in [5].

One of the objectives of the study was to analyze the influence of the shape and structure of particles to conditions and characteristics of ignition of mazut. Therefore, in the experiments used , steel particles in the form of a disc (diameter of d_p=6 · 10^{-3} m, the thickness is h_p=3 · 10^{-3} m), spheres (d_p=6 · 10^{-3} m), and formed after welding.

3 The discussion of experiments' results

Fig. 1 shows the experimental dependence $\tau_{ign}(T_p)$ for the three possible configurations of particles - a heating source (disc, sphere, drop, formed after welding). The solidified porous particles formed after welding, selected (sorted) before experiments so that their area of the mid-section was as close as possible to the cross-sectional area of the particle in the form of a sphere.

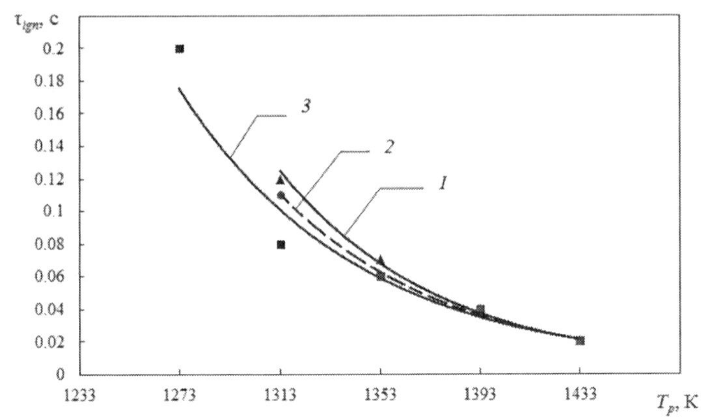

Fig. 1. Experimental dependences of mazut ignition time delay from the initial temperature of the particle: 1 – spheres d_p=6 · 10^{-3} m; 2 – disc d_p=6 · 10^{-3} m, h_p=3 · 10^{-3} м; 3 – particle after welding.

The maximum error in determining the delay time in all the performed experiments did not exceed 15% (small T_p). A comparison of the values of τ_{ign} obtained for typical mazut and distillate fuels – kerosene (Fig. 2). It may be noted that in all the range of initial temperatures particle heat sources, the time delay ignition of the mazut was less than that obtained for kerosene.

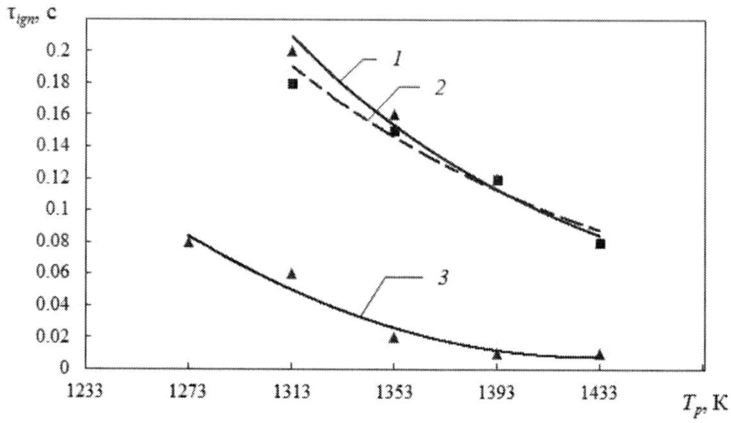

Fig. 2. Experimental dependences of kerosene ignition time delay from the initial temperature of the particle: 1– spheres d_p=6 · 10^{-3} m; 2 – disc d_p=6 · 10^{-3} m, h_p=3 · 10^{-3} м; 3 – particle after welding.

It is not obvious at first glance, the ratio of the values τ_{ign}, probably due to the peculiarities of the formation of flammable vapors of these fuels. Fuel oil (mazut) refers to liquid fuels with a certain share of the solid residue (coke) of the pyrolysis process. At the stage of vaporization of the mazut consumed is significantly less compared to any distillate fuels (including kerosene) heat of phase transition per unit mass of fuel.

It should be noted that the analysis video frames experiments showed that the mazut vapor ignition occurs at a small distance from the surface of the fuel film. But the mechanism of mazut ignition quite significantly different from the corresponding mechanisms of distillate fuels. As a result, separation into fractions of low-temperature gasification residue mazut temperature increases, and the thermal decomposition starts. The mass of the nonvolatile residue amounts to 60 % of the initial mass of the fuel [18]. Therefore, it can be divided into two stages of mazut ignition process. In the first step (at relatively low temperatures (1313-1400 K)) the ignition of vapor occurs. In a further ignited products of thermal decomposition the second part, which is formed during heating of high-molecular fractions of the fuel. In Fig. 3 shows a typical video frames obtained in the experiment when the ignition of the film of mazut spherical particle.

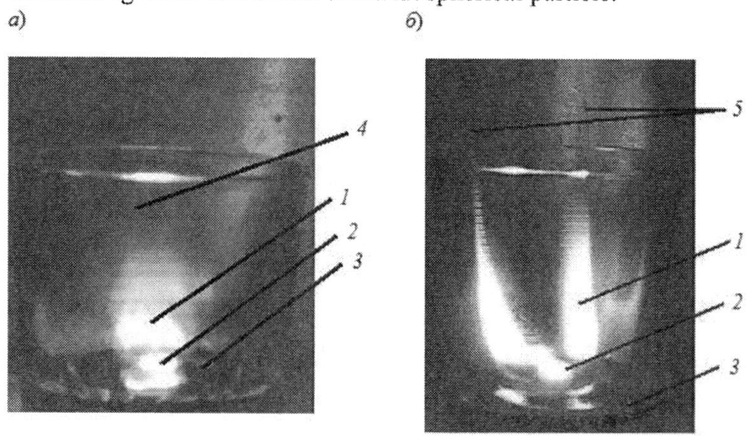

Fig. 3. Video frames of the experiment with ignition of mazut by a steel hot particle: 1 – flame; 2 – hot particle; 3 – mazut; 4 – glass vessel; 5 – products of the combustion of light mazut fractions vapor.

In experiments also found that when exposed to mazut porous particles (formed after weld) ignition occurs at temperatures almost 40 degrees lower than for particles in the form of a disk or sphere. This deviation to the marginal conditions of ignition due to the influence of the open porosity of the particles (formed after welding) on the intensity of the heat transfer to the vaporization surface. Increase heat to surface evaporation rate leads to an increase in the phase boundary temperature, and accordingly, the evaporation rate and temperature of the vapor.

Furthermore, in the pores, the conditions for the ignition of air mixture local (located at the initial instant in the pores of the particle) and products of vaporization process. Perhaps the ignition of this mixture occurs in the cross section of the particle at the "oil-to-air" section.

Also the duration of induction period and conditions ignition is affected by the low (compared to solid particles), the density of the porous particles. For this reason, the deposition rate of the porous particles, much less than monolithic. Accordingly, the duration of the period of heating the fuel in the first case, significantly more.

4 Conclusions

In conclusion, it should be noted that the results of the experiments illustrate the high fire hazard of mazut films on a solid (e.g. concrete or metal) surfaces, and the ability to forecast conditions and characteristics of ignition of that liquid fuel.

This study was financially supported by the Russian Science Foundation (project no. 14-39-00003).

References

1. The technical operation of thermal power plants, Ministry of Energy **115** (2003)
2. RD-153-34.0-03.301-00. Guidance document “Fire safety rules for power plants”. M .: JSC “Energy technologies” (2000)
3. N.I. Anikin, N.N. Bulxov, V.A. Gerish, *Pozharovzrivobezopasnoct* **19**, 53 (2010)
4. V.D. Zahmatov, *Pozharovzrivobezopasnoct* **22**, 63 (2013)
5. A.V. Zakharevich, G.V. Kuznetsov, V.I. Maksimov, V.F. Panin and D.S. Ravdin, Bulletin of the Tomsk Polytechnic University **313**, 25 (2008)
6. N.S. Belkov, A.V. Zakharevich, G.V. Kuznetsov, Chem. Petrol. Eng. **50**, 424 (2014)
7. A. V. Zakharevich, G. V. Kuznetsov, V. I. Maksimov and V. T. Kuznetsov, Combust. Expl. Shock. **44**, 543 (2008)
8. A.V. Zakharevich, J. Eng. Phys. Thermophys. **87**, 98 (2014)
9. O.V. Vysokomornaya, G.V. Kuznetsov, P.A. Strizhak, Russ J.Phys Chem. B **5**, 668 (2011)
10. G.V. Kuznetsov, P.A. Strizhak, J. Phys Chem. B **3**, 441 (2009)
11. D.O. Glushkov, G.V. Kuznetsov, P.A. Strizhak, Russ J. Phys. Chem. B **5**, 1000 (2011)
12. G.V. Kuznetsov, P.A. Strizhak, J. Eng.Thermophysics **17**, 224 (2008)
13. G.V. Kuznetsov, P.A. Strizhak, Combust. Explo. Shock. **46**, 683 (2010)
14. G.V. Kuznetsov, G. Ya. Mamontov, G. V. Taratushkina, Combust. Explo. Shock. **40**, 70 (2004)
15. G.V. Kuznetsov, G. Ya. Mamontov, G. V. Taratushkina, Khim. Fiz. **23**, 67 (2004)

16. G.V. Kuznetsov, P.A. Strizhak, J. Eng.Thermophys. **18,** 162 (2009)
17. G.V. Kuznetsov, P.A. Strizhak, J. Eng.Thermophys. **18,** 72 (2009)
18. B. S. Beloselsky, *Fuel oil and energy technology* (Power Engineering Institute, Moscow, 2005)
19. GOST 10585-99 Fuel oil, Fuel oil, Specifications (changes with 1, 2, 3) (2012)

Analytical estimation of particle shape formation parameters in a plasma-chemical reactor

Ilya A. Zhukov[2], *Sergei* S. Bondarchuk[1,2], *Alexander* S. Zhukov[2], *Ivan* S. Bondarchuk[2], *Boris* V. Borisov[3*], and *Tatiana* N. Ayeshina[3]

[1] Institute for Problems of Chemical and Energetic Technologies of the Siberian Branch of the Russian Academy of Sciences, 659322 Biysk, Russia
[2] National Research Tomsk State University, 634050 Tomsk, Russia
[3] National Research Tomsk Polytechnic University, 634050 Tomsk, Russia

Abstract. Analytical estimation of particle shape formation parameters in a plasma-chemical reactor implementing the process of thermochemical decomposition of liquid droplet agents (precursors) in the flow of a high-temperature gaseous heat-transfer medium was obtained. The basic factor which determines the process is the increase of concentration of a dissolved salt precursor component at the surface of a liquid particle due to solvent evaporation. According to the physical concept of the method of integral balance the diffusion process of concentration change is divided into two stages: the first stage is when the size of gradient layer does not reach the center of a spherical droplet and the second stage when the concentration at the center of a liquid droplet begins to change. The solutions for concentration fields were found for each stage using the method of integral balance taking into account the formation of salt precipitate when the concentration at the surface of the droplet reaches certain equilibrium value. The results of estimation of the influence of various reactor operation parameters and characteristics of initial solution (precursor) on the morphology of particles formed – mass fraction and localization of salt precipitate for various levels of evaporation.

1 Introduction

The use of ultra-fine powders of metal oxides for production of ceramic composite materials makes it possible to synthesize multipurpose products with unique strength characteristics, create construction materials replacing metals as well as protective coatings, filters, membranes, heat-insulation and catalytic materials with specified pore size. Plasma-chemical synthesis is one of the most promising techniques with controlled parameters, including thermochemical decomposition of liquid-droplet agents (precursors) in the flow of a high-temperature gaseous heat-transfer medium [1-3].

* Corresponding author: bvborisov@tpu.ru

The basic process parameters which determine the structure of obtained final product are represented by reactor operation characteristics (droplet speed, time inside the reactor, heat-transfer medium temperature), as well as relationship K between characteristic times of evaporation t_{vap} and intra-droplet diffusion t_{dif}:

$$K = \frac{t_{vap}}{t_{dif}} .$$

It is shown in papers [3-5] that in the process of precursor droplet movement in the flow of heat-transfer medium the temperature of a liquid particle increases up to the point where solvent evaporation begins at its surface relatively fast and the temperature profile becomes homogeneous. Then the concentration of salt near the surface increases and salt diffuses to the center of the droplet and at certain concentration value a condensed precipitate is formed.

2 Numerical Simulation

The rate u of change of radius R of the droplet due to solvent evaporation can be found in the framework of reduced film model [4, 6, 7]:

$$u = \frac{dR}{dt} = \frac{\lambda_g}{2\varrho} \frac{\text{Nu}}{R c_V} \ln\left(1 + \text{B}\right) = K\frac{D}{R} , \quad (1)$$

where B – Spalding number; Nu – Nusselt number; D – diffusion coefficient; λ_g – thermal conductivity of gas at boiling point; ϱ – gas density; c_V – specific heat capacity of solvent vapors at constant volume.

As the temperature of the surface of precursor droplet reaches the equilibrium point of evaporation (boiling) its radius *R*starts to reduce. Then it is assumed that the temperature of the surface of the droplet remains constant and equal to the temperature of equilibrium evaporation. The concentration C of salt at the surface increases due to solvent evaporation and diffusion process described by the following equation begins.

$$\frac{\partial C}{\partial t} = \frac{D}{r^2} \frac{\partial}{\partial r}\left(r^2 \frac{\partial C}{\partial r}\right), \quad r \in \left[0; R\right], \quad (2)$$

where r – coordinate oriented from the center of the droplet.

The initial condition (moment of time t=0) for equation (2) is.

Boundary conditions in the center of the droplet:

$$\left.\frac{\partial C}{\partial r}\right|_{r=0} = 0 . \quad (3)$$

Boundary conditions on the surface of the droplet in the regime of solvent evaporation:

$$D\left.\frac{\partial C}{\partial r}\right|_{r=R} = -C_S \frac{dR}{dt} = uC_S , \quad (4)$$

where $C_s = C(R)$ – concentration on the surface of the liquid particle.

Using the analogy of heat and diffusion processes the method of integral (in this case mass) balance [8, 9] can be used to solve equation (2). The method assumes the following concentration profile for considered conditions in the gradient zone with the thickness δ (Fig. 1, *a*):

$$C(r) = C_0 + \frac{\Delta C}{\delta^2}\frac{R}{r}\left(r - r_*\right)^2, \; r_* \le r \le R, \tag{5}$$

where δ – thickness of gradient zone of concentration; $r_* = R - \delta$, $\Delta C = C_s - C_0$.

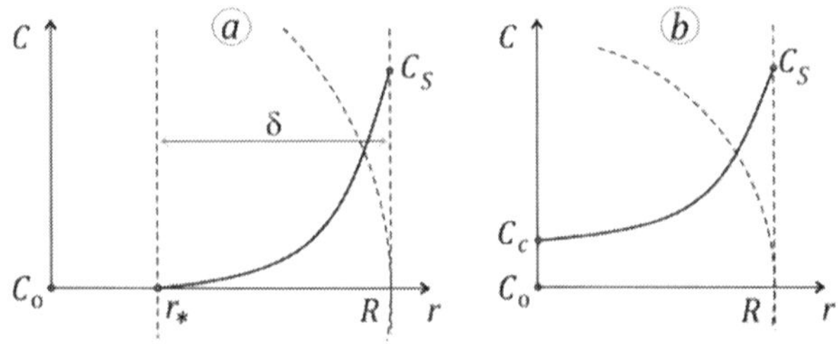

Fig. 1. Solution region scheme.

Integrating the equation (2) over the coordinate r

$$\int\limits_{r_* = R-\delta}^{R} r^2 \frac{\partial C}{\partial t} dr = \int\limits_{r_* = R-\delta}^{R} D\frac{\partial}{\partial r}\left(r^2 \frac{\partial C}{\partial r}\right) dr,$$

Taking into account relationship (5) in the limits of variable droplet radius R and length $\delta$of the gradient zone according to Leibniz's rule

$$\frac{d}{dt}\left[\frac{C_0}{3}R^3 - \frac{C_0}{3}R_0^3 + \Delta C R^3\left[\psi(R) - \psi(r_*)\right]\right] = 0,$$

where $\psi(r) = \left(\frac{r}{\delta}\right)^2\left[\frac{1}{4}\left(\frac{r}{R}\right)^2 + \left(1 - \frac{\delta}{R}\right)\left[\frac{1}{2}\left(1 - \frac{\delta}{R}\right) - \frac{2}{3}\frac{r}{R}\right]\right]$.

Integrating the obtained ordinary differential equation over time we can obtain the final relationship between the concentration at the droplet surface and the size of the gradient zone

$$C_s = \frac{C_0}{3}\left[3 + \frac{(1-\varphi)\left(\frac{R_0}{R}\right)^3 - 1}{\psi(R) - \psi(r_*)}\right]. \tag{6}$$

Another relationship between C_s and δ can be obtained from differentiation of the temperature profile (5), boundary conditions (4) taking into account the relation (1)

$$C_S = C_0 \frac{2-\frac{\delta}{R}}{\left(2-\frac{\delta}{R}\right)-\frac{\delta}{R}K} . \tag{7}$$

Thus, from expressions (6) and (7) for any current droplet radius Rsurface concentration C_s and gradient layer thickness δ can be determined if condition $R \geq \delta$ is satisfied.

In case when concentration disturbance reaches the center of the droplet the following relationship can be used [8] instead of the profile (fig. 1, *b*)

$$C(r) = C_C + (C_S - C_C)\left(\frac{r}{R}\right)^2 , \quad 0 \leq r \leq R ,$$

where C_c – concentration in the center of a liquid particle.

After the transformations analogous to the ones described above an ordinary differential equation can be obtained for this case

$$\frac{d}{dt}\left[C_c \frac{R^3}{3} + \frac{C_s - C_c}{5} R^3\right] = 0 ,$$

solving which we can determine the concentration C_c at the center of a liquid particle and at the droplet surface C_s:

$$C_c = \frac{5(2-K)}{2(5-K)}(1-\varphi)C_0\left(\frac{R_0}{R}\right)^3 , \quad C_s = \frac{2}{2-K}C_c . \tag{8}$$

Parameter φ, from equations (6) and (8), determines the fraction of precipitating salt mass (regarding the initial mass).

3 Analysis of the results

The algorithm for finding concentration distribution parameters and the fraction of precipitated salt is as follows.

1. On the basis of thermophysical properties of materials and reactor's geometry characteristics using averaged heat-transfer medium flow parameters from formula (1) the following values are found: a dimensionless criterion K – ration between characteristic times of evaporation and diffusion, evaporation rate u, the time of a liquid particle inside the reactor and radius R of a particle leaving the reactor.
2. Concentration C_s on the droplet surface is calculated for φ=0 using expressions (6), (7) or (8).
3. If $C_s > C_*$ (C_* – equilibrium concentration after which salt formation starts), then we choose the value of parameter φ, providing $C_s = C_*$. Thus, we can determine all characteristics of a liquid-droplet particle leaving the reactor including the coordinates of salt precipitation.

The evolution of the droplet of low-concentration (C_0=0.04) water solution of the salt with the radius R_0=20 μm in the process of diffusion $D=2\cdot10^{-9}$ m^2/s and conditional value C_*=1. is considered as an example.

Figure 2 shows the dependence of concentration C_s on evaporation level R_0-R for various values of Kcriterion. As it can be seen the higher the value of K the sooner the concentration at the surface reaches the equilibrium point C_*, which determines the beginning of salt precipitation process. The dependence of mass fraction of salt precipitate on evaporation level for various values of Kcriterion is shown Figure 3. The data correspond to development of physical process shown in Figure 2 and illustrate the size and location of the area (in the particle volume) of salt precipitation. Figure 4 shows the diagram for localizations of 80% salt mass precipitation characterizing the condensate area after precursor solvent evaporation.

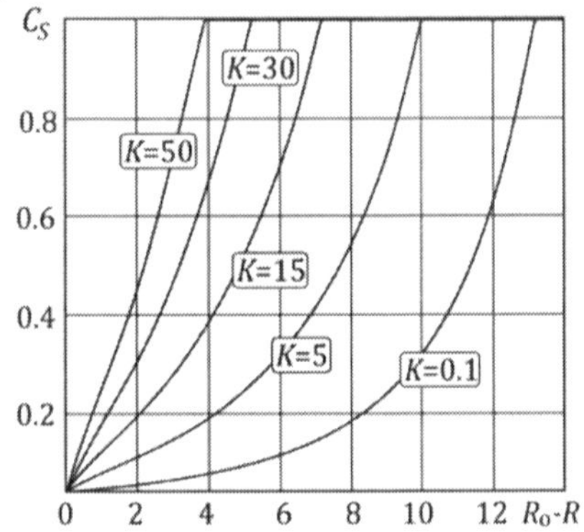

Fig. 2. Dependence of concentration C_S on evaporation level.

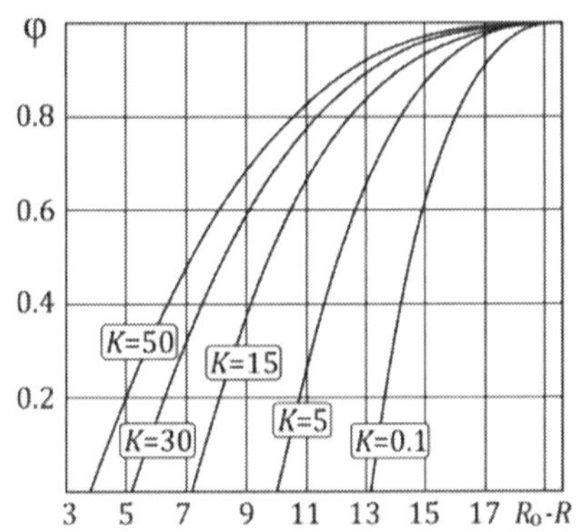

Fig. 3. Dependence of mass fraction of salt precipitate on evaporation level.

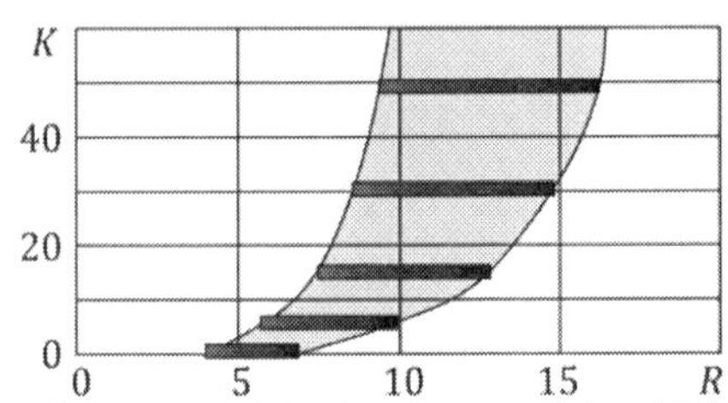

Fig. 4. Localizations of 80 % salt mass precipitation for a particle with initial size of 20 μm.

Thus, the algorithm for analytical estimation of parameters of particle shape formation in the plasma-chemical reactor based on solving the problem of diffusion in the intra-droplet precursor volume in the process of solvent evaporation is described in the paper. The offered approach makes it possible to determine mass fraction and localization of salt precipitate for various evaporation levels.

The reported study was funded by RFBR, according to the research project No. 16-38-60031 mol_a_dk.

References

1. A.S. Zhukov, V.A. Arkhipov, S.S. Bondarchuk, Izv. vuz. Phys. **53**, 12 (2010)
2. V.A. Arkhipov, A.S. Zhukov, S.S. Bondarchuk, Tomsk state pedagogical university bulletin **8** (2011)
3. V.A. Arkhipov, A.S. Zhukov, S.S. Bondarchuk , J. Eng. Phys. Thermophys. **86**, 775 (2013)
4. A.S. Zhukov, V.D. Goldin, V.A. Arkhipov, S.S. Bondarchuk, Russ. J. Phys Chem. **7** (2013)
5. A.A. Brin, S.P. Fisenko, Yu.A. Khodyko , J. Eng. Phys. Thermophys. **84**, 274 (2011)
6. A.S. Zhukov, V.A. Arkhipov, S.S. Bondarchuk, T.V. Zhukova, Butlerov communications **28** (2011)
7. V.V. Pomerantsev, K.M. Arefyev, D.B. Akhmedov, *Fundamentals of practical combustion theory* (Energoatomizdat, Leningrad, 1986)
8. T. R. Goodman, Advances in Heat Transfer **1**, 51 (1964)
9. S.S. Bondarchuk, A.S. Zhukov, P.A. Pesterev, *Tomsk state pedagogical university bulletin* **8,** 49 (2011)

Comparative analysis of simulated kinetic schemes when calculating the nitrogen oxide formation in solid fuel flaring

Elena Zueva[1], *Natalia*Vizgavlyust[1*], *Lyudmila* Vorobjeva[1], and *Alexander* Starchenko[2]

[1] National Research Tomsk Polytechnic University, 634050 Tomsk, Russia
[2] National Research Tomsk State University, 634050 Tomsk, Russia

Abstract. This paper presents a comparative analysis of the kinetic schemes with the numerical analysis of the generation of nitrogen oxides during the combustion of the pulverized fuel. The distributions of the concentration of nitrogen oxides of two studied kinetic schemes with height of the combustion chamber are obtained. The conclusions about the accuracy of numerical researches and the choice of optimal kinetic scheme are made.

1 Introduction

Protection of atmosphere from pollutant emissions is one of crucial current problems. Rapid growth of energy consumption is known to be accompanied by the increase of emissions polluting atmosphere.

Thermal stations consuming big quantities of organic fuel emit combustion products containing solid particles, sulfur, nitrogen and carbon oxides into atmosphere.

Every year requirements to environment protection become stricter. Due to the above one of the problem of designers is to adhere to the requirements on ultimate permissible concentrations set up by sanitary bodies (first of all - nitrogen oxide) both for new thermal plants and for the ones already operating.

In order to define the most efficient way to control NO_x for each specific boiler it is necessary to perform some field experiments. However it is not always possible to run such field experiments.

In order to optimize designing and to reduce the experiment costs the numerical simulation will be used. It would allow analyzing the nitrogen oxide emissions during designing of new plants and working out several options for upgrade of the existing boilers choosing the best option for solution of this problem.

For introduction of effective techniques for suppression of nitrogen oxide formation it is necessary to develop numerical simulations of combustion of nitrogen containing fluids in the furnaces and combustion chambers. When creating the numerical model describing the formation of nitrogen oxide it is important to consider all specific processes going on in the furnace: complicated spacial mechanics, combustion of coal duster etc.

* Corresponding author: vizgavljust@tpu.ru

2 Description of the model and the object of research

At present the mechanism for thermal wastes of nitrogen oxide is well studied and described by the diagram of Ya.B. Zeldovich [1]. Simulation of formation of fuel and rapid nitrogen oxides is more complicated. As a rule the offered kinetic models require accounting the significant number of chemical reactions accompanying this condition. It significantly limits their use when performing numerical investigations for field facilities.

This article presents the numerical simulation of nitrogen oxide formation in combustion of powdered nitrogen containing fuel in the furnaces of industrial boilers basing on the kinetic model for NO formation developed by Mitchell and Tarbell [2], and with use of the model developed by Gusev, Zaychik nd Kudryavtsev [3].

Mitchel-Tarbell model is fairly convenient for using; its kinetic scheme includes only 13 reactions. Kinetic scheme of this model is described by the system consisting of seven equations simulating chemical reactions of combustion of hydrocarbons, coke, emission and combustion of volatile substances, formation of thermal nitrogen oxides, heterogeneous reaction of nitrogen oxide conversion in the result of their interaction with coke carbon [2, 4, 5].

Gusev-Zaychik-Kudryavtsev model supposes that all fuel nitrogen oxides are formed from nitrogen fuels transferring into gas form along with volatile substances and the impact of nitrogen remaining in the coke is neglected. Formation of fuel nitrogen oxide is calculated supposing that during emission and combustion of volatile substances the nitrogen containing components of fuel are decomposed into active atomic nitrogen N or hydrogen cyanide HCN. Further the process is supposed to go into two directions: oxidation of nitrogen to nitrogen oxide and formation of molecular nitrogen due to recombination of atomic nitrogen or as a result of nitrogen oxide reduction. The estimated scheme for fuel oxides generation is described by the system consisting of three equations [2, 4].

The above numerical models for NO_x generation were applied to the operating boiler of BKZ-220-100ZHSH, which configuration is presented on Fig.1. Fig. 2 shows distribution and concentration of nitrogen oxide in the furnace of BKZ-220-100ZHSH boiler with abundant air α=1.2.

BKZ-220-100ZHSH boiler has prismatic furnace. The boiler is 24.9 m high; the furnace is 8.64 m wide and 7.74 m deep. The furnace is completely shielded by pipes of 60 mm in diameter. Slightly sloping symmetric hearth (α=15°) formed by pipes of the front and back shields complies with the system of fluid slag removal and the lower part of the furnace is refractory-faced and covered with chrome plastic up to 5.750 m high. Lighting-up burners are installed to the walls at 9.6 m high and built in into special burners with secondary air supply [6].

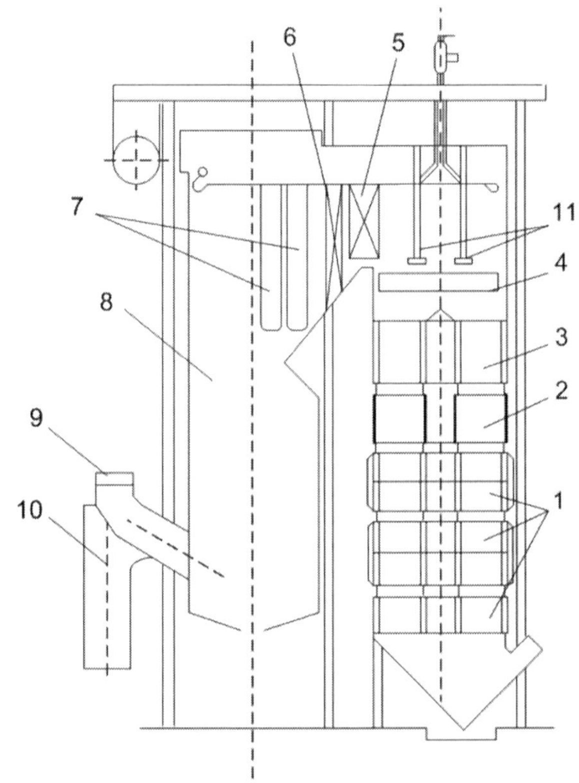

Fig. 1. BKZ-220-100ZHSH boiler: 1, 3–air heater stages; 2, 4–economizer stages; 5–convention stage; 6, 7–outlet and furnace shields; 8–furnace; 9– secondary air nozzle; 10–miller separator; 11–shot cleaning system.

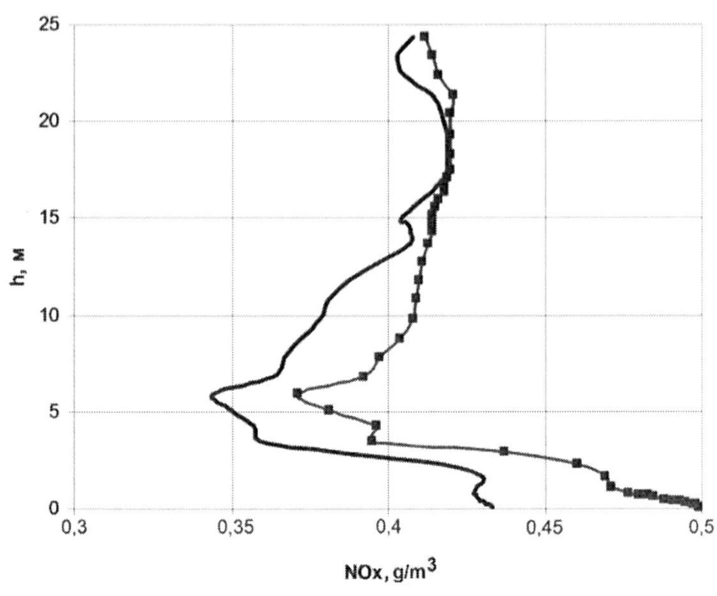

Fig. 2. Formation of nitrogen oxides respective the furnace height of BKZ-220-100ZHSH boiler: ——— – model [2], –■–■–■–■– – – model [3].

3 Results and discussion

Fig. 2 shows that the selected numerical models for investigation of nitrogen oxide formation provide fairly similar results. The accuracy of the forecast NO_x formation greatly depends on the part of nitrogen fuel emitted together with volatile substances or during the combustion of coke residue. Model [2] describes more accurately and fully the process of nitrogen oxide formation since it includes the combustion of coke residue, turbulent diffusion and transfer of furnace medium components by averaged movement. Fig.3 shows the typical of all models [2, 3] character for NO_x concentration distribution in the vertical section of the furnace between direct-injection burners and 3D model.

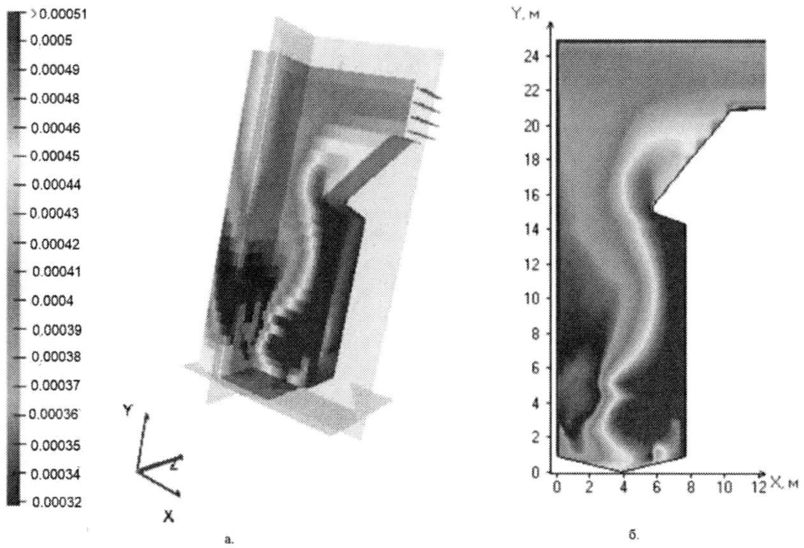

Fig. 3. Distribution of NO_x weight concentration in the furnace of BKZ-220-100ZHSH boiler: a – 3D model, b – vertical section.

Fig. 3 shows that the maximum quantity of NO_x nitrogen oxide is formed in the upper and lower parts of active burning area which is characterized by the increased values of flow temperature. NO_x concentration changes to very insignificant extend in the part of the furnace from 10 to 25 mm high where combustion is already over. The area of maximum temperature values (up to 1900 K) is located in the part of the furnace opposite to burners. The furnace outlet temperature is 1250 K.

4 Conclusions

Model [2] is an effective tool for forecasting nitrogen oxide generation in coal-dust furnaces compared to model [3], which is confirmed by the results of calculation experiments and tests showing fair convergence with the data of field measurements. Model [2] considers the data on outlet process and combustion of fuel volatile components, after-burning of coke residue, non-isothermality, turbulent diffusion and transfer of furnace medium components by the averaged movement, and using model [2] for NO formation during coal combustion will allow optimizing combustion process for powdered fuel in boilers of the thermal plants in order to increase the efficiency of fuel burning out and reduction of NO_x emissions.

References

1. V.R. Kotler, *Nitrogen oxides in the flue gas boilers* (Energoatomizdat, Moscow 1987)
2. J.W. Mitchell, J.M. Tarbell, AIChE Journal **28**, 302 (1982)
3. I.N. Gusev, L.I. Zaytchik, N.Yu. Kudryavtsev, Therm. Eng. **32**, 1 (1993)
4. N.V. Vizgavljust, A.V. Starchenko, A.V. Gil, T.S. Taylasheva, EPJ Web of Conferences **82**, 01040 (2015)
5. E.A. Zueva, N.V. Vizgavlyust, L.V. Vorobyeva, A.V. Starchenko, MATEC Web of Conferences **72,** 01124 (2016)
6. I.I. Fedetsky, *Issledovanie protsessov v mineralnoy chasti nazarovskogo uglya v topke s ploskimi parallelnymi struyami* (Diss. kand. med. nauk., Tomsk, 1980) [in Russian]

Peculiarities of teaching translation of scientific and technical papers to engineering students

Dmitry L. Matukihn[1,*], *Nikolay* A. Kachalov[1], and *Roman* M. Fedorenko[1]

[1]National Research Tomsk Polytechnic University, 634050 Tomsk, Russia

Abstract. The article deals with the issues and methods of teaching translation and interpretation of scientific and technical papers to engineering students from a linguistic point of view. The authors describe the concept and special features of technical translation as a professional activity compared to different kinds of translation. The paper analyzes stylistic, lexical and phraseological, and grammatical peculiarities of scientific and technical papers. Moreover, the authors suggest the main techniques of translating scientific and technical papers such as direct and indirect translation methods.

1 Introduction

Recently, a large number of state-of-art machinery, equipment and technology have been imported in the Russian Federation. The rate of exchange of scientific and technical information with other countries is increasing rapidly. As a result of these intrinsic reasons, the ability to translate a foreign language text and interpret speech is one of the major goals of teaching foreign language to engineering students. However, the translation is the most reliable means of verification of understanding a foreign language, its semantic and stylistic features. Special attention shall be drawn to teaching translation under the circumstances of limited contact hours in the engineering universities, as the translation is one of the most important elements of performance monitoring at different stages of mastering a foreign language. In the course of training a future expert it is required to provide the development of overall necessary professional and personal qualities, the leading of which is cognitive independence. Linguists note that the importance of the ability to organize independent activities is most clearly defined in the process of spoken and written foreign language communication that requires constant analysis of linguistic facts. Translation skills of technical documentation, manuals for industrial and household equipment are necessary for senior students of engineering specialties, doing practical training in various enterprises and institutions [1, 2].

Translation is an exact reproduction of the original source by means of another language while preserving the unity of content and style. It is believed that technical translation is less challenging than fiction translation, because the technical translator needn't look for equivalents of various metaphors and other stylistic means, which are inherent in fiction and journalistic style. Technical translation difficulty lies elsewhere. The basis of technical

* Corresponding author: mdlbuddy@mail.ru

translation is formal and logical style, which is characterized by precision, unemotionality and impersonality of information transfer. Technical translation shall be as close to the original source.

A slight deviation from the meaning may have unintended consequences (for example, inaccurate translation of technical equipment manuals can lead to an accident on the site). Therefore, in contrast to the fiction translation, technical translation does not allow any inaccuracies or deviation from the original source. In technical translation, each word and definition is translated precisely, eliminating ambiguity of understanding. Technical translation is saturated with various highly-specialized terms. Grammar of technical translations is characterized by the use of specific and firmly established grammar rules, e.g., *indefinite-personal and impersonal constructions, passive voice, and non-finite forms of the verb* [3]. Logical highlights are presented by means of inversion. Whatever the complexity of the information in a document to be translated is: description of technological processes, calculations of any parameters, principles of operation of equipment and its operating procedures, etc., the translator is required to understand the meaning of the text, and then skillfully and accurately convey it to the target language. Thus, technical translation is a special kind of translation, the quality of which depends primarily on the personality of a translator.

If one needs to quickly get acquainted with the simple technical documentation, it is sufficient to use computer-based translators. To provide in-depth study of scientific literature, translation must be performed by a professional who commands the foreign language and obtains technical knowledge in a particular area, e.g., electrical power engineering, heat power engineering, etc. The goal of training technical translation skills is to teach very accurately and clearly to convey in another language any technical information. No interpretation, no ambiguity, no distortion of the meaning is permitted. This paper dwells on the issues and methods of teaching professional translation of scientific and technical papers to engineering students from a linguistic point of view.

2 Peculiarities of translation of scientific and technical papers

The peculiarity of technical translation of scientific documentation is an excessive amount of specialized terms, service and introductory words, frequent use of abbreviations, lexical neologisms, and realias. A significant role in the scientific and technical papers is played by prepositions and complex conjunctions (e.g., *in order that, provided that, as though, etc.*) which create logical connections between the individual elements of sentences. In addition, the scientific and technical papers frequently use adverbs, which are integral parts of the development of logical thinking. Under the realias of scientific and technical papers we understand the titles, the brands of equipment and materials.

Specialized terms are to be translated using special dictionaries and glossaries. The realias, as a rule, are not translated but transliterated or given in the text in their original writing (e.g., the Russian *GOST, Act, sputnik, etc.*). Optimal translation solution for the phraseology is a search for identical phraseological items in the target language. In the absence of direct correspondence, the original idiom can be translated by searching for similar phraseology items, having a similar meaning with the original one, but built on a different verbal and image basis. It should be remembered that idioms, often having similar meaning but different form in various languages, are uniquely emotionally and associatively colored, and not always interchangeable. Geographical names and well-known proper names are given in Russian transcription.

The translation process always takes place strictly within the logical thinking, i.e., those who are capable of thinking logically are likely to make the proper translation. Experts in logic say that to understand an unfamiliar term, we need to analyze the context. In the course of this analysis, we set different semantic connections and relationships between the unknown term and other words, the meaning of which is well known. Such cases are very common in the translation from the foreign language into the native language [4].

Stylistic peculiarities of the translation of scientific and technical papers are language clarity, lack of expressive, emotional and imaginative patterns. The texts of technical manuals and specifications often contain sentences without a predicate or a subject, as well as sentences, consisting of recitations only. The accuracy of the scientific and technical translation is determined by how well the translator commands the source language and the subject of discussion. The clarity of the text can be defined as its semantic unambiguity. In [5] the authors identified three main categories of stylistic defects that deny the uniqueness of the text:

- sentence amorphousness;
- logical stress shift in the sentence;
- "parasitic" relationships between words.

The authors suggest that semantic unambiguity of the text can be achieved by overcoming such defects of style as the excessive use of possessive case, the repetition of the same single-root words in a sentence, agglomeration of long words, pseudo-scientific patterns, bureaucratic style, etc. It is necessary to avoid literal translation which always leads to the distortion of an original meaning, or violation of the rules of the Russian language. The literal translation is based either on formal resemblance of the English word with the Russian one, or the use of the main or most common meaning of the English word without considering the context as a whole, e.g.,

plant – a herb not a station,
machinist – an engine driver not an operator,
package – wrapping not software, etc.

Knowledge of the typical translation errors and their causes contribute to their prevention. Translation errors are divided into two main types: speech, and functional and informative. Speech errors, in turn, are subdivided into system-speech and normative speech errors. They feature a violation of the norms of language and speech standards (usage). Functional and informative errors are functionally unreasonable deviation from the original text, adversely affecting the transfer of its meaning. In terms of its negative impact on the quality of the original source translated, they are divided into the "distortions" and "inaccuracies".' The distortions are errors in the translation, misleading the translation recipient about the content of a source text. Inaccuracy features lesser degree of equivalence violations, and less misleading impact, e.g.,

provide a means – give a chance not ensure a measure,
to process – for the procedure not to produce,
silicon – graphite not carbonium, etc.

Grammatical peculiarities of the translation of scientific and technical papers are characterized by frequent use of participial and verbal adverb phrases, simple, expended and complex sentences, passive voice, infinitive, participial and gerund phrases.

The great danger for the translator lies in the fact that both English and Russian feature a lot of similar grammatical forms and syntactic structures, the functions and meanings of which do not match in the languages. Similar names of such forms and constructions, borrowed both in Russian and in English from Latin, should not be misleading. English infinitive and participle functionally and meaningfully differ from the corresponding Russian forms. In the course of translation, it is required to use grammatical transformations. Functional correspondence is achieved by a complete change of structures, word order and by replacing parts of speech and sentence, adding and omitting words. The ability to adequately translate foreign text is one of the goals of training, even though it is not aimed at training a highly-qualified translator. The essence of adequate translation is not the transfer of letters but the spirit of the text. Each language has its own unique means, characteristics and properties, so to faithfully convey an image or phrase, sometimes a translator shall completely change them [6].

3 Translation methods of scientific and technical papers

The translation process features a number of special methods aimed at creating adequate translation. Experts distinguish two translation methods: direct and indirect translation. Indeed, the message in the original language may properly be translated in the target language, since it is based on the parallel categories, either the parallel concepts. Although, sometimes due to structural or metalinguistic distinctions, some stylistic effects can't be transferred to the target language, without changing in varying degrees, the order of the elements or even lexical items [7, 8, 9]. It is clear that in the second case, it is necessary to resort to more sophisticated methods – indirect methods of translation. Borrowing, calquing and literal translation are methods of direct translation. Transposition, modulation, equivalence and adaptation – indirect translation.

Direct translation methods

Borrowing is the simplest method of translation. Borrowed approximately from 50 languages, lexical items account for almost 75% of the vocabulary of the English language and include layers of vocabulary borrowed from various historical periods and under the influence of different conditions of existence and development. Many borrowings under the influence of the system, into which they have entered, undergo significant phonetic, grammatical, and even semantic changes, adapting to the phonetic, grammatical and semantic laws of the system. The process of assimilation can be so deep that the foreign origin of such words is not obvious for native English speakers and is exposed only with the help of etymological analysis. This is most true for the Scandinavian and Latin borrowings, e.g., engineer, figure, get, skill, etc. It should be noted that the borrowings are often introduced to language through translation, among them are semantic borrowings, or "false friends", which should be avoided. For example:

fixed – repaired not mounted,
current – electricity not present/existing,
realize – work out not understand,
accurate – tidy not precise,
conductor – a cable not a fare taker,
mark – a stamp not a label, or a sign, etc.

Calquing is a special kind of borrowing: we borrow from a foreign language a syntagm and literally translate the elements that compose it. For example: *the word 'superpower' is translated, using the method of calquing as a 'power house'*, other examples are as follows *point of view, high voltage, air conditioned, etc.*

Literal translation is to transfer the structure of the sentence without changing the construction and without substantial changing the word order. If the Russian sentence has a structure similar to the English one and it can be used without violating the grammar rules and reasoning, the literal translation is acceptable. For example:

bolt of lightning – lightning strike not lightning flash,
main – primary not manifold,
line – a string not pipe, etc.

Indirect translation methods

Transposition refers to the method, which consists in replacing the words in one language without changing the meaning of the sentence in the second language. Replacement is often necessary because of the different grammatical structures in the language. The translator shall use the method of transposition, if the resulting pattern matches better the whole phrase or allows you to restore the stylistic features. For example:

"...to enhance the international nuclear power safety measures for civic scientific reactors." – "...совершенствование международных усилий по обеспечению атомной безопасности относительно гражданских научно-исследовательских ядерных реакторов."

Modulation is a variation of the message, using a phrase with the different meaning in the original language and the target language for the transfer of the same idea. This method can be resorted when it is clear that literal translation or even transposition results in a grammatically correct statement, which is not contrary to the spirit of the target language. Inadequate translation can be avoided through modulation. For instance:

"...which provide the extensive foundation for assuring reactor safety." – "... формирующие исключительную основу по обеспечению безопасности функционирования реактора."

In addition to modulation, *concretization* method is widely used in translation, i.e., the translator uses a narrow meaning, belonging to a word, but does not provide a new, as in modulation. For example:

"...according to the operable codes." – "...согласно действующим сводам положений." The word *'codes'* stand for *'кодекс'*, but it has got another meaning such as *'свод'*, for example, *'свод законов.'*

Equivalence often occurs that two texts describe the same situation trough very different stylistic and structural means. Most of equivalences are stable and are part of the idiomatic phraseology, including clichés, sayings, adjectival or substantive fixed patterns and so on, e.g.,

dog bolt – откидной болт,
cast iron – чугун,
pin – шпилька,
liquid rockets – ракеты на жидком топливе, etc.

Quite often is used a translation method called *grammatical replacement*, which can be seen at the level of word forms or the whole sentences, e.g.,

"...that would clearly set the preferable attributes for safety control of a scientific reactor." – "...однозначно определяющего необходимые атрибуты для котроля за безопасностью научно-исследовательского реактора." A verb *'would set'* is translated as a participial phrase *'определяющий'*.

Compensation is used when some semantic elements and stylistic peculiarities can't be transferred reliably. The translator shall remember that the translation is not just the transfer of words from one language to another, it's also transfer of cultures. For example:

a bogthrotting son of a bloody noun – такой-рассякой,
clunker – развалюха, etc.

Adaptation. When the situation in the original language does not exist in the target language, it shall be transferred through a different situation, which we consider an equivalent, e.g.,

"Nature asked both candidates" – "Журнал "Nature" задал вопросы обоим кандидатам",

"Department of Energy people" – "представители министерства энергетики США",

"Safety posters and signs are displayed in the territory of Power Station" – "На территории ТЭЦ развернута наглядная агитация по технике безопасности", etc.

4 Conclusions

This paper discusses the issues and methods of professional translation of scientific and technical papers by the example of power engineering from a linguistic point of view. Translation of technical documents is the field of professional translation activity. The professional translation shall be performed by highly qualified personnel, commanding a foreign language, professional knowledge, specific terminology, being able to adequately and competently express their thoughts in the target language. phrases.

Specific requirements for the scientific and technical translation are the ability to suggest accurate terminology equivalents, which is a necessity for adequate translation. Rather serious issue in the translation of scientific and technical texts is the problem of non-equivalent terms. The origin of non-equivalent vocabulary is differences in the realias of scientific environment. In this case, most of the non-equivalent terms can be translated by calquing, using the lexical and grammatical transformations. Another translation issue is when one English term corresponds to several terms of the Russian language. In this case, the translator requires not only a high enough level of native language and standard of speech, but also deep knowledge in the field the document to be translated refers to.

References

1. E.S. Tarasova, MJSS **6,** 350 (2015)
2. Yu.I. Lashkevich, M.D. Grozdova, *On the translation of scientific and technical texts* (Practica, Moskva, 2016)
3. V.S. Gorbunova, S.V. Sbotova, XXI century: the results of past and present problems **2,** 194 (2015)
4. A.V. Kartashova, Bulletin of Siberian Science **1,** 617 (2011)
5. D.L. Matukhin, E.N. Gorkaltseva, MJSS **6,** 525 (2015)
6. A.V. Obskov, S.I. Pozdeeva, D.L. Matukhin, G.A. Nizkodubov, MJSS **6,** 319 (2015)
7. Yu. Kobenko, E. Tarasova, Procedia Soc. Behav.Sci. **206,** 3 (2015)
8. Ju. Kobenko, A. Ptashkin, MJSS **5,** 1694 (2014)
9. Y.V. Kobenko,. N.N. Zyablova, S.I. Gorbachevskaya, MJSS **6,** 494 (2015)

MATEC Web of Conferences **92**, 01042 (2017) DOI: 10.1051/matecconf/20179201042
Thermophysical Basis of Energy Technologies - 2016

Measurements of the contact angle of Noa81 photoresist for different temperatures

Ekaterina O. Kirichenko[1,2,*], and *Elizaveta* Ya. Gatapova[1]

[1]Kutateladze Institute of Thermophysics, 630090 Novosibirsk, Russia
[2]Novosibirsk State University, 630090 Novosibirsk, Russia

Abstract. In this work we study the properties of the surface, obtained by applying Noa 81 photoresist at different temperatures of the substrate. The measurement of free energy was conducted by the Owens, Wendt, Rabel and Kaelble method under isothermal conditions. The paper presents the obtained data on contact angle for different temperatures.

1 Introduction

Intensive development of electronic technology is one of the main reasons for the continuous improvement of the existing cooling systems as well as designing brand new heat exchangers. This needs availability of advanced surfaces capable to resist thermal and mechanical stress [1]. Evaluating the possibility of using such surfaces in various devices, as well as studying physical phenomena near the three phases contact line [2, 3] plays a crucial role in designing advanced cooling systems.

This article examines the surface obtained by applying the Noa 81 photoresist. Using the Drop Shape Analyzer-E-100 produced by KRUSS, we investigated surface properties such as surface free energy and contact angle. By applying add-on TC40 module, which is a chamber for rapid change of temperature, the contact angle was measured within the substrate temperature range from 13 to 88°C.

2 Experimental setup and technique

Studies of the contact angle and surface free energy were performed with the use of DSA 100 instrument produced by KRUSS. This device consists of three systems: high-precision dosing system with the dosing step of 0.1 μl, motorized object table for sample with program driven axes, and the optical system, which includes 50 W light source and a CCD camera with a resolution of 780x580 pixels (viewing field from 3.7x2.7 mm to 23.2x17.2 mm).

Free surface energy was measured using Owens, Wendt, Rabel and Kaelble method. Measurement technique is as follows. A few drops of water and diiodomethane are deposited on the surface. The surface tension is considered from the viewpoint of polar and disperse components. The method is based on the equation for the calculation of surface

* Corresponding author: kirichenko_ekaterina@outlook.com

tension combined with Young equation. The system is solved based on contact angles of two liquids with known polar and disperse components of the surface tension. Dispersion component includes van der Waals forces and other non-specific interactions, while polar component is responsible for strong interaction and hydrogen bonds. These considerations laid basis for derived equation:

$$\frac{\sigma_L(\cos\theta+1)}{2\sqrt{\sigma_L^D}} = \frac{\sqrt{\sigma_S^P}\cdot\sqrt{\sigma_L^P}}{\sqrt{\sigma_L^D}} + \sqrt{\sigma_S^D}\,, \tag{1}$$

The contact angle is investigated by forming a drop with a volume of 5 μl deposited on the surface. The drop is formed by means of the syringe pump. Measurement of contact angle on the Noa 81 surface is shown in Fig. 1.

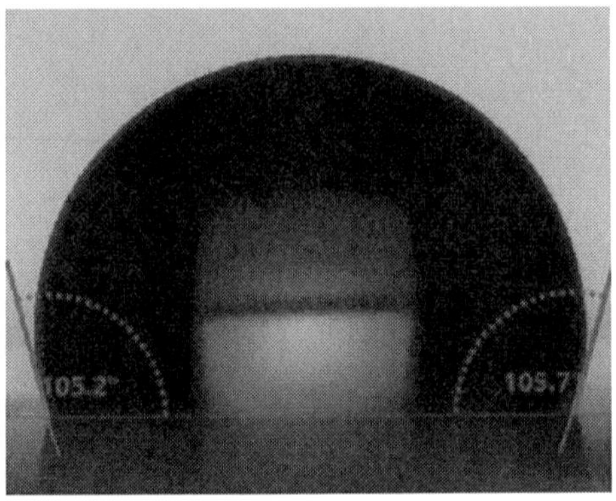

Fig. 1. Measurement of the contact angle on the Noa 81 surface.

In this article, the measurements were conducted within the temperature range from 13 to 88°C with the application of TC40 supplementary module with Peltier elements (Fig. 2).

Fig. 2. The TC40 thermostatic chamber with Peltier elements.

The substrate is placed into a thermostatic chamber. A thermocouple is installed at a distance of 1 mm from the investigated surface to control ambient temperature. The liquid is fed through the hole from the top by means of the syringe pump.

3 Results

The free energy of the Noa 81 surface substrate amounts to 23.24±0.66 mN/m, at that, a dispersive component is equal to 22.83±0.46 mN/m, while polar component is equal to 0.42±0.19 mN/m. For example, for pure Teflon this value amounts to 18 mN/m while the polar component is missing.

Fig. 3 shows the dependence of contact angle on the surface temperature. Studies performed in a thermostatic chamber revealed that with increasing temperature the contact angle varies within the hysteresis. The contact angle hysteresis under isothermal conditions on a given surface was investigated in [4]. At an ambient substrate temperature, the contact angle hysteresis was 20°.

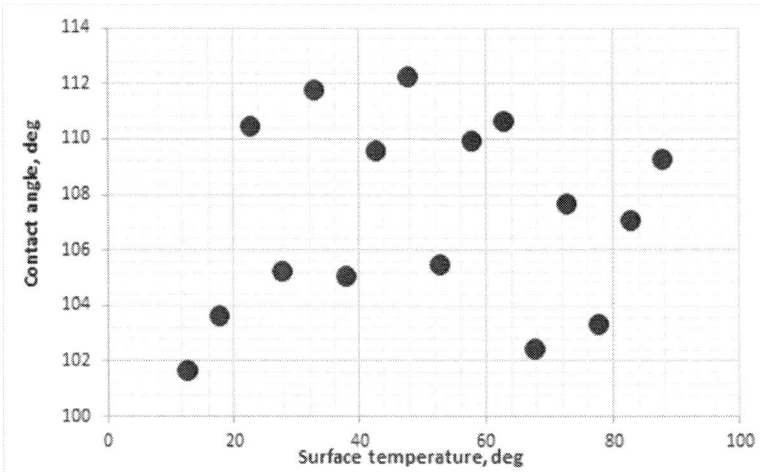

Fig. 3. The contact angle versus the surface temperature.

4 Conclusions

We have studied basic properties of the surface, obtained by applying the Noa 81 photoresist, carrying out measurements of a free surface energy and contact angle at different surface temperatures. It is shown that the contact angle varies within the hysteresis when changing surface temperature.

The authors gratefully acknowledge the support of this work by the Ministry of Education and Science of the Russian Federation (Agreement 14.616.21.0016, project identifier RFMEFI61614X0016).

References

1. S.R. Rao, Y. Peles, Int. J. Heat Mass Tran. **89**, 482 (2015)
2. E.Y. Gatapova, O.A. Kabov, Microgravity Sci. Technol. **19**, 132 (2007)
3. A. Zou, A. Chanana, A. Agrawal, P.C. Wayner Jr., S.C. Marool Sci. Rep. **6**, 20240 (2016)
4. E.O. Kirichenko, E.Y. Gatapova, J. Phys. Conf. Ser. **754**, 032009 (2016)

An experimental study of thermocapillary rupture of a locally heated horizontal liquid layer on various substrates

Dmitry Kochkin[1,2,*]

[1]Kutateladze Institute of Thermophysics, 630090 Novosibirsk, Russia
[2]Novosibirsk State Technical University, 630073 Novosibirsk, Russia

Abstract. This article experimentally studies a thermocapillary rupture of a locally heated horizontal liquid film on substrates of different materials and different thickness. In the course of work the dependence of the critical heat flux on the initial and final film thickness was determined. Upon reaching the heat flux and film thickness at which the liquid film boils up, the film disrupts at almost constant finite thickness. The research has shown that with an increase in thermal conductivity and thickness of the substrate the critical heat flux increases. On the substrates of a material with high thermal conductivity (copper) the heat flux is redistributed on the substrate; at that the localization of heating is lost, and the rupture is not observed.

1 Introduction

The devices using liquid film flows are widely used in industry. Thin liquid films are promising for cooling equipment with high local heat generation, in particular, high-performance electronics [1-3]. Such equipment requires removal of high local heat fluxes from electronic components such as computer chips and power electronics (transistors and thyristors).

The film cooling systems are efficient since they provide high intensity of heat and mass transfer at relatively low coolant flow rates. The decrease in film thickness allows enhancing heat and mass transfer; however, thin films are susceptible to ruptures, which leads to a drastic increase in the temperature of the cooled section and its failure.

Thermocapillary rupture of flowing liquid films under local heating were studied in [4-6]. Influence of substrate wettability and liquid properties on thermocapillary rupture of flowing films was studied in [7-9]. The paper [10] investigated the influence of substrate structuring on the rupture of flowing liquid films. Authors of [11, 12] were first to study the thermocapillary rupture of a stationary horizontal layer of fluid. The works [7, 13] showed that the critical heat flux for horizontal liquid films may several times exceed the heat flux for the films flowing under the action of gravity. This explains the motivation of our work.

* Corresponding author: kochkin1995@mail.ru

The main goal of this research is to determine the dependence of the critical heat flux on the thickness of the horizontal liquid film using substrates of different materials and different thicknesses.

2 Methods

The experiments were conducted on locally heated substrates of copper, brass, stainless steel and glass with a diameter of 50.8 mm and thickness of 1 and 3 mm. The substrate was mounted on a textolite base with a round copper rod with a diameter of 12.7 mm (Fig. 1). A ceramic heater was attached to the bottom of the rod. Heat flux was determined on electric power, dissipated on the ceramic heater. The working fluid was distilled water with an initial temperature of 25°C. The liquid was supplied to the substrate by using a syringe pump. The substrate was cooled at the perimeter of the substrate. The water temperature in the cooling circuit was 10°C.

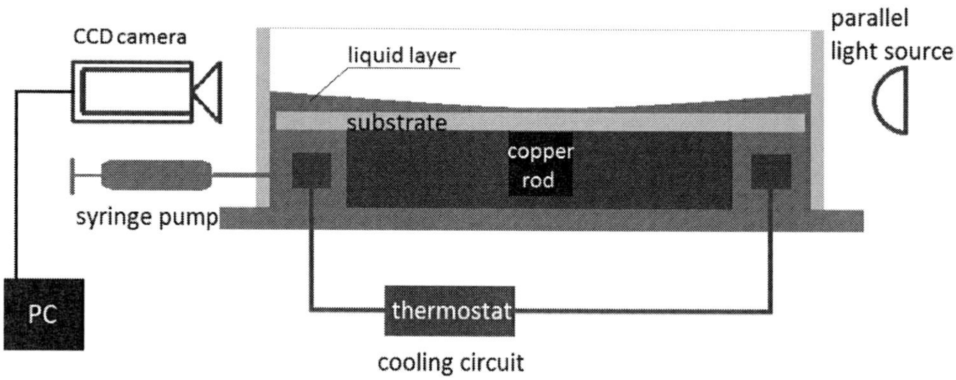

Fig. 1. Schematic diagram of the stand (the power source of the heater is not specified).

The use of replaceable substrates in this work allows studying the rupture of a horizontal liquid layer at different thermal conductivity and wettability of the substrate. To change wettability the substrate may be applied with various coatings that do not affect its thermal conductivity. In this work, coatings were not applied to the substrate.

To determine the layer thickness the optical method using a CCD camera DMK 23GP031 and source of parallel light MI-150 was applied (Fig. 2). This method caused a problem in determining the true liquid layer because of the meniscus preventing the observation. This was surmounted by mounting the parallel light source on the other side of the work section on the same axis with the camera. The light was reflected from the meniscus and did not enter the camera lens; however through the liquid layer the light freely penetrated, resulting in the light-struck band on the screen, being the image of the film thickness. Having obtained the film of the desired thickness, we gradually increased the heater power. When the film was disrupted, we fixed the critical heat flux (at the rupture) and the final (before rupture) thickness of the liquid layer.

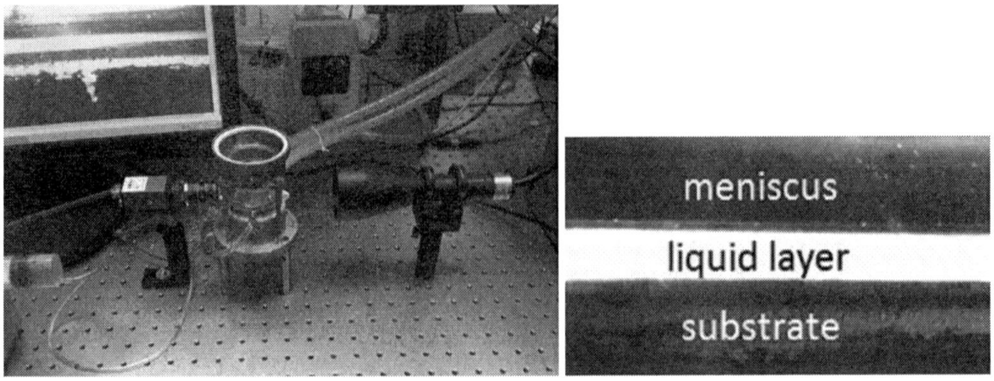

(*a*) (*b*)

Fig. 2. Photo of experimental stand (*a*) and liquid layer (*b*).

3 Results

According to the results of the experiments conducted on the glass substrate of 3 mm thickness, the graph of dependences of the critical heat flux (at film rupture) on the initial and final (before rupture) thicknesses of the fluid layer was built (Fig. 3). From its analysis it may be inferred that the larger the heat flux, the more intense liquid evaporation (the difference between the initial and final film thicknesses increases). Upon reaching the initial thickness ~ 660 μm and higher, the liquid starts boiling, until the film thickness becomes critical and the film disrupts; at that the value of finite thickness practically does not change. At small film thicknesses of ~ 300–400 microns the final film thickness slightly decreases, as the evaporation is insignificant, and the graphs in this section are practically the same. When the thickness is up to ~ 300 μm the film disrupts without turning on the heater, which is apparently due to the fact that the air temperature is 15◦C higher than the temperature of water in a cooling loop; hereupon the thermocapillary thinning of the film takes place.

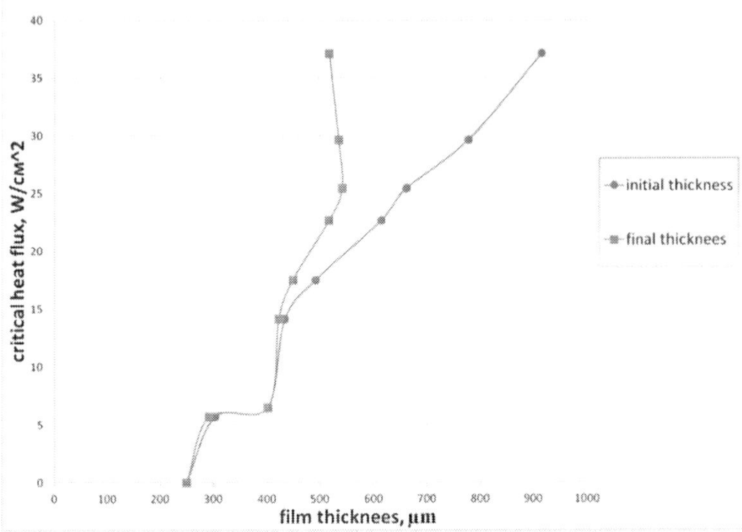

Fig. 3. Dependence of the critical heat flux on film thickness.

Figure 4 shows the dependence of the critical heat flux on the initial thickness of liquid film on stainless steel substrates of 1 and 3 mm thickness, and glass one with a thickness of

3 mm. It is seen that on stainless steel substrates of 1 mm thickness the critical heat flux is several times lower than on substrates of the same material with a thickness of 3 mm. This allows concluding that with increasing thickness of the substrate its heat resistance increases, hereupon the surface temperature of the substrate decreases. Apparently because of an increase in heat spreading in the substrate, the local heat flux density decreases. Because of the influence of these two factors the critical heat flux increases. From the comparison of dependencies for 3 mm thick stainless steel and glass substrates it is seen that the critical heat flux on the substrate of stainless steel is several times larger than on the glass one. It is apparently due to the fact that the thermal conductivity of stainless steel is ~ 20 times higher than the one of glass that causes much larger spread of heat flux in the substrate. This decreases the locality of heating, and the critical heat flux increases.

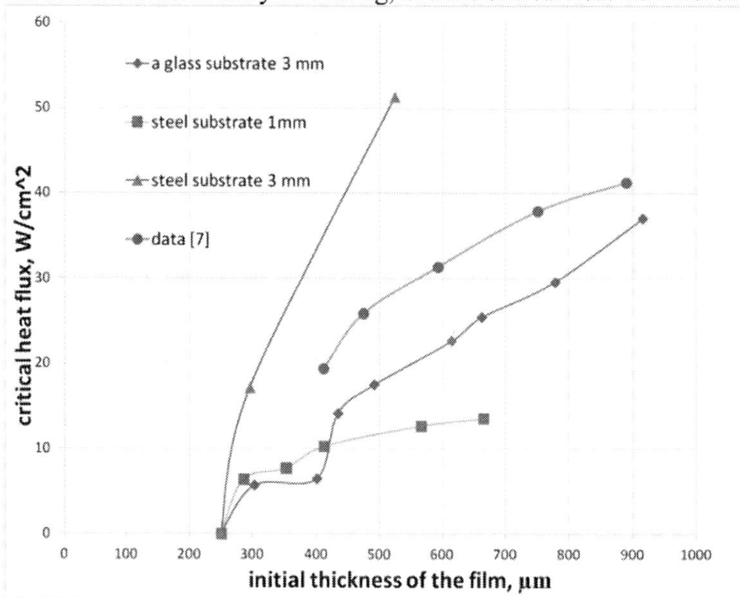

Fig. 4. Dependence of the critical heat flux on the initial film thickness.

4 Conclusion

It is established that with an increase in thermal conductivity and thickness of the substrate the critical heat flux for film rupture increases. When using substrates made of materials with high thermal conductivity (copper and brass) the heat flux is redistributed over the substrate and heat is removed to the cooling circuit; at that the film rupture is not observed. It is also worth noting that the final film thickness before rupture in the boiling liquid does not change.

The work was supported by the RF Ministry of Education and Science (Agreement No 14.604.21.0053, project identifier RFMEFI60414X0053).

References

1. D.V. Zaitsev, D.A. Rodionov, O.A. Kabov, Tech. Phys. Lett.+ **35**, 680 (2009)
2. O.A. Kabov, D.V. Zaitsev, Multiphase Science and Technology **21**, 249 (2009)
3. E.A. Chinnov, F.V. Ron'shin, O. A. Kabov, Interfacial Phenomena and Heat Transfer **3**, 243 (2015)

4. O.A. Kabov, Thermophys. Aerodyn. **7**, 513 (2000)
5. E.A. Chinnov, O.A. Kabov, A.V. Muzykantov, D.V. Zaitsev, International Journal of Heat and Technology **19**, 31 (2001)
6. D.V. Zaitsev, D.A. Rodionov, O.A. Kabov, Microgravity Sci. Technol. **19**, 100 (2007)
7. D.V. Zaitsev, D.P. Kirichenko, and O.A. Kabov, Tech. Phys. Lett.+ **41**, 551 (2015)
8. D.V. Zaitsev, A.A. Semenov, O.A. Kabov, Thermophys. Aeromech. **23**, 625 (2016)
9. D.V. Zaitsev, O.A. Kabov, V.V. Cheverda, N.S. Bufetov, High Temp. **42**, 450 (2004)
10. D.V. Zaitsev, A.M. Lozano, H. Auracher, O.A. Kabov, Microgravity Sci. Technol. **19**, 71 (2007)
11. A. Orell, S.G. Bankoff, Int. J. Heat Mass Tran.**14**, 1835 (1971)
12. J.P. Burelbach, S.G. Bankoff, S.H. Davis, Phys. Fluid. A **2**, 321 (1990)
13. D.V. Zaitsev, O.A. Kabov, Microgravity Sci. Technol **19**, 174 (2007)

Numerical methods application to study processes in the CFB boilers combustion chambers

Oleg M. Koksharev[1,*], and *Andrey* V. Gil[1]

[1]National Research Tomsk Polytechnic University, 634050 Tomsk, Russia

Abstract. This paper presents the application of numerical calculation of a hybrid Euler–Lagrange approach to model the gas–solid flow combined with a combustion process in the CFB boiler that has been resolved by applying the ANSYS FLUENT 14.0. In this work, the numerical modelling of furnace processes at various speeds of airflow supply from below was estimated. For the design speed adopted values of 0.5, 1.5, 2.5 and 5 m/s. The research object is the combustion chamber of boiler unit with CFB, with steam capacity of 230 t/h.

1 Introduction

The constantly developing fluidized combustion technology has become competitive with a conventional pulverized coal (PC) combustion. Circulating fluidized bed (CFB) boilers can be a good alternative to PC boilers due to their robustness and lower sensitivity to the fuel quality. However, appropriate engineering tools that can be used to model and optimize the construction and operating parameters of a CFB boiler still require development. To achieve these objectives, considering the level of development of modern computer technology, it is prospective to use the methods of numerical analysis and simulation with a number of well-known advantages over experimental methods [1].

2 Problem statement

Currently, there is a lack of information about experimental results, and knowledge of various technological aspects of the use of fluidized bed for the implementation of any heat and mass transfer processes. To cover the needs of the different furnaces creation with a fluidized bed, a substantial progress is necessary in the field of mathematical and numerical methods modelling of various physical and chemical processes in fluidized bed. However, the process of modelling is complicated by a number of problems [2, 3], such as the requirement of the boundary conditions correctness, which is often almost impossible to explore theoretically. The validity of the model is most often confirmed by the practice of numerical modelling.

* Corresponding author: omk2@tpu.ru

The research object is the combustion chamber of boiler unit with CFB, with steam capacity of 230 t/h [4, 5] (Fig.1). The height of the furnace reaches 28 m. The boiler has 4 burners on the front for the supply of coal fuel in the furnace and 6 burners embrasure on the back side that intended to return the inert particles and unburned organic part of the fuel. Leaving the furnace particles are separated into two streams and fed into the cyclone for separation.

(a) *(b)*

Fig. 1. The research object: *(a)* - 3-D model; *(b)* – mesh.

3 Mathematical Model

In this study, the Euler-Lagrange aerothermochemical technique of describing processes was used. Turbulent gas properties are calculated using the «k-ε» two-parameter turbulence model. Radiative heat transfer is represented within the P1 method approximation of spherical harmonics [6, 7].

Tasks are solved numerically on non-staggered grids in the Cartesian coordinate system using the relaxation method. Discretization of initial differential equations is performed by using the finite volume method. Diffusion flow on the facets of the control volume are approximated by a central difference scheme, which has the second order of accuracy. To determine the convective flows Van Leer monotonic upwind scheme of the second-order was used [8, 9].

4 Results and discussion

Figure 2 shows the concentration of substances in the furnace volume, (concentration in fractions) at a 0.5 m/s of airflow supply from below.

In the lower part of the furnace there is a layer of 4 m (Fig. 2, a), where the concentration of fuel reaches the maximum value. By supplying air from below the particles are distributed throughout the volume of the furnace chamber. The concentration of fuel particles near the walls and in the central part of the furnace is uniform.

At 1.5 m/s, the height of the layer at the bottom of the furnace is lower than at a speed of 0.5 m/s, and is about 2 m. At 2.5 m/s of airflow supply, the height of the layer at the bottom of the furnace is lower than at a speed of 1.5 m/s, and is about 1.7 m (Fig. 3).

Analyzing simulation results at 5 m/s of airflow supply from below, it can be noted that the height of dense layer of solid particles is about 0.4 m (Fig. 4). In addition, it has the highest distribution density of particles in the upper part of the furnace in contrast to the results at a low speed (0.5 m/s).

DOI: 10.1051/matecconf/20179201044

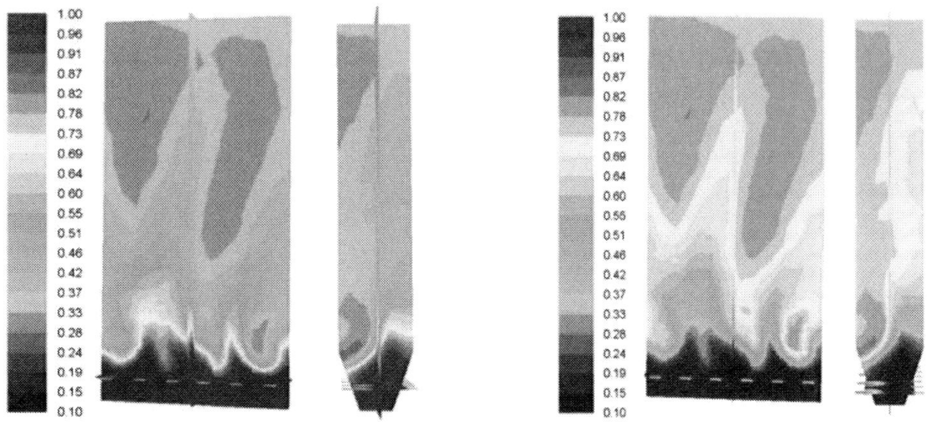

(a) *(b)*

Fig. 2. Height variation of the furnace estimated phases concentration values in vertical cross-section at 0.5 m/s of airflow supply from below: *(a)* particles; *(b)* air.

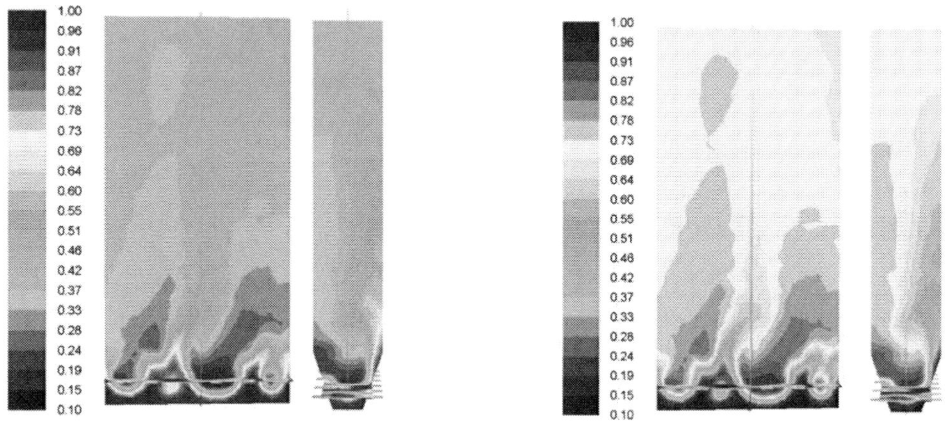

(a) *(b)*

Fig. 3. Height variation of the furnace estimated phases concentration values in vertical cross-section at 2.5 m/s of airflow supply from below: *(a)* particles; *(b)* air.

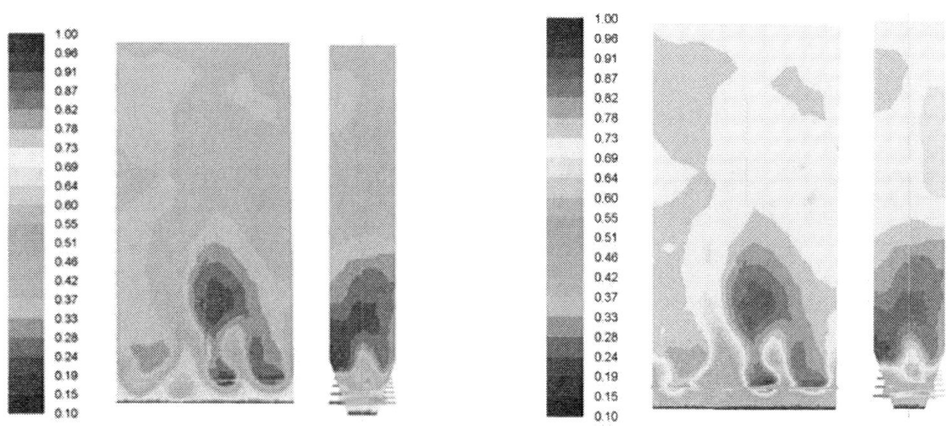

(a) *(b)*

Fig. 4. Height variation of the furnace estimated phases concentration values in vertical cross-section at 5 m/s of airflow supply from below: *(a)* particles; *(b)* air

Figure 5 shows the change in the concentration (in fractions) of fuel in the furnace chamber height at different rates of airflow supply from below. According to the obtained results by increasing the feed rate of airflow supply the fractions volume at the exit of the furnace will increase, and in the lower part of the combustion chamber will decrease.

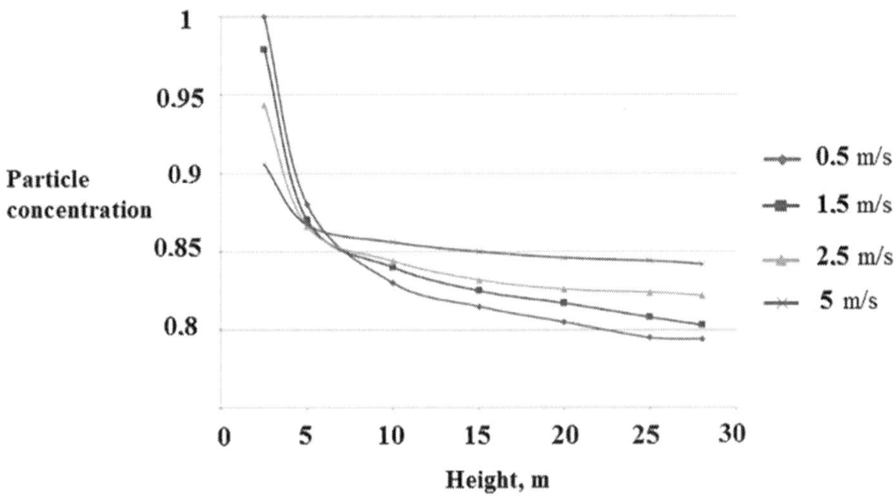

Fig. 5. Height variation of the fuel particle concentration.

5 Conclusions

In conclusion, in accordance with the task it can be stated that the most efficient pattern is observed at the height of the layer that is not exceed the level of the burners, not more than 2.1 m. Consequently, the rate of airflow supply for the investigated furnace should be not less than 2 m/s.

References

1. A. Tourunen, J. Saastamoinen, H. Nevalainen, Fuel **88**, 1333 (2009)
2. P. Basu, *Combustion and gasification in fluidized beds* (CRC Press, Boca Raton, 2006)
3. A.V. Gil, A.Y. Gil, EPJ Web Conf. **82**, 01044 (2015)
4. A.V. Gil, D.A. Baturin, MATEC Web Conf. **72**, 01009 (2016)
5. A.V. Gil, D.A. Baturin, IOP Conf. Ser. Mater. Sci. Eng. **93**, 012003 (2015)
6. J. Xie, W. Zhong, B. Jin, Y. Shao, Y. Huang, Adv. Powder Technol. **24**, 382 (2013)
7. A.V. Zavorin, A.V. Gil, P.S. Khaustov, R.B. Tabakaev, D.A. Buslov, , IOP Conf. Ser. Mater. Sci. Eng. **66**, 012038 (2014)
8. M.H. Bordbar, K. Myohanen, T. Hyppanen, Appl. Therm. Eng. **76**, 344 (2015)
9. A.S. Zavorin, A.V. Starchenko, S.V. Obukhov, Power Techn. Eng. **45**, 42 (2011)

Measurement of the temperature profile during evaporation of water and ethanol

Ekaterina G. Korbanova[1,2,*]

[1]Kutateladze Institute of Thermophysics, 630090 Novosibirsk, Russia
[2]Novosibirsk State University, 630090 Novosibirsk, Russia

Abstract. The temperature profiles across a liquid–gas layers at normal atmospheric conditions are measured for water and ethanol. A thin liquid layer is locally heated from the bottom and evaporates from the liquid–gas interface. Micro-thermocouple with the sensor element thickness of 3 μm is used for measurements. It is shown that the temperature profile has a different character for different liquids.

1 Introduction

Investigations of phase transition phenomena are important for microelectronics cooling applications [1]. The liquid–gas interface is difficult to define for micro-systems, and the effects arising in the Knudsen layer become significant.

The existence of the temperature jump in the Knudsen layer during evaporation has been known in kinetic theory [2,3]. The approach to the description of heat and mass transfer based on the Navier–Stokes equations with temperature and pressure jumps conditions at liquid–vapor system was suggested in [4]. Analysis showed a good agreement with the temperature and pressure profiles calculated basing on the Boltzmann kinetic equations. However, most of the experimental data on the temperatures jumps at the liquid-gas interface were available only for reduced pressure conditions [5,6]. Recently, authors of [7,8] performed precise experiments on the temperature jump measurements using micro-thermocouple with the sensor thickness of 3 μm for normal atmospheric conditions. A temperature jump at the liquid–gas interface has been clearly detected even for small evaporation rate and the temperature jump value is found to increase with increasing the temperature difference between heater and ambient gas [8].

2 Experimental setup

Results of an experimental investigation of the temperature field of liquid–gas layers are presented. Experiments are carried out for two types of liquids at local heating and evaporation. Ultrapure Water (MilliQ) and 99% ethanol are used as working fluid for experiments. The experiments are performed at quasi-stationary heating mode. The temperature of the heater is controlled by three thermocouples. The humidity and

* Corresponding author: korbanova20@gmail.com

DOI: 10.1051/matecconf/20179201045

temperature of the ambient air inside the box are monitored by Testo device. The pressure is measured by barometer. The temperature profiles across the liquid–gas interface are measured using the specially fabricated micro-thermocouple with the sensor thickness of less than 3 μm (Fig. 1). The experimental setup presented in [8] is used.

Fig. 1. SEM image of micro-thermocouple by Hitachi S340.

Calibrations of all thermocouples are carried out in the temperature range 5–100 °C. Two reference thermometers, ETS-100, are used. Each one has uncertainty of 0.02 °C in the tested temperature range. Calibration is performed using calibrator KS-100-1. Data registration is carried out by means of temperature control and measurement system consisting of data collection (NI 9214) and the corresponding software. The precise miniature motorized linear stage (Zaber), used as micro-positioner. It displaces the micro-thermocouple and determines its position. The range of the micro-positioner' movement is 25 mm with the minimal step of 0.05 μm. The micro-positioner is connected to a personal computer and it is controlled by specially developed software. The shadow technique is used for visualization with video camera providing a resolution of 3 μm per pixel.

2 Results

Synchronization of the micro-thermocouple readings and its position via time is made by specially developed software. The measurements allow us to obtain detailed information on the temperature profile across the liquid-gas layers. The existence of the temperature jump at the liquid-gas interface at atmospheric conditions has been established in [8] for water–air system. Fig. 2a shows the temperature profile for water–air system. The temperature jump is clearly indicated. Fig. 2b presents the temperature profile for ethanol–air system. One can see a different temperature curves for water and ethanol. The evaporated liquid flow rate is measured to be of 0.215 μl/s for water – air system at the heating power of 1.07 W.

 DOI: 10.1051/matecconf/20179201045

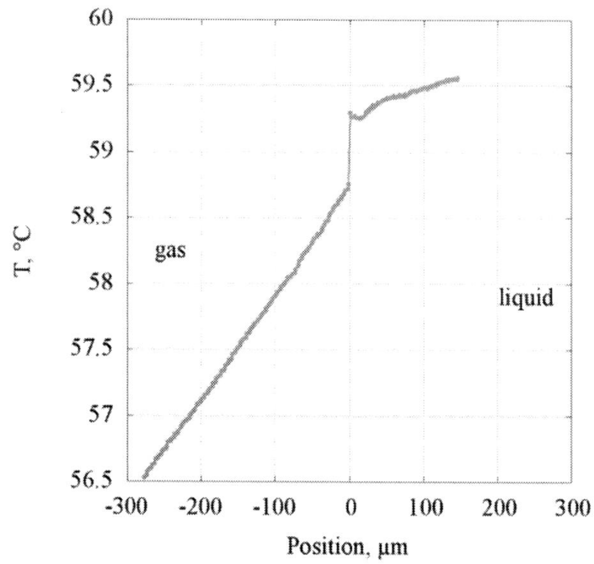

a

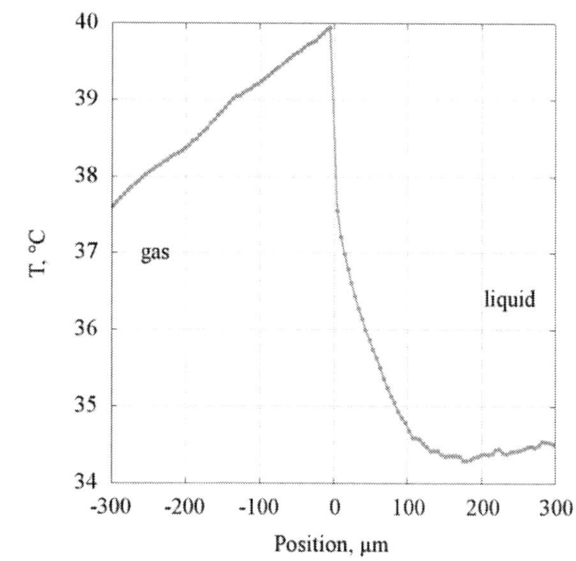

b

Fig. 2. Temperature profile: *a* - for air – water system, heating power is 1.07 W, average temperature of the heater is 88.2 °C, air temperature is 32.3 °C, air humidity is 30.2 %, pressure is 99.46 kPa; *b* – air – ethanol system, heating power is 1.11 W, average temperature of the heater is 84.7 °C, air temperature is 24 °C, air humidity 42.5 %, pressure is equal to 99.59 kPa.

4 Conclusions

The temperature profiles across a liquid - gas layers are measured at normal atmospheric conditions for water and ethanol. A thin liquid layer is locally heated from the bottom substrate and it evaporates from the liquid-gas interface. Micro-thermocouple with the sensor element thickness of 3 μm is used for measurements. It is shown that the temperature profile may have a different character for water and ethanol. The evaporated liquid flow rate is measured to be of 0.215 μl/s for water – air system at the heating power of 1.07 W.

The authors gratefully acknowledge the support of this work by the Ministry of Education and Science of the Russian Federation (Agreement 14.616.21.0016, project identifier RFMEFI61614X0016).

References

1. O.A. Kabov, E.Ya. Gatapova, D.V. Zaitsev, *Thermal and Thermomechanical Phenomena in Electronic Systems, 2008. ITHERM 2008. 11th Intersociety Conference on*, 520 (2008)
2. Y.P. Pao, Phys. Fluids **14**, 1340 (1971)
3. R.Ya. Kucherov, L.E. Rikenglaz, J. Exp. Theor. Phys.+ **37**, 88 (1960)
4. E.Y. Gatapova, I.A. Graur, F. Sharipov, O.A. Kabov, Int. J. Heat Mass Tran. **83**, 235 (2015)
5. P.N. Shankar, M.D. Deshpande, Phys. Fluids A **2**, 1030 (1990)
6. G. Fang, C.A. Ward, Phys. Rev. E **59**, 417 (1999)
7. E.Y. Gatapova, R.A. Filipenko, Y.V. Lyulin, I.A. Graur, I.V. Marchuk, O.A. Kabov, Thermophys. Aeromech. **22**, 701 (2015)
8. E.Y. Gatapova, I.A. Graur, O.A. Kabov, V.M. Aniskin, M.A. Filipenko, F. Sharipov, L. Tadrist, Int. J. Heat Mass Tran. **104**, 800 (2017)

Experimental study into the evolution of pulse atomization-generated aerosol systems in confined space

Natalya Korovina[1], *Irina* Zharova[2,*], and *Evgeny* Muravlyov[1]

[1]Siberian Branch of the Russian Academy of Sciences, 659332 Biysk, Russia
[2]National Research Tomsk State University, 634050 Tomsk, Russia

Abstract. Basic characteristics of the powder aerosol generated by the shockwave method using a pyrotechnic atomizer were experimentally studied.

1 Introduction

The study of mechanisms and laws of the formation and propagation of aerosol media is among the fundamental problems of two-phase flow hydrodynamics, the solution of which is the basis for aerosol practical application in different industries. For instance, with developing chemical technologies and dangerous emissions of noxious and toxic agents resulting from the manufacturing process or man-caused emergency events, the fast and effective decontamination of toxicants is a hot topic.

The most commonly used method to neutralize toxic agents is their sorption by sorbent-containing filters. With this, the contaminated air must be allowed to have completely passed through the said filters, which is not always technically feasible under conditions of a specific production. From this point of view, the most promising is the other technique of purifying the air from toxic gases – the adsorption by fine aerosols atomized in contaminated indoor space. The adsorption conceptually allows the extraction of any contaminant in a wide concentration range from the emissions.

The main indicator of the decontamination efficiency of a toxicant is the high dispersiveness of atomized aerosols because adsorption rate increases with increasing specific-mass surface of the sorbent particles. Under real conditions, the effective use of powders having high sorption activity to decontaminate confined indoor spaces or deactivate the surfaces depends on an atomization technique used and a device implementing the same. Today, the following techniques are commonly used: airstream atomization with an injector sand spray gun, pulsed dispersion with a shock-wave atomizer, pulsed dispersion with a pyrotechnic atomizer, and electrostatic atomization.

The pulsed dispersion techniques [1, 2] (using shock-wave or pyrotechnic atomizers) has the advantage of generating aerosols at a high rate – for fractions of a second or for seconds—by the pulse action from the triggered pyrotechnic charge of an explosive or a high-energy material (HEM). Despite the similarity of these atomization techniques (the

* Corresponding author: zharova@niipmm.tsu.ru

use of a pyrotechnic charge), there are differences between them. The shock-wave atomizer uses a more powerful charge with less gas formation. The combustion of a HEM produces a shock wave propagating through the layer of the atomized substance to break down agglomerates and additionally crush the particles. The atomization is due to the shock pulse from the triggered pyrotechnic charge. In the pyrotechnic atomizer, the HEM charge is a source of cold gas that pulls the powder out of the atomizer vessel. In this respect, the processes of formation and propagation of the aerosol cloud, when the said atomization techniques are used, take place in accordance with different regularities.

The present work reports the results of the experimental investigation into the formation and propagation laws of fine aerosols generated by pulsed atomization techniques.

2 Experimental test procedure for pulsed powder atomization

To study regularities of the dispersion process of powdery materials, experiments were run with shock-wave and pyrotechnic spray generators. The experiments with the shock-wave atomizer to generate fine aerosol were performed on a test bench rigged with an optical diagnostics system [3]. A schematic of the test bench with a measurement box of 1 m^3 in volume is illustrated in Fig. 1.

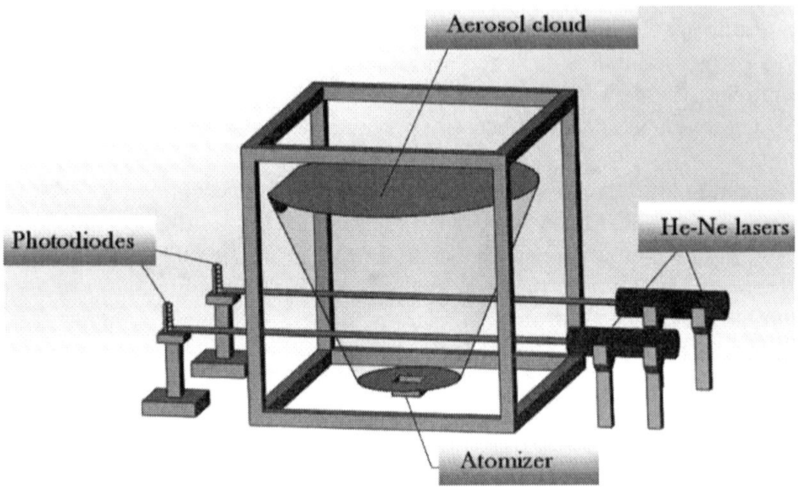

Fig. 1. A schematic of the test bench.

For the aerosol cloud to be uniformly distributed across the measurement box volume, the shock-wave atomizer was placed at the center of the measurement box (Fig. 1).

The initial dispersiveness of the powder was measured with a Pip9.0 optical particle-size analyzer in a measurement range of 0.5-3000 μm. The variation of the powder dispersiveness parameters in the flow was monitored by the laser radiation small-angle scattering method using a two-channel LID-2M laser instrument. The instrument allows continuous, remote, contactless measurements of particle size and concentration directly in the aerosol.

The high-rate generation of fine aerosol was carried out by the shock-wave atomizer. A conceptual diagram of the atomizer is displayed in Fig. 2.

DOI: 10.1051/matecconf/20179201046

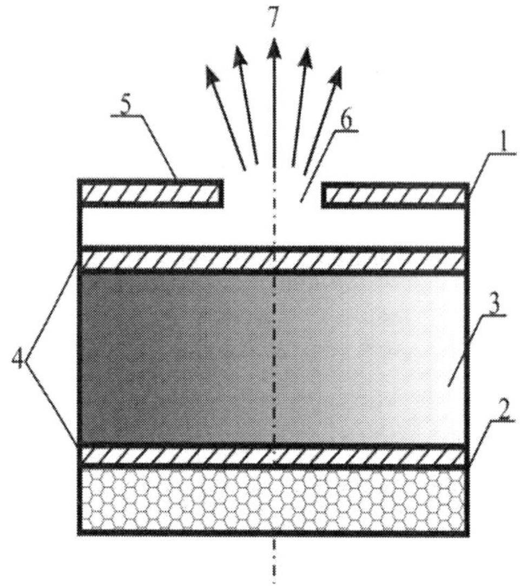

Fig. 2. A diagram of the shock-wave atomizer: 1 – casing, 2 – pyrotechnic charge compartment, 3 – atomized powder, 4 – limiting membranes, 5 – limiters, 6 – nozzle, 7 – dispersed flow.

To determine spatial-time parameters of the aerosol cloud, a VideoSprint G4/NG high-speed camera was used that is designed for recording high-speed processes at up to 250 000 fps. The high-speed videorecording of the aerosol cloud is able to identify basic parameters (geometry and velocity) and uncover peculiar features of aerosol formation and propagation.

The following requirements were followed to assure safety while experimenting:

– the weight of the HEM used in a separate trial must be as little as 0.5 g;

– the weight of the atomized powder should be no more than 7 g.

The trials on examining the propagation of the aerosol cloud produced by the pyrotechnic atomizer were run in a measurement box of 4 m^3 in volume. With that, the atomizer was accommodated at the bottom of the box at 30° angle to the horizontal surface for the aerosol cloud to be uniformly distributed in the confined space. The appearance of the measurement box is depicted in Fig. 3.

Fig. 3. Measurement box appearance.

Figure 4 illustrates a diagram of the pyrotechnic atomizer.

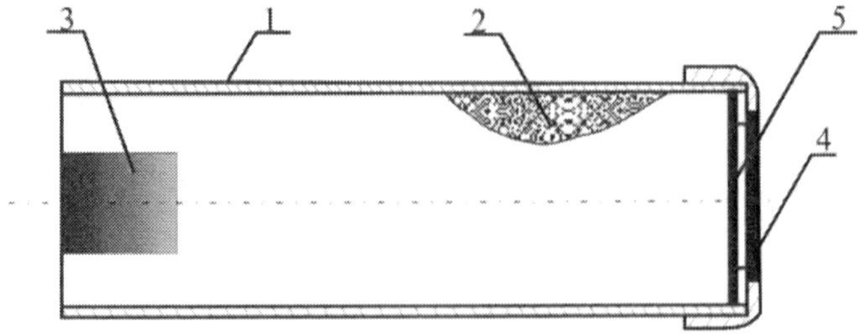

Fig. 4. A schematic of the pyrotechnic atomizer: 1 – casing, 2 – atomized powder, 3 – cold-gas source, 4 – spray nozzle, 5 – membrane.

In the course of the experiments, a Canon XA20 video camera at a frame frequency of 25 fps was used.

3 Experimental results analysis

The experimental study into the aerosol cloud formation and propagation was conducted by the pulse atomization of inorganic powders: silicon oxide (SiO2), aluminum (ASD-6), calcium (chalk), and pseudoboehmite (AlO(OH)). Images of the original pseudoboehmite powder obtained on a JSM-840 scanning electron microscope are shown in Fig. 5.

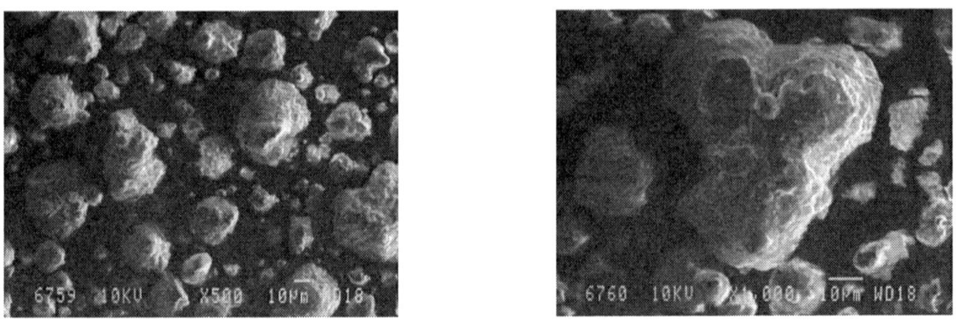

(a) *(b)*

Fig. 5. SEM images of the original powder (pseudoboehmite): *(a)* x500 zoom and *(b)* x1000 zoom.

Figure 6 shows time profiles for the propagation velocity of the powder aerosol cloud generated by the shock-wave atomization in the confined space. Figure 7 displays video frames of the propagation of the aerosol cloud generated by atomizing the aluminum powder by the shock-wave dispersion method (each snapshot has a fixed moment of time).

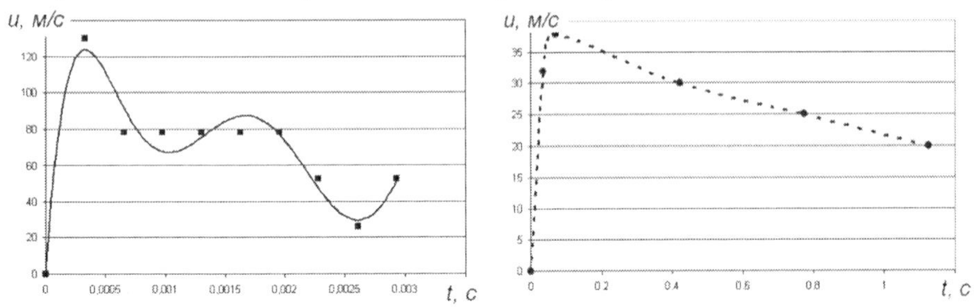

(a) *(b)*

Fig. 6. Time profile for the aerosol cloud propagation: *(a)* aluminum and *(b)* silicon oxide.

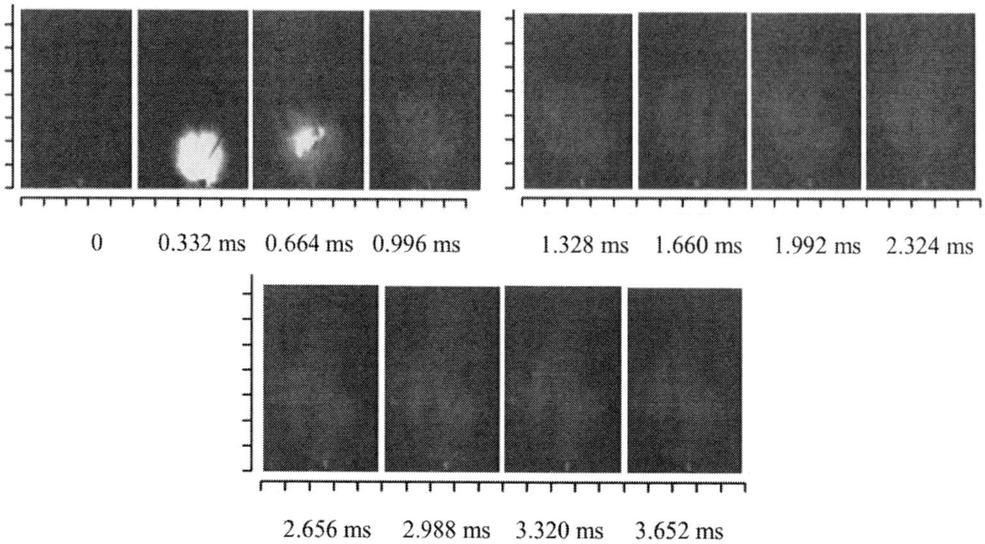

Fig. 7. Video frames of the shock-wave atomization of auminum powder.

Figure 8 depicts an image of the pseudoboehmite powder dispersed by the shock-wave atomizer.

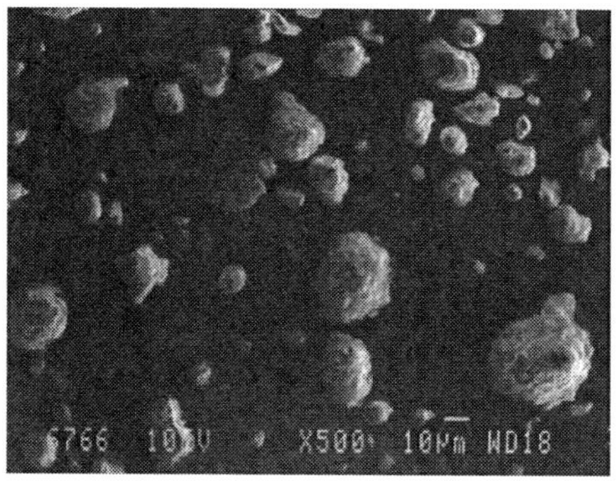

Fig. 8. Pseudoboehmite powder atomized by the shock-wave atomizer.

Figure 9 illustrates video frames of the propagation of the powder aerosol generated by the pyrotechnic atomizer.

Fig. 9. Snapshots of the pseudoboehmite powder atomization.

Characteristics of the original and pulse-atomized powders (D_{43} –s mass median particle diameter) are listed in Table 1.

Table 1. Characteristics of pulse-atomized powders.

Powder	D_{43}, μm		
	Original	Atomization	
		Shock-wave	Pyrotechnic
Calcium (chalk)	49.0	20.0	
SiO_2	63 ÷ 100	16.5	
ASD-6	8.3	7.5	
Pseudoboehmite (AlO(OH))	53.9	11.3	13.9

Analysis of the experimental findings has revealed the following.
In the shock-wave atomization:
– the spray plume is conical and symmetric in shape;
– the powder ejection from the atomizer is completed in 3.6 ms;
– the travel velocity of the flow front is ~120 m/sec;
– the aerosol cloud is formed within ~8 msec.
In the pyrotechnic atomization:
– the propagation velocity of the powder aerosol is ~3.2 m/sec;
– the aerosol cloud is formed for (1 ÷ 3) sec.

4 Conclusions

The findings from the experimental study have comparatively been analyzed and demonstrate that the shock-wave atomizer provides higher dispersiveness of the powder aerosol under the conditions examined, which can be explained by the destruction effect of the shock wave resulted from the HEM combustion. Herewith, it takes considerably less time for the aerosol cloud formation compared to the pyrotechnic atomization. However, it is more appropriate to use the pyrotechnic atomizer when it comes to large confined spaces requiring large amounts of the powder (~300 g) to generate fine aerosol cloud.

The work was accomplished with support from the RFBR grant No. 16-38-50168 (mol_nr).

References

1. B.I. Vorozhtsov, O.B. Kudryashova, A.N. Ishmatov, I.R. Akhmadeev, G.V. Sakovich, J. Eng. Phys. Thermophys. **83**, 1149 (2010)
2. O.B. Kudryashova, M.Yu. Styopkina, N.V. Korovina, A.A. Antonnikova, E.V. Muravlyov, A.A. Pavlenko, J. Eng. Phys. Thermophys. **88**, 833 (2015)
3. N.V. Korovina, I.K. Zharova, O.B. Kudryashova, S.S. Titov, EPJ Web Conf. **110**, 01029(2016)

Numerical analysis of influence of heat load on temperature of battery surface with cooling by a two-phase closed thermosyphon

Alexander Krasnoshlykov[1,*]

[1]National Research Tomsk Polytechnic University, 634050 Tomsk, Russia

Abstract. Numerical analysis of thermal conditions of a two-phase closed thermosyphon using the software package ANSYS FLUENT has been carried out. Time dependence of temperature of heat source surface, which characterize the efficiency of thermosyphon at critical temperatures of batteries have been obtained.

1 Introduction

Problem with aircraft operation are the system thermal control heat-loaded aircraft equipment (batteries, controls, electronics) [1-2]. For example, when charging the lithium-ion battery of the aircraft there is a possibility of ignition due to its thermal runaway effect [3].

Two-phase closed thermosyphon are promising devices cooling and thermal control of various energy and technical systems [1, 4, 5].

Application thermosyphons allows increase cooling efficiency by reducing temperature gradients in the system.

Thermosyphons not been widely used, because so far not developed a general theory of heat transfer processes in such devices.

Modern studies of heat transfer through a two-phase closed thermosyphon mainly presented experimental data on heat and mass transfer processes [6-8]. At the same time, the number of works dealing with mathematical modeling of such processes is insignificant (for example, [9, 10]). Model [9, 10] describe only some particular cases of operation of such heat transfer devices, and do not consider the whole set of heat transfer processes in the vapor and liquid phases of the working fluid.

Purpose of the work - defining surface temperature of the battery when using a two-phase closed thermosyphon for cooling.

2 Formulation of the problem

Chart of the closed thermosyphon rectangular cross section considered in figure1.

* Corresponding author: krasnas@tpu.ru

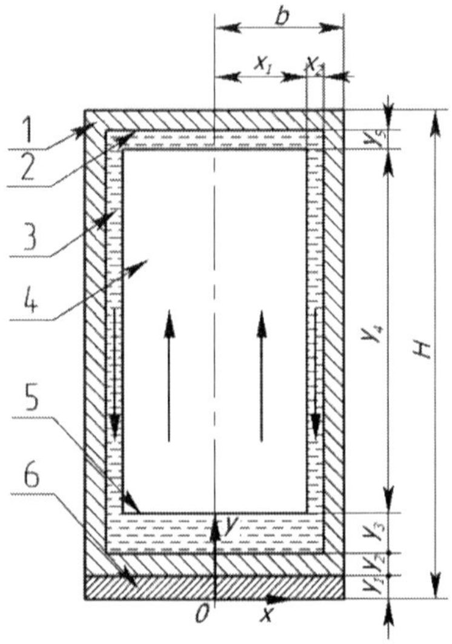

Fig. 1. Schematic diagram of the thermosyphon. 1 – body of thermosyphon; 2 – surface of condensation; 3 – liquid film; 4 – vapor flow; 5 – surface of evaporation; 6 – surface of heating source.

Mathematical model of thermal processes occurring in the study area is as follows:

$$\frac{\partial \rho}{\partial t}+\frac{\partial(\rho u)}{\partial x}+\frac{\partial(\rho \upsilon)}{\partial y}=0; \quad (1)$$

$$\frac{\partial \rho_1 u_1}{\partial t}+\frac{\partial(\rho_1 u_1 u_1)}{\partial x}+\frac{\partial(\rho_1 \upsilon_1 u_1)}{\partial y}=\rho_1 g_x-\frac{\partial P_1}{\partial x}+\frac{\partial}{\partial x}\left(\mu_1\frac{\partial u_1}{\partial x}\right)+\frac{\partial}{\partial y}\left(\mu_1\frac{\partial u_1}{\partial y}\right); \quad (2)$$

$$\frac{\partial \rho_1 \upsilon_1}{\partial t}+\frac{\partial(\rho_1 u_1 \upsilon_1)}{\partial x}+\frac{\partial(\rho_1 \upsilon_1 \upsilon_1)}{\partial y}=\frac{\partial P_1}{\partial y}+\frac{\partial}{\partial x}\left(\mu_1\frac{\partial \upsilon_1}{\partial x}\right)+\frac{\partial}{\partial y}\left(\mu_1\frac{\partial \upsilon_1}{\partial y}\right); \quad (3)$$

$$\frac{\partial \rho_2 u_2}{\partial t}+\frac{\partial(\rho_2 u_2 u_2)}{\partial x}+\frac{\partial(\rho_2 \upsilon_2 u_2)}{\partial y}=\rho_2 g_x-\frac{\partial P_2}{\partial x}+\frac{\partial}{\partial x}\left(\mu_2\frac{\partial u_2}{\partial x}\right)+\frac{\partial}{\partial y}\left(\mu_2\frac{\partial u_2}{\partial y}\right); \quad (4)$$

$$\frac{\partial \rho_2 \upsilon_2}{\partial t}+\frac{\partial(\rho_2 u_2 \upsilon_2)}{\partial x}+\frac{\partial(\rho_2 \upsilon_2 \upsilon_2)}{\partial y}=\frac{\partial P_2}{\partial y}+\frac{\partial}{\partial x}\left(\mu_2\frac{\partial \upsilon_2}{\partial x}\right)+\frac{\partial}{\partial y}\left(\mu_2\frac{\partial \upsilon_2}{\partial y}\right); \quad (5)$$

$$\rho_1 C_{p1}\left(\frac{\partial T_1}{\partial t}+u_1\frac{\partial T_1}{\partial x}+\upsilon_1\frac{\partial T_1}{\partial y}\right)=\lambda_1\left(\frac{\partial^2 T_1}{\partial x^2}+\frac{\partial^2 T_1}{\partial y^2}\right); \quad (6)$$

$$\rho_2 C_{p2}\left(\frac{\partial T_2}{\partial t}+u_2\frac{\partial T_2}{\partial x}+\upsilon_2\frac{\partial T_2}{\partial y}\right)=\lambda_2\left(\frac{\partial^2 T_2}{\partial x^2}+\frac{\partial^2 T_2}{\partial y^2}\right). \quad (7)$$

The heat equation for a solid wall thermosyphon:

$$\rho_3 C_{p3}\frac{\partial T_3}{\partial t}=\lambda_3\left(\frac{\partial^2 T_3}{\partial x^2}+\frac{\partial^2 T_3}{\partial y^2}\right), \quad (8)$$

where u, υ – the velocity components in the projection on the axis x, y, respectively; P – pressure; T – temperature; ρ – density; x, y – Cartesian coordinates; t – time; C_p – heat capacity; g – acceleration of gravity; λ – coefficient of thermal conductivity; μ – dynamic viscosity; indices *1*, *2*, *3* – properties of the liquid, vapor and the solid material wall of thermosyphon.

The initial conditions for the system of equations (1–8) are as follows:

$$u(x,y)=0;\ T(x,y)=T_0;\ P(x,y)=P_0.$$

The boundary conditions for the equations (1–8) have the form:

$$x=0, 0<y<H,\quad \frac{\partial T_2}{\partial x}=0;\quad \frac{\partial u_2}{\partial x}=0,\ \frac{\partial \upsilon_2}{\partial x}=0,$$

$$x=b,\ 0\le y\le H,\quad \lambda_3\frac{\partial T_3}{\partial x}=0,$$

$$x=x_1, y_1+y_2+y_3\le y\le y_1+y_2+y_3+y_4,\quad \begin{cases} T_1=T_2 \\ \lambda_1\dfrac{\partial T_1}{\partial x}=\lambda_2\dfrac{\partial T_2}{\partial x}\end{cases};\quad \begin{cases} u_1=u_2=0 \\ \upsilon_2=0\end{cases},\ \frac{\partial \upsilon_1}{\partial x}=0,$$

$$x=x_1+x_2,\ y_1+y_2\le y\le y_1+y_2+y_3+y_4+y_5,\quad \begin{cases} T_1=T_3 \\ \lambda_1\dfrac{\partial T_1}{\partial x}=\lambda_3\dfrac{\partial T_3}{\partial x}\end{cases};\quad \begin{cases} u_1=0 \\ \upsilon_1=0\end{cases},$$

$$y=y_1+y_2,\quad 0\le x\le x_1+x_2,\quad \begin{cases} T_1=T_3 \\ \lambda_1\dfrac{\partial T_1}{\partial x}=\lambda_3\dfrac{\partial T_3}{\partial x}\end{cases};\quad \begin{cases} u_1=0 \\ \upsilon_1=0\end{cases},$$

$$y=y_1+y_2+y_3+y_4+y_5,\ 0\le x\le x_1+x_2,\quad \begin{cases} T_1=T_3 \\ \lambda_1\dfrac{\partial T_1}{\partial x}=\lambda_3\dfrac{\partial T_3}{\partial x}\end{cases};\quad \begin{cases} u_1=0 \\ \upsilon_1=0\end{cases},$$

$$y=y_1+y_2+y_3, 0\le x\le x_1,\quad \begin{cases} T_1=T_2 \\ \lambda_1\dfrac{\partial T_1}{\partial y}=\lambda_2\dfrac{\partial T_2}{\partial y}-Q_e w_e-\upsilon_2 C_p\rho(T_2-T_0)\end{cases};\quad \begin{cases} \upsilon_2=\dfrac{w_e}{\rho_2}, \\ \upsilon_1=\dfrac{w_e}{\rho_1},\end{cases}$$

$$y = y_1 + y_2 + y_3 + y_4,\ 0 \le x \le x_1,\ \begin{cases} T_2 = T_1 \\ \lambda_2 \dfrac{\partial T_2}{\partial y} = \lambda_1 \dfrac{\partial T_1}{\partial y} + Q_c w_c + \upsilon_2 C_p \rho (T_1 - T_0) \end{cases};\ \begin{cases} \upsilon_1 = \dfrac{w_c}{\rho_1}, \\ \upsilon_2 = 0, \end{cases}$$

$$y = 0,\ 0 \le x \le b,\ \lambda_3 \frac{\partial T_3}{\partial y} = q_h,$$

$$y = H,\ 0 \le x \le b,\ \lambda_3 \frac{\partial T_3}{\partial y} = \alpha (T_3 - T_{oc}).$$

where – Q_e, Q_c – heat of condensation and evaporation, respectively; w_e, w_c – mass rate of evaporation and condensation, respectively;

Mathematical modeling of the task carried out in the package ANSYS FLUENT software. [11].

Mass rate of evaporation and condensation were calculated according to the formula:

$$w_c = \beta \sqrt{\frac{M}{2\pi R T_H}} (P_2 - P_H),$$

where $P_н$ – saturation pressure; $T_н$ – saturation temperature; R – universal gas constant; M – molar weight; β – accommodation coefficient.

3 Results and Discussion

Numerical study of thermal processes in the two-phase closed thermosyphon rectangular performed under the following geometrical parameters (Figure 1): height H = 300 mm, the transverse dimension L = 70 mm. The water is considered as a working fluid.

It was assumed that the lower cover of thermosyphon can reach a temperature corresponding to the critical thermal regime of a typical lithium-ion battery [3]. Operating temperature of a battery operation ability is from -60 to +60°C [3].

In accordance with this range, the following variations the heat flux from the heat source: $1.1 \cdot 10^4$ W/m^2; $1.7 \cdot 10^4$ W/m^2; $2.3 \cdot 10^4$ W/m^2.

Figure 2 shows time dependence of temperature of surface of heat (battery) at different heat flux densities.

It can be seen that when the heat load q_h=$1.1 \cdot 10^4$ W/m^2, the temperature of the battery surface for 800 seconds, reaches a constant value of 340 *K*. By increasing the thermal load to $1.7 \cdot 10^4$ W/m^2 is the reduced time to 600 seconds. However, the surface temperature of the heat source is not raised above 341 *K*. Further increase in heat flux from the heat source to $2.3 \cdot 10^4$ W/m^2 leads to rapid having linear, increase T (Figure 2).

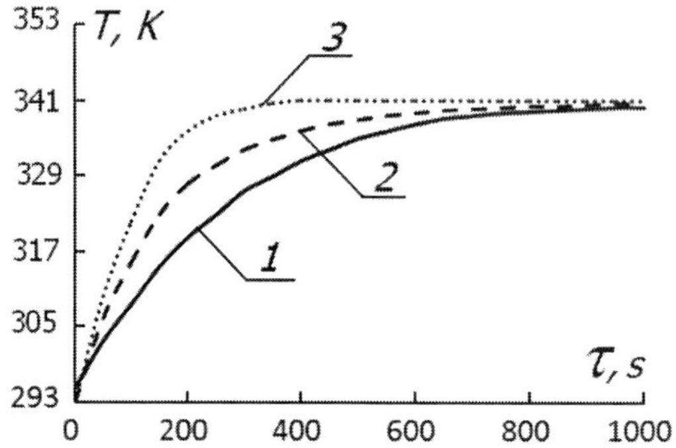

Fig. 2. Temperature on the lower cover of the thermosyphon depending on time at various heat flux density. 1 – q_h=1.1·10^4 W/m^2; 2 – q_h=1.7·10^4 W/m^2; 3 – q_h=2.3·10^4 W/m^2.

4 Conclusion

The results of numerical studies characterize the battery temperature at various thermal loads.

At temperatures, corresponding to the critical temperature batteries (over 60 °C) aircraft equipment in the heating zone is intensification of the evaporation process.

It was found that in the range of heat flux densities of 1.1·10^4 Вт/м^2 to 2.3·10^4 Вт/м^2 the boiling conditions of crisis and emergency operation of lithium-ion batteries are not reached.

The reported research was supported by Russian Federation President Grant for state support of the Russian Federation leading scientific schools SS-7538.2016.8.

References

1. C.H. Lee, S.J. Bae, M. Jang, J. Power Sources **293,** 498 (2015)
2. R. Saada, D. Patel, B. Saha, Process Saf. Environ. **97,** 109 (2015)
3. D.A. Khrustalyov, *Tehnologiya, oborudovanie i podgotovka kadrov v svarochnom proizvodstve*, 183 (2003) [in Russian]
4. N.I. Klyuev, V.A. Bruyaka, Izvestiya Vysshikh Uchebnykh Zavedenij. Aviatsionnaya Tekhnika **1,** 58 (2003) [in Russian]
5. M.K. Bezrodnyi, I.L. Pioro, T.S. Kostyuk, *Protsessyi Perenosa V Dvuhfaznyih Termosifonnyih Sistemah* (Fakt, Kiev, 2005) [in Russian]
6. H. Jouhara, A.J. Robinson, Appl. Therm. Eng. **30**, 201 (2010)
7. V.Y. Kravetz, V.A. Chernobay, A.A. Nikitenko, East Europe Journal **50**,40 (2011)
8. A.B. Solomon, R. Roshan, Int. J. Heat Mass Tran. **82**, 521 (2015)
9. G.V. Kuznetsov, A.E. Sitnikov, High Temp. **40**, 898 (2002)
10. G.V. Kuznetsov, M.A. Al-Ani, M.A. Sheremet, J. Eng. Thermophys. **20**, 201 (2011)
11. Ansys Help. FLUENT Theory Guide

Thermographic investigation of surface temperature of the evaporating liquid layer under the action of gas flow

Aleksei Kreta[1,2*], and *Yuriy* Lyulin[2,3]

[1]Novosibirsk State University, 630090 Novosibirsk, Russia
[2]Kutateladze Institute of Thermophysics, 630090 Novosibirsk, Russia
[3]National Research Tomsk Polytechnic University, 634050 Tomsk, Russia

Abstract. An experimental study of the temperature field on the surface of horizontal liquid layer (Ethanol) evaporating into gas flow (Air) has been performed. Temperature gradient of the gas-liquid interface has been measured with the help of Titanium 570M IR camera. Shear stresses on gas-liquid interface induced by thermocapillary effect and inert gas flow have been defined.

1 Introduction

Study of heat and mass transfer through the gas-liquid interface is one of the most important problems in the present time. The intensive evaporation from the liquid surface into the gas flow induces various convective flows within the liquid. Two mechanisms have been spotted to be responsible for the convection. One of them is the gravity that acts essentially on density differences due to temperature gradients, the other one is the thermo-capillarity originated by the temperature dependence of the surface tension at the liquid interface. The convective motions caused by buoyancy are known as the Rayleigh-Bénard convection. When the convection is induced by the surface tension, it is referred to as Marangoni-Bénard convection. In the most of investigations the fluid layers are steadily heated from below with a static temperature gradient and without evaporation. The evaporating fluid layers with nonlinear temperature gradient have much less been studied due to the complexity of the problem. The works [1] report the experimental results concerning the effects of evaporation on the thermocapillary convection in a thin liquid layer. It was found that evaporation is not only a means of cooling the layer surface, but also a driving power for the development of the convective instabilities.

The problem of convection of the evaporating liquid layer under the action of the inert gas flow is much more complicated than described above, because includes also the shear-stress effect induced by gas flow [2]. The coupling between evaporation and convection cased by a shear-stress, thermocapillary and buoyancy effects in the liquid layer, has a direct effect on the evaporation flow rate, being relevant, for example, for heat-pipe grooves and thin-film evaporators. The increased interest to the problem is caused by

* Corresponding author: alexy.slav@yandex.ru

experiments under microgravity conditions [3–5], some experiments under normal gravity and the preparation of experiments on the International Space Station in the framework of the scientific project "Convection and Interfacial Mass Exchange" (CIMEX) of the European Space Agency [6]. These experiments are aimed to investigate the features of convective flows of a fluid in an open cavity and in an open horizontal layer under conditions of a co-current gas flow and evaporation. The processes of convection, accompanied by evaporation at the interface, are actively studied experimentally [7], numerically [8, 9] and theoretically in the present time [9, 10].

The aim of this work is to experimentally study the structure of convective flows and to measure temperature distribution on the surface of horizontal liquid layer under the action of the gas flow using IR camera.

2 Experimental setup

The experimental investigations were conducted on the setup shown in Fig. 1. The setup consists of the following components: test section, gas and liquid loops, data acquisition system, thermal stabilization system, and optical techniques.

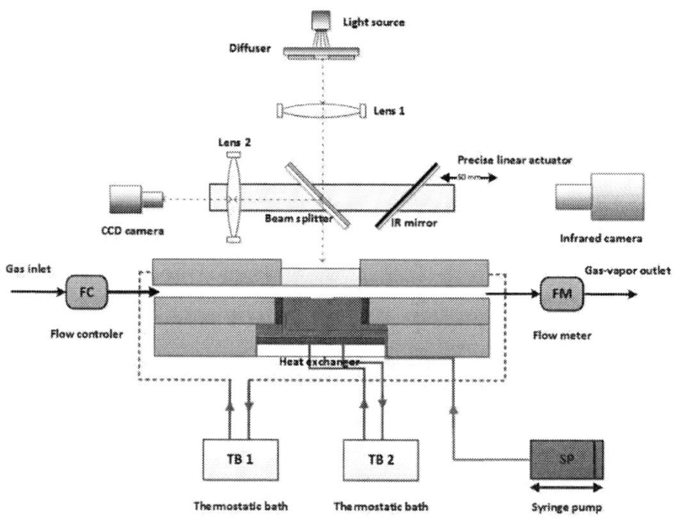

Fig. 1. Scheme of the experimental setup.

Pure gas enters to the test section from the compressor. The mass flow controller sets flow rate of the gas at the inlet of the test section. The liquid is injected into the test section by a precision syringe pump. The working liquid evaporates under the gas flow and the vapor-gas mixture flows to the outlet of the gas channel in test section. The vapor-gas mixture flow at the outlet is measured by a mass flow meter. The liquid temperature in the test section is determined by the temperature of the substrate base, which controlled with the help of thermoelectric module. The position of the interface level is controlled with an accuracy of 10 micrometres using a Schlieren technique and the syringe pump [7].

Measurements of the temperature distribution on gas-liquid interface were conducted with the help with of Titanium 570M IR camera. An infrared wavelength region is 3-5 micrometres. Infrared imaging of liquid surface was performing through quartz optical window that implemented in the top cover of the test cell. IR camera was placed on the optical tabletop horizontally and measuring of the surface temperature of the liquid was contacted with the help of the IR mirror. The IR mirror was inclined to 450 to infrared camera and liquid surface. Optical components of the Schlieren technique and IR mirror were mounted on an optical platform of the linear actuator. Linear actuator is controlled from personal computer and by special software. The optical platform moved in the range of 50 mm and speed of 104 mm/s in horizontal direction. Initially, the platform was set in position where optical axes of Schlieren technique coincide and the interface visualized. When a flatness of interface and the stationary regime of the experiment were established the linear actuator moved the platform in position where optical axes of the IR camera and mirror also coincide and measurement of surface temperature was running.

3 Results and discussion

The experiments were conducted at atmospheric pressure in a test section, and a 5 mm thick liquid layer. Ethanol liquid was used as the working liquid. The evaporation surface area was 100 mm^2 with the corresponding 10x10 mm cut-out in the plate. The gas flow rate was in the range 100–1000 ml/min that corresponds to gas average velocities for 0.0138 to 0.138 m/s (Re=2.8–28), respectively. The temperature of the "liquid-gas" system has changed from 20 °C to 40 °C. The difference between the temperature of the gas and liquid does not exceed 0.1 °C.

Fig. 2 shows a thermogram of the liquid surface at the intensive evaporation. The IR image snapshot does not explain all dynamic of the process. Analyzing of the video the movement of gas-liquid interface in counter-current direction to the gas flow is observed.

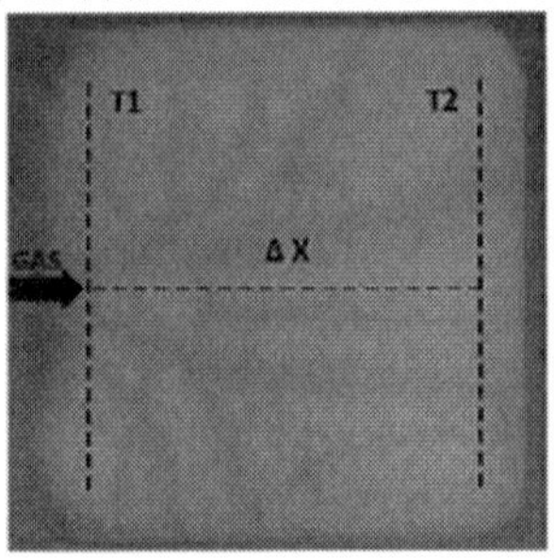

Fig. 2. The thermogram of the evaporating liquid layer.

In Fig. 3 a typical temperature distributions of the interface in the direction of gas flow are plotted for the various temperatures. It is shown the differences in the interfacial temperature. Intensive evaporation induces temperature decreasing in initial of the evaporation surface and temperature gradient is appeared. In this case the thermocapillary effect on the gas-liquid interface takes place.

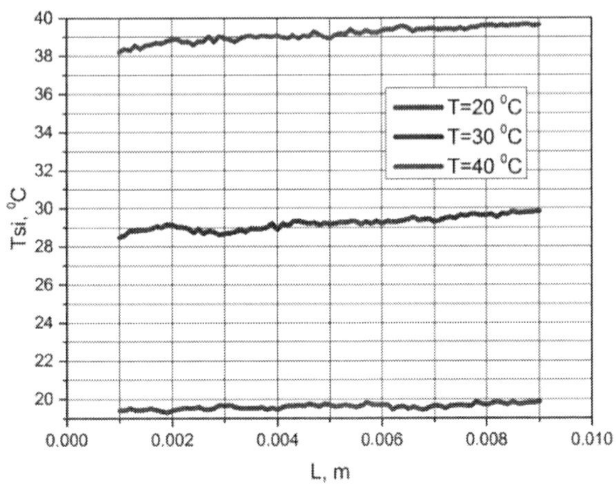

Fig. 3. The temperature distribution on the surface of the evaporating liquid layer.

The shear stress on the liquid surface τ_{sur} caused by the thermocapillary effect can be estimated as:

$$\tau_{sur} = \frac{\delta\sigma}{\delta T}\frac{\delta T}{\delta x}, \tag{1}$$

where, $\sigma_T = \frac{\delta\sigma}{\delta T}$ - temperature coefficient of surface tension is equal 1,14 x 10^{-4} N/mK.

The temperature gradient is defined as: $\frac{\delta T}{\delta x} = \frac{T_2 - T_1}{\Delta x}$. Points of the temperature measuring and length x=0,008 m are shown in Fig. 2.

Shear stresses on the liquid surface induced by gas flow can be estimated as [11]:

$$\tau_g = \eta\frac{dV}{dy} = -\frac{H}{2}\frac{\delta p}{\delta x}, \tag{2}$$

where, H= 0,005 m is height of gas channel; η – dynamic viscosity.

Pressure gradient is calculated from the formula: $V_g = -\frac{H^2}{12\eta}\frac{\delta P}{\delta x}$, where average gas velocity V_g, m/c is known value. It is assumed that in the gas channel the Poiseuille flow is realised. The gas-liquid interface is considered as a rigid surface.

Fig. 4 shows the relationship of forces shear stresses associated with the temperature gradient at the surface of the liquid layer to the forces of shear stresses caused by friction on the surface of the gas flow at different temperatures system "liquid - gas" and at different Reynolds numbers.

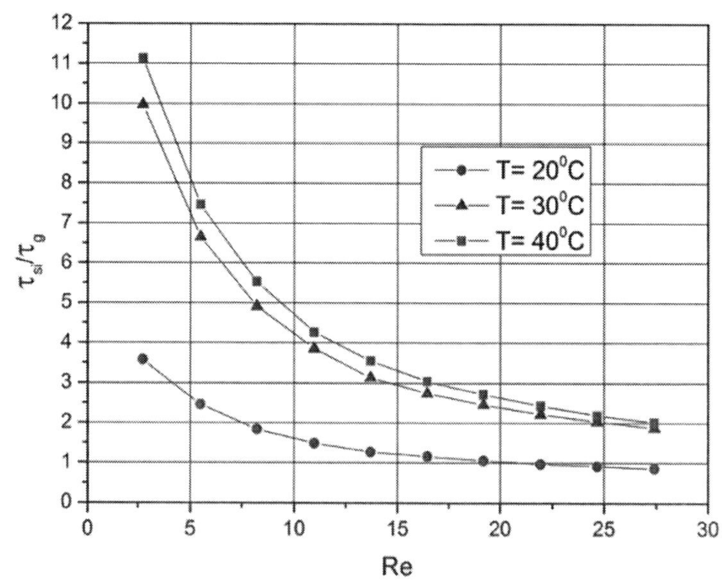

Fig. 4. Shear stresses.

The plot show that with the increase of the gas flow relationship the thermo-capillary shear stress to the gas shear stress are decreased. It can be seen that when the small gas

Reynolds the thermocapillary shear stresses are bigger than gas shear stresses in several times. This means that thermocapillary forces dominate at low gas velocities. When the maximum gas flow rate relationship the thermo-capillary shear stress to the gas shear stress are significantly decreased. However, thermocapillary forces all the same give a decisive contribution to the development of convection currents in the liquid layer.

4 Conclusions

Experiments were carried out to study the influence of the gas flow velocity and temperature system liquid-gas on the temperature distribution on the interface. Infrared study and measurements of the temperature field on the surface of horizontal liquid layer evaporating into gas flow has been performed. Temperature gradient of the gas-liquid interface has been measured with the help of Titanium 570M IR camera. Shear stresses on gas-liquid interface induced by thermocapillary effect and inert gas flow have been defined. It is shown that the main factor influencing on the development of convection in the liquid layer under the action of gas flow is thermocapillary effect.

The work was financially supported by the Russian Ministry of Education and Science (Project identifier RFMEFI61314X0011).

References

1. N. Zhang, D.F. Chao, Int. Commun. Heat and Mass **26,** 1069 (1999)
2. O.A. Kabov, Int. J. Emerging Multidiscip. Fluid Sci. **2**,161 (2010)
3. L. Joannes, O. Dupont, S. Gorza, P. Colinet, J. Legros, T. Lanen, H. Oksuzoglu, *Proceedings of Fringe 2001, 4th international workshop on automatic processing of fringe patterns*, 506 (2002)
4. K. Löth, B. Larsson, H. Schneider, O. Janson, Y. Houltz, P. Colinet, C. Iorio, L. Joannes, O. Dupont, *Proceedings of the 53rd international astronautical congress, Houston, Texas (The World Space Congress*, (2002)
5. P. Colinet, L. Joannes, C.S. Iorio, B. Haut, M. Bestehorn, G. Lebon, J.C. Legros, Adv. Space Res. **32,** 119, (2003)
6. T. Balázs, Microgravity Sci. Technol. **24**, 189 (2012)
7. Yu.V. Lyulin, O.A. Kabov, Int. J. Heat Mass Tran. **70,** 599 (2014).
8. H. Machrafi, C.S. Iorio, P.C. Dauby, Interfacial Phenomena and Heat Transfer **2**, 1 (2014)
9. O.N. Goncharova, O.A. Kabov, Int. J. Heat Mass Tran. **53,** 2795 (2010)
10. O.N. Goncharova, M. Hennenberg, E.V. Rezanova, O.A. Kabov, Interfacial phenomena and heat transfer **1,** 317 (2013)
11. L. Landau, E. Lifshitz, *Hydrodynamics* (Science, Moscow, 1986)

Airflow mixing augmentation device for hot-air heating systems in modular boilers

Nikolay I. Kurilenko[1], *Pavel A.* Artamonov[1], *Elena Yu.* Kurilenko[1], and *Gennady Ya.* Mamontov[2,*]

[1]Industrial University of Tyumen, 625001 Tomsk, Russia
[2]National Research Tomsk Polytechnic University, 634050 Tomsk, Russia

Abstract. The article demonstrates the results of the theoretical research devoted to the study of air flow interaction in hot-air heating systems of automatic modular boilers involving the use of fan heaters. The work quotes the results of mathematic simulation of various density air flows that are vertical to each other.

1 Introduction

In recent times, a great attention has been paid to energy-saving issues in various production spheres. This article studies one of the techniques to optimize the hot-air heating system of automatic modular boilers. Air thermal conditions have a direct control over the boiler efficiency factor [1].

The use of water fan heater-based hot-air heating in automatic modular boilers is one of the efficient solutions of the air thermal conditions issue. The fan heater is set at right angle to a supply grill. An incoming stream of the cold outer air is thrown off by the hot air stream from the fan heater and is heated [2]. This heating and ventilation system (HV) compared to air handling units is characterized by lower capital and maintenance costs, a higher reliability (warm-air furnace freezing is excluded) and ease of automatic control. Despite its advantages, this system efficiency reduces together with the decrease in the outer air temperature. It is related to a higher flow velocity (1–1.5 m/s) and a different flow density of hot and cold air streams (the cold air density can exceed the hot air density by more than 50 %).

2 Problem statement and solution procedure

We simulate the boiler block box room with the fan heater-based heating and ventilation system. The room dimensions are 3.0x4.0x8.0 m with coordinates X, Z and Y respectively (fig.1). The cold air flows through the supply grill with a fixed-rate mass flow G equal to 1.37 kg/s and a temperature of 223K (-50 °C). The fan heater soaks up the inner air at a rate of G and this air is supplied at right angles to the inflowing air stream at 323K. All specified area is filled with the air whose characteristics change at the ideal gas law. We use

* Corresponding author: gmamontov@tpu.ru

the finite-volume difference scheme [3] and solve the following equations [4] to determine the temperature distribution by the hole cross section that the heated air is evacuated through.

– equation of continuity:

$$\frac{\partial \rho}{\partial t} + \nabla \bullet (U\rho) = 0 \tag{1}$$

where ρ – density, kg/m^3; t – time, s; U – velocity vector, m/s;

– impulse law:

$$\frac{\partial (U\rho)}{\partial t} + \nabla \bullet (U\rho \otimes U) = -\nabla p + \nabla \bullet \tau + S_M \tag{2}$$

where p – thermodynamic pressure, Pa; S_M – volume mass source, kg/m^2·s^3; τ – stress tensor is described by (3):

$$\tau = \mu(\nabla U + (\nabla U)^T - \frac{2}{3}\delta \, \nabla \bullet U) \tag{3}$$

where μ – dynamic coefficient of viscosity, Pa·s; δ – Kronecker function; T – temperature, K;

– energy conservation law:

$$\frac{\partial (\rho e)}{\partial t} + \nabla \bullet (\rho U e) = \nabla \bullet (\lambda T) + p\nabla \bullet U + \tau : \nabla U + S_E \tag{4}$$

where e – internal energy; $\tau : \nabla U$ – viscous dissipation; S_E – volume energy source, kg/m·s^3.

The gravity differential air pressure is described by the buoyancy model based on Boussinesq thermal convection equation:

$$\rho = \rho_{Bref}(1 - \beta(T - T_{Bref})) \tag{5}$$

where ρ_{Bref} – substance density at the temperature of the problem initial data, T_{Bref}, kg/m^3, T – temperature at the reference point, K, β – coefficient of volume expansion, that can be determines as:

$$\beta = \frac{1}{\rho} \cdot \frac{\partial \rho}{\partial T} \tag{6}$$

At all geometry boundaries, we use a nonslip wall model and as a turbulence model, we take Menter's shear stress transport model [5].

The (1)–(6) equations solution is a temperature distribution by the specified geometry cross section that the heated air is evacuated through (fig.2). As the figure 2 illustrates, there is a high thermal block box height gradient (in this case it is equal to 8.4 °C/m) when the fan heater-based heating and ventilation system that operates as a hot air curtain under inflowing air low temperatures, is used and it proves the low efficiency of the heating and ventilation system under these conditions.

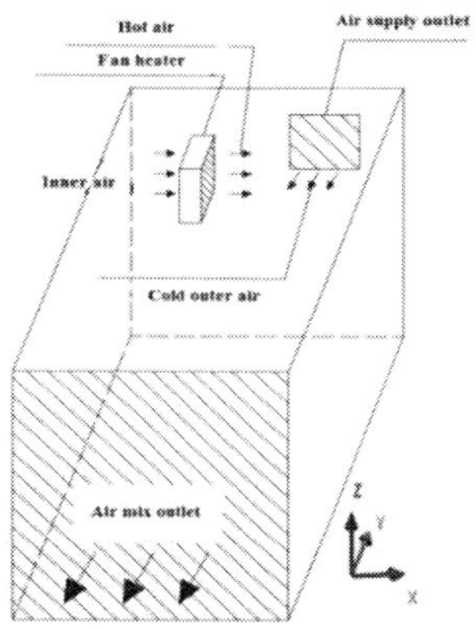

Fig. 1. Analytic model.

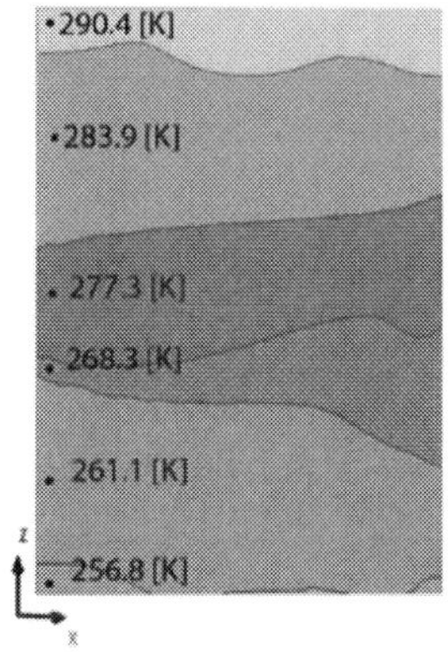

Fig. 2. Temperature field distribution at the exit area.

When the hot air from the fan heater makes contact to the cold inflowing air, there is not an overall mixing and a part of the heat goes up under gravity forces, and the cold inflowing air goes down creating a high thermal gradient. This effect can be observed in figure 3 (B). The numerical simulation result illustrates that the major part of cold and hot air flows does not mix. As it can be seen in figure 3 (C), the mixing occurs only in the contact area of two jets. Despite the fan heater in service, the cold air with a temperature lower than -30 °C descends to the room floor and drifts low over it (fig. 3 B, C).

Due to the high thermal gradient, we can observe the cold air flow formation over the block box room floor that can result in equipment freezing. The use of modular boilers shows that the fan heater-based heating and ventilation system can be operated without any additional changes at the outer air temperature up to - 30°C. Under these conditions, we observe a high value of the thermal gradient, but if the supply grills are at the top of the room, there are not negative temperature streams over the floor. If the inflowing air temperature is below 30°C, there is a need in an engineering solution that will allow to heat the cold air coming to the room through the fan heater and to decrease the thermal gradient room heightwise.

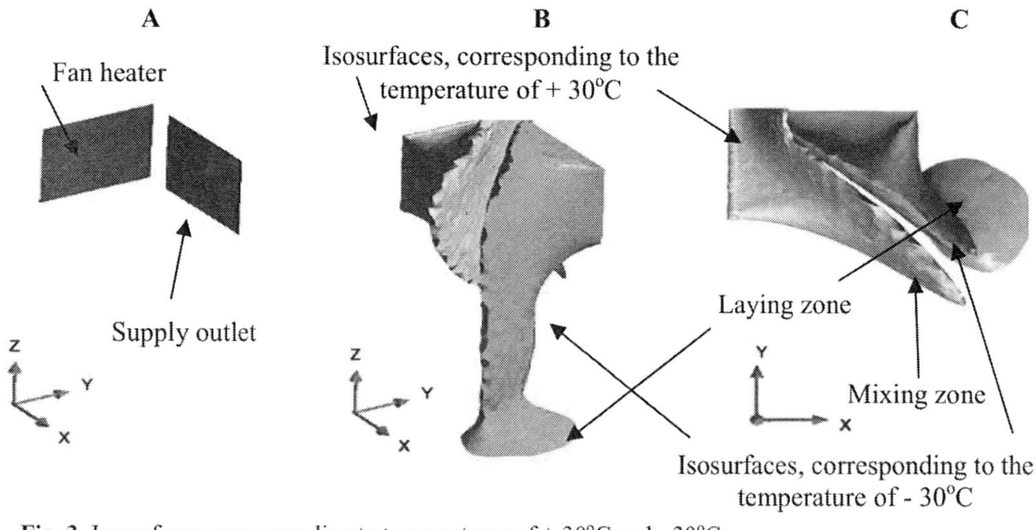

Fig. 3. Isosurfaces, corresponding to temperatures of + 30°C and - 30°C.

The energy efficiency is based on the minimization of capital costs to obtain a useful effect [6] so the optimization task of the fan heater-based heating and ventilation system involves the development of the device that will allow providing an efficient inflowing air heating at low capital and maintenance costs.

We have analyzed the cold and hot air streams distribution (fig.3) and have supposed that one of the ways to increase the mixing efficiency is a flow motion limitation in space.

Let us study the box that is made of a sheet metal. The box is supposed to have holes for the fan heater, the supply grill and the outlet (fig.4). This design will allow limiting the stream distribution at Z axis and keep the cold air from going downwards, and the hot air – upwards. The outlet that is smaller than the total area of the air supply outlet and the fan heater can direct the air mixture upwards at the velocity that is two times higher than in the air supply outlet.

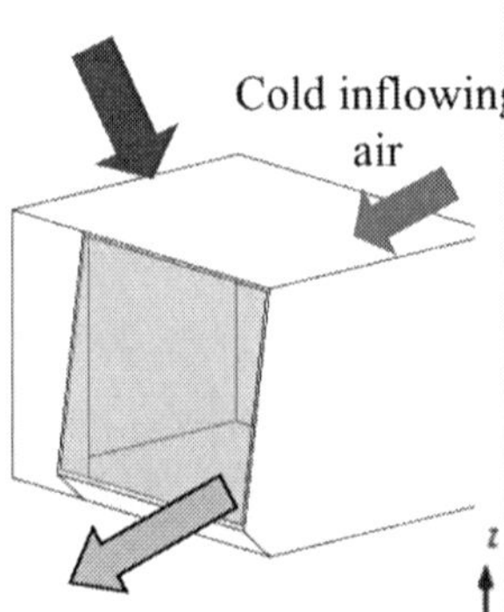

Fig. 4. General view of the space box to mix air flows.

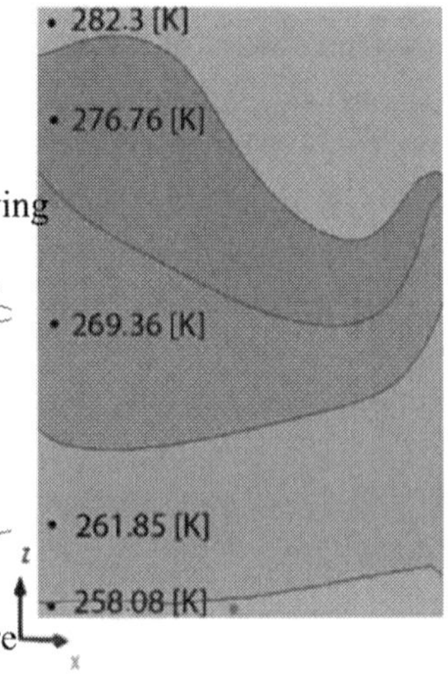

Fig. 5. Temperature field distribution at the exit area when the mixing box is used.

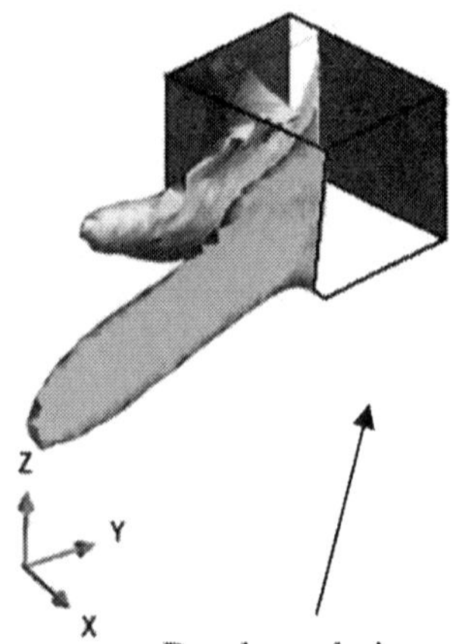

Fig. 6. Isosurfaces, corresponding to temperatures of + 30°C and - 30°C when the mixing box is used.

We have carried on modeling based on the box geometry and we have had the temperature field distribution at the specified geometry exit area (fig. 5), and isosurfaces with the temperatures of + 30°C and - 30°C (fig. 6). As it can be seen from these figures, the box use decreases the thermal gradient up to 6 °C/m. If we take the thermal gradient as a quality criteria of the hot and cold air stream mixing, we can assume that the box use in this design allows to increase the fan heater-based heating and ventilation system efficiency by 28.6 % without any maintenance cost increase. The figure 6 shows that the cold air stream with the temperature of - 30°C does not come to the simulated room floor.

Despite the fact that the box use has increased the mixing efficiency almost by a third, this system is not still efficient enough for heating and ventilation systems at inflowing air low temperatures. The box concept of this design has a range of disadvantages. Firstly, the cross section area is reduced, thus it increases the airflow resistance (the resistance has increased from 1 Pa to 14.5 Pa), so the discharge characteristic of the fan heater changes. Secondly, even if the box has limited the stream distribution along Z-axis, this design has not allowed the stream mixing that gives a higher cold air stream penetration due to the speed increase. This system can cause the equipment freezing at the central part of the room.

The jet theory of G. Abramovich is studied in detail in [4], [5], [6] and [7], and these works reproduce the calculation of hot air curtains bearing in mind various factors and operation conditions. Based on the results of these works and concepts described in inventions No. 290154 of 23.11.1971 and No. 688786 of 30.09.1979, we have developed the space structure to enhance the cold and hot air stream mixing. The general view of this structure is presented in figure 7. The device operation is based on the flow distribution principle. The flows are divided into two zones by plates. In the first zone only the hot air moves. The second zone is characterized by the contact of the heated air and the air under heating. Thus, the total flow is divided into sandwiched layers of the hot and cold air that mix as they leave the flow separation device.

3 Conclusion

We have carried on the mathematic simulation with the flow separation device and have got the temperature distribution at the specified geometry exit area (fig. 8) and isosurfaces that correspond to the temperatures of +30°C and - 30 °C (fig. 9).

As it can be seen in figures 7, 8 and 9 the flow separation device provides the decrease of thermal gradient room heightwise from 8.4 °C/m до 1.4 °C/m. The this design does not oversize the separation device, but it allows to reduce the resistance up to 7.4 Pa due to the cross section area extension.

Thus, the developed design helps increase the air heating system efficiency at low temperatures of the inflowing air a sixfold without any maintenance cost increase. In addition, the cold air flow jet penetration is four times lower.

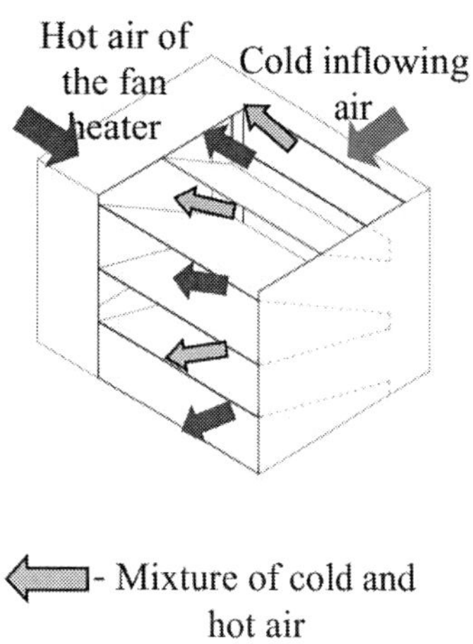

Fig. 7. General view of the flow separation device.

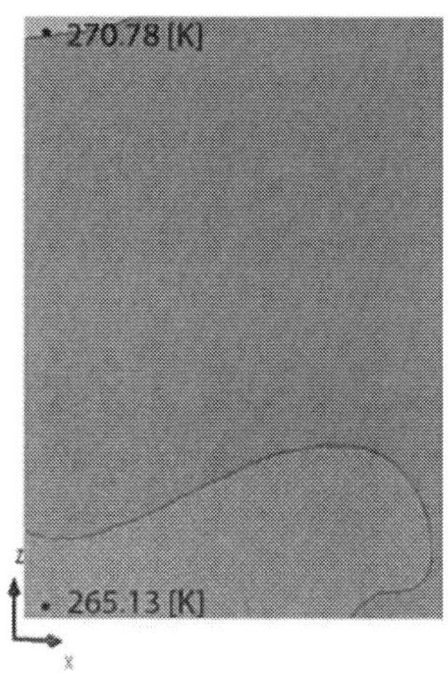

Fig. 8. Temperature field distribution at the exit area when the flow separation device is used.

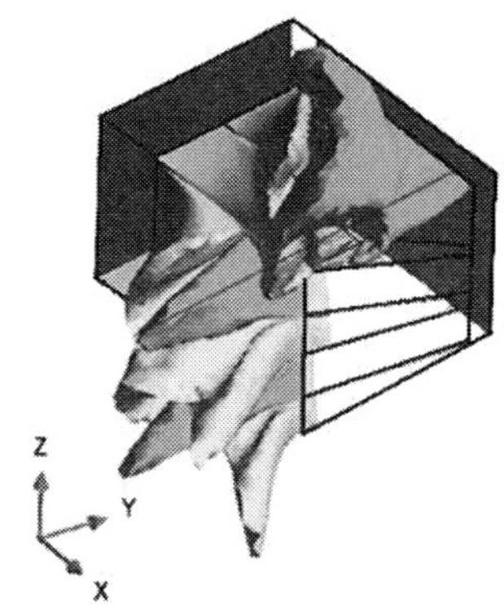

Fig. 9. Isosurfaces, that correspond to the temperatures of + 30°C and - 30°C when the flow separation device is used.

References

1. G.N. Abramovitch, Theory of turbulent jets. Reprint version of 1960. //ECOLIT, 720 (2011)
2. N.I. Kurilemko, Privolzhsky academic journal **2**, 116 (2014)

3. N.I. Kurilenko, L.Yu. Mikhaylova, P.A. Artamonov, Bulletin of TSUAB **4**. 171 (2012)
4. A.B. Kaplun, E.M. Morosov, M.A. Olferieva, *Editorial UMSS*, 272 (2003)
5. ANSYS HELP / CFX / Modeling Guide http://orange.engr.ucdavis.edu/Documentation12.0/120/CFX/xmod.pdf
6. The Menter Shear Stress Transport Turbulence Model/ Turbulence Modeling Resource / - NASA. Langley Research Center - http://turbmodels.larc.nasa.gov/ sst.html
7. D. Macklin, A. Yashina, Energy bulletin **4**, 43 (2009)
8. V.N. Bogoslovsky, V.I. Novozhilov, B.D. Simakov, V.P. Titov, *Heating and ventilation* (Stroyizdat, Moscow, 1976)
9. V.V. Baturin, *Fundamentals of industrial ventilation, Fourth edition* (Profizdat, Moscow, 1990)
10. V.N. Taliev, *Ventilation aerodynamics, Manual for graduate students* (Stroyizdat, Moscow, 1979)
11. G.N. Abramovitch, *Theory of turbulent jets. Reprint version of 1960* (ECOLIT, Chicago, 2011)

Thermal conductivity coefficient UO_2 of theoretical density and regular stoichiometry

A V. Kuzmin[1,*], and *M* M. Yurkov[1]

[1]National Research Tomsk Polytechnic University, 634050 Tomsk, Russia

Abstract. The main methodological research results of the thermal conductivity of uranium dioxide (UO_2) were considered. UO_2 is the main fuel of modern nuclear energetic reactors. The assessment of efficiency of the proposed approximations compared to Fink-Ronchi formula at the theoretical values of UO_2 density is proposed. According to this methodology we conducted the analysis of the experimental dependences of the thermal conductivity obtained by domestic authors.

1 Introduction

Nowadays uranium dioxide is the main fuel of nuclear energetic reactors owing to its chemical stability and radioactive resistance. The first accessible reference data on thermal and physical characteristics UO_2 were presented in research work [1] at the end of 60s.

Operating experience and further researches of oxide fuel in 1968-1998 showed significant differences in dependence of thermal conductivity coefficient on temperature λ(T), both fuel production technology and further influence on its meaning during reactor work of fission products, porosity changes and stoichiometry composition of fuel [2-4].

In our opinion, important methodical specifics of the research stage of thermal and physical properties UO_2 are:

- introduction of idea of theoretical density UO_2;
- detailed understanding of thermal transfer mechanism to thermal conductivity models of crystal sampling λ(T) in the form of sum of phonon or lattice conductivity λ_{ph}, photon or radiation λ_r and electronic conductivity, including ambipolar conductivity, λ_e;
- receiving of the first recommended Harding-Martin's formula [5] for thermal conductivity design UO_2 of theoretical density at temperature ranges 773–3120K which included two main components of thermal conductivity mechanism (pic.1);
- use of integral thermal conductivity for temperature changes design in fuel pellet.

On the whole, at this stage of research insufficiency of experimental data on thermal conductivity UO_2 at high temperatures conditioned by numerous amount of influencing factors and difficulty of making experiments was identified.

There are ranges of temperature changes of fuel pellets in cassettes of medium and maximum reactor power VVER-1000 [6] shown on the graph of thermal conductivity

* Corresponding author: kuzminav@tpu.ru

coefficient dependence UO_2 on temperature (pic.1) which underline the role of each component of thermal transfer mechanism.

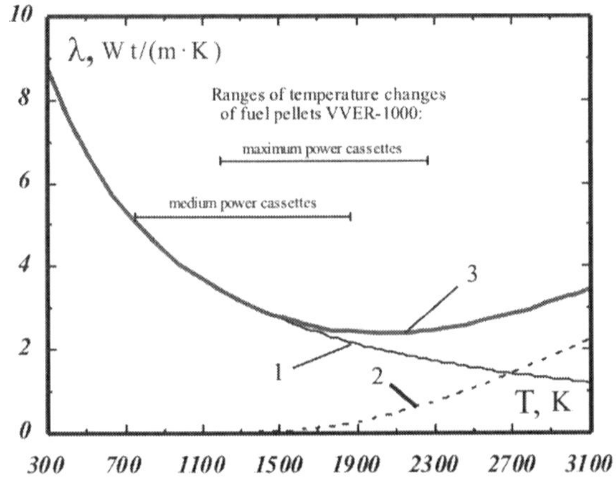

Fig. 1. Thermal conductivity UO_2: 1 – phonon component of thermal conductivity; 2 – electronic component; 3 – full thermal.

Thereafter in 1999-2008 as a result of generalization of more extensive experimental data Ronchi [7] offered a new dependence for ambipolar component. Later Fink generalized other researcher's results and found modified expression for phonon component. He used Ronchi's expression for ambipolar component and received a design formula of thermal conductivity coefficient UO_2 [8,9] which inaccuracy is +10% till the limit of 2000 K, and +20% if the limit is higher. Coefficient 1.158 was introduced in the Fink-Ronchi's formula below for recomputation of the meaning of thermal conductivity samples UO_2 coefficient from 95 to 100% theoretical density [10]:

$$\lambda = 1.158 \cdot \left(\frac{100}{7.5408 + 17.692 \cdot \tau + 3.6142 \cdot \tau^2} + \frac{6400}{\tau^{5/2}} \exp\left(\frac{-16.35}{\tau} \right) \right), \quad (1)$$

where $\tau = (T/1000)K$.

Introduction of theoretical density UO_2 allows detecting efficiency of offered approximations for thermal conductivity coefficient UO_2 design of theoretical density – λ_0. Let us compare design dependences made by Russian scientists in different years with Fink-Ronchi's formula.

2 Comparison of the approximations on the thermal conductivity of UO_2 at the theoretical values of density

An experimental work [11] proved a theoretical dependence on thermal conductivity coefficient UO_2 design presented in a reference book [12]:

$$\lambda_0 = 1.158 \cdot 10^2 \cdot \left(\frac{1}{3.77 + 0.0258 \cdot T} + 1.1 \cdot 10^{-6} \cdot T + 1.01 \cdot 10^{-13} \cdot T^3 \cdot \exp\left(7.2 \cdot 10^{-4} \cdot T\right) \right). \quad (2)$$

There is also an empiric formula there [12]:

$$\lambda_0 = 1.158 \cdot \left(11.5 - 1.14 \cdot 10^{-2} \cdot T + 4.4 \cdot 10^{-6} T^2 - 5 \cdot 10^{-10} \cdot T^3\right). \quad (3)$$

Influence of inaccuracy from stoichiometry composition UO_{2+x} on thermal conductivity was researched in work [13]. As a consequence, the following dependence led to theoretical density of uranium dioxide was received in the range of temperatures 300-2000K.

$$\lambda_0(x,T)=1.158\cdot 10^2\cdot\left(4.5+43\cdot x^{1/2}+0.0219\cdot T-0.033\cdot x\cdot T\right)^{-1}. \tag{4}$$

Approximation formula [14] given below is shown without specified conditions of application.

$$\lambda=\frac{5500}{560+T}+0.942\cdot 10^{-10}\cdot T^3. \tag{5}$$

Comparison of formulas (2)–(5) received before 2003 and led to theoretical density UO_2 with Fink-Ronchi's dependence is presented in picture 2; it shows significant differences.

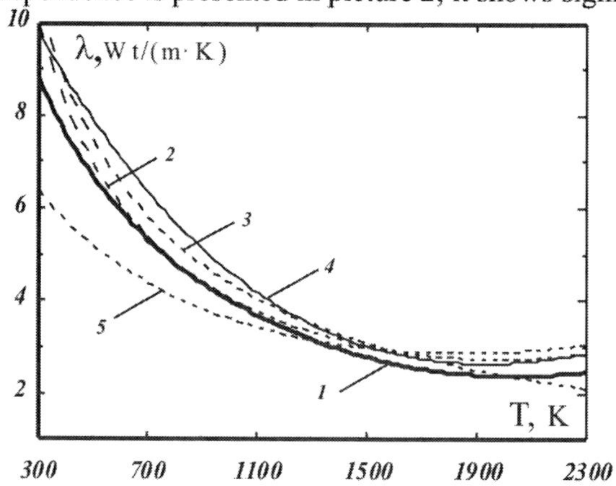

Fig. 2. Comparison of approximations on thermal conductivity UO_2: 1 – Fink-Ronchi's formula (1) [8]; 2 – formula (2), data [11,12]; 3 – formula (3) [13]; 4 – empiric dependence (4) [12]; 5 – formula (5) [14].

In the succeeding reference books and books after 2003 there is move towards standard formula (1). There is empiric dependence in work [15]

$$\lambda_{norm}=\frac{4820}{351+T}+2.434\cdot 10^{-14}\cdot T^4, \tag{6}$$

in which it is easy to detect a significant difference from Fink-Ronchi's formula led to theoretical density UO_2 (1) (see fig.3, curve 2) while initial conditions of its use were not stipulated. Dependence (1) becomes general in work [16] and further a design formula becomes based on it [17] which includes medium fuel burning-out B, MW·day/kgU:

$$\lambda(B,T)=1.158\cdot\left\{\begin{array}{l}\left[\begin{array}{l}\left(0.1148+3.0864\cdot 10^{-3}\cdot B\right)+\\ 2.475\cdot 10^{-4}\cdot\left(1-2.94\cdot 10^{-3}\cdot B\right)\cdot\left(T-273.15\right)\end{array}\right]^{-1}+\\ +0.0132\cdot\exp\left(1.88\cdot 10^{-3}\cdot\left(T-273.15\right)\right)\end{array}\right\}. \tag{7}$$

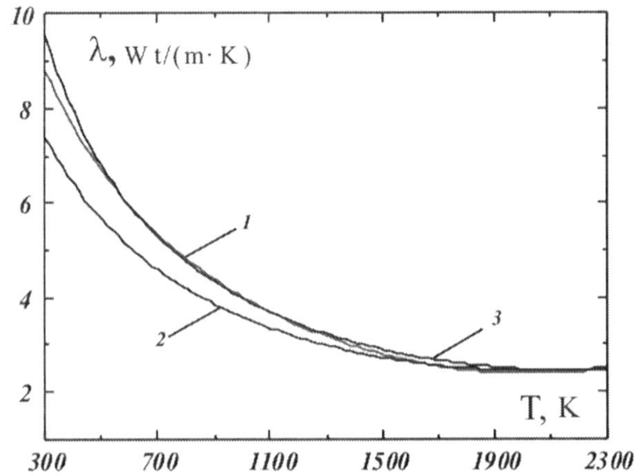

Fig. 3. Comparison of design formulas on thermal conductivity UO_2: 1 – Fink-Ronchi's formula (1) [8,16]; 2 –formula (6), [15]; 3 – formula (7) [17].

Thus, analysis of comparison of experimental dependences of thermal conductivity coefficient UO_2 made by different authors and Fink-Ronchi's formula at theoretical density of fuel allowed:

- to identify significant differences in input of some mechanisms of thermal conductivity assessment with the growth of fuel temperature;
- to detect a base limit of uranium dioxide thermal conductivity change making it possible to clearly distinguish porosity influence, nonstoichiometry and burning-out at present.

References

1. V.S. Chirkin *Teplofizicheskie svojstva materialov jadernoj tehniki* (Atomizdat. Moscow, 1968) [in Russian]
2. Thermophysical Properties of Materials for Water Cooled Reactors: Report IAEA-TECDOC-949. Vienna: IAEA. (1997)
3. P.L. Kirillov, Atomnaja tehnika za rubezhom **6**, 3 (1998) [in Russian]
4. E.Ju.Sinjatkin, A.V. Kuzmin, Sovremennye tehnika i tehnologii: sbornik trudov XVII mezhdunarodnoj nauchno-prakticheskoj konferencii studentov, aspirantov i molodyh uchenyh **3** 18 (2011) [in Russian]
5. J.H. Harding, D.G. Martin , J. Nuclear Mater. **166** 223 (1989)
6. A.M. Afrov, S.A. Andrushechko, V.F. Ukraincev, *Universitetskaja kniga* (Logos, Moscow, 2006) [in Russian]
7. Thermal Conductivity and Thermal Diffusivity of Solid UO2: 'Internet' site: htpp: // www.insc. anl.gov./matprop/uo2/thcsuo2. pdf
8. Thermophysical Properties Database of Materials for Light Water Reactors and Heavy Water Reactors: Final Report of a Coordinated Research Project 1999-2005; IAEA-TECDOC-1496. Vienna: IAEA. (2006)
9. P.L. Kirillov, M.I. Terent'eva, N.B. *Deniskina Teplofizicheskie svojstva materialov jadernoj tehniki: Uchebnoe spravochnoe posobie dlja studentov special'nostej: 14.03.05 – Jadernye reaktory i jenergeticheskie ustanovki, 14.04.04. – Atomnye jelektricheskie stancii i ustanovki* (IzdAt, Moscow, 2007) [in Russian]
10. Ju.G. Godin, A.V. Tenishev, V.V. Novikov *Fizicheskoe materialovedenie: Uchebnik dlja vuzov* (MIFI, Moscow, 2008) [in Russian]

11. V.I. Koljadin, Je.P. Il'in, A.G. Harlamov, V.V. Jakovlev, Atomnaja jenergija **36**, 59 (1974) [in Russian]
12. P.L.Kirillov, Ju.S. Jur'ev, V.P. Bobkov *Spravochnik po teplo-gidravlicheskim raschetam (jadernye reaktory, teploobmenniki, paroge-neratory)* (Jenergoatomizdat, Moscow, 1990) [in Russian]
13. A.N. Andrianov, V.G. Baranov, Ju.G. Godin, V.B. Kruglov, A.V. Tenishev, Nauchnaja sessija MIFI-2003. Perspektivnye naukoemkie tehnologii. Fizika, himija i komp'juternaja razrabotka materialov **9**, 21 (2003) [in Russian]
14. B.A. Dement'ev *Jadernye jenergeticheskie reaktory: Uchebnik dlja vuzov* (Jenergoatomizdat, Moscow, 1990) [in Russian]
15. B.S. Petuhov, L.G. Genin, S.A. Kovalev, S.L. Solov'ev *Teploobmen v jadernyh jenergeticheskih ustanovkah: Uchebnoe poso-bie dlja vuzov* (MJeI, Moscow,2003) [in Russian]
16. P.L. Kirillov, G.P.Bogoslovskaja *Teplomassoobmen v jadernyh jenergeticheskih ustanovkah: Uchebnoe posobie dlja vuzov* (IzdAT, Moscow, 2008) [in Russian]
17. P.L. Kirillov, A.V. Zhukov, N.I. Loginov, V.M. Mahin, I.L. Pioro, Ju.S. Jur'ev, *Spravochnik po teplogidravlicheskim raschetam v jadernoj jenerge-tike. Tom 2. Jadernye reaktory, teploobmenniki, parogeneratory* (IzdAt, Moscow, 2003) [in Russian]

MATEC Web of Conferences 92, 01051 (2017) DOI: 10.1051/matecconf/20179201051
Thermophysical Basis of Energy Technologies - 2016

Power plant ash and slag waste management technological direction when Kansk-Achinsk brown coal is burned

Snejana A. Lihach[1*], *Roman* N. Kulesh[1], *Valentina* I. Nikolaeva[1], *Ksenia* Y. Orlova[1], and *Aygul* S. Ilyasova[1]

[1]National Research Tomsk Polytechnic University, 634050 Tomsk, Russia

Abstract. Today resource efficiency technology development in all industries where conventional raw material is being replaced by local natural resources and industrial waste is an essential matter. Along with that most producing operations are overload with wide range of waste produced during technological process. Thermal power stations are real world evidence. Their main waste is ash and slag which accumulated in great amounts in often overfull ash dumps. The goal of present work is to find perspective ash dump waste utilization methods. The study will be based on experimentally obtained data: elementary compound and properties of Kansk-Achinsk brown coal. Research methods: experimental, chemical silicate analysis, mineralogical forms identification within samples by using ASM X-ray diffraction analysis. Experiments resulted with the following conclusions: silica is ash main component, and ash has the form of ore concentrate analogy in a number of elements. We think that ASM main properties which make it useful for utilization are: high content of calcium oxide; high ash sorption properties; ASM radiation safety class which makes them safe to be used in materials, goods, and structures production for residence and public buildings construction and reconstruction; sufficiently high content of individual elements.

1 Introduction

Yearly Russian power industry produces not only electric energy, but also about 27 million tons of ash and slag waste (ASW). Just a small part of it, less than 15 percent, is reprocessed in other industries [1]. The main part of the waste is stored for years in ash dumps which take prime quality lands and deteriorate ecology. In past years several cities start really feel the problem: power plants built outside cities in due course time appeared to be within the city limits (Tomsk, Kemerovo, Belovo, Novosibirsk).

Meanwhile coal solid part and its components might become a huge source of raw materials and be widely applied in some industries [2]. Therefore problems of ASW common usage in construction materials manufacturing, construction and other industries are essential.

* Corresponding author: korotkikh@tpu.ru

2 Ash and slag research

In order to determine ASW line of utilization we need to perform a number of studies. An ash and slag materials (ASM) passport is put together based on the studies as a product of a power plant. Then the decision is made how an ASM will be utilized. A difference between ASM and ash dump is that an ASM is a classified product with physical-chemical and granulometric properties which concur with specifications. ASM elementary and compound constitutions are needed to be specified in order to determine whether or not ASM could be used as a raw material. The constitutions are based on many factors which determined by coal burning conditions and boiler operating regime. Consequently ASM passport should contain a possible range of products desired parameters [3].

In line with the above ash and slag elementary composition and properties study was per-formed for Kansk-Achinsk brown coal burning. The study was pursued in several stages: first ash and slag sampling from a boiler and ASM sampling from the ash dump were made; then laboratory studies were performed. Ash and slag samples were taken from working boilers with bottom-ash and slag-tap removal. All sampling work, samples preparation and study were performed in compliance with standard technique.

We obtained percentage ranges of elements content calculated using oxides by elementary composition research. The results are presented in table 1.

Table 1. Percentage ranges of elements content

Elements content calculated using oxides	Range, %		
	Ash	Slag	Ash and slag mixture
Ignition loss	0.07-0.18	-	below 10
SiO_2	32.97-48.01	41.05-49.16	49.23-65.79
TiO_2	0.27-0.48	0.27-0.31	0.32-0.55
AL_2O_3	6.25-9.01	5.89-7.41	7.08-10.14
Fe_2O_3	10.78-15.94	15.23-19.05	6.29-18.1
CaO	26.7-33.9	20.6-25.07	10.2-21.4
MgO	5.15-7.09	3.39-4.29	2.32-6.45
K_2O	0.2-0.3	0.2-0.5	0.2-0.42
Na_2O	0.3-0.6	0.35-0.6	0.2-0.4
SO_3	1.31-2.12	0.48-2.12	0.48-1.47
CaO_{free}	6.5-9.7	нет	0.14-0.98

We paid attention to the fact that ash and slag mixture from ash dump has large weight loss value. It might be explained by intrusion of coal dust into an ash sluiceway in a boiler shop. Besides, supply of organic substance into an ash dump from the outside is possible as well. Also, we should take into consideration chemical reactions which take place when ash and slag waste is stored in a waterlogged ash dump.

For the purpose of ASM content research we performed chemical silicate analysis and ASM X-ray diffraction analysis. The last one let us to identify mineralogical forms within samples.

3 Results and discussion

Based on the research findings we obtained percentage ranges of elements content calculated using oxides with Kansk-Achinsk coal burning based on fly ash, slag and stored ash and slag mixture. Using the obtained results we made the fol-lowing conclusion:

1. The ash samples from under multicyclones appear to be amorphized textural features with finely dispersed physical mixture of some crystalline phases. Low temperature β–quartz (β–SiO2) of hexagonal syngony is the largest comparing masses. Overall re-searched ash samples contain from 33 to 10 weight % of β–SiO_2. The second crystalline structural constituent of ash samples is caustic limestone (CaO). The samples contain from 8 to 18 weight % of CaO as well as from 4 to 16 weight % of anhydrite (anhydrous plaster, CaSO4). All ash samples are magnetic and contain cryptocrystalline magnetic fraction in form of spinel magnetite (Fe3O4).There are from 11 to 22 weight % of free iron crystalline oxide in the form of hematite (α–Fe_2O_3) and magnetite.
2. Slag samples in case of wet bottom operation are appeared as glass-like amorphized mass. X-ray diagrams show geometric feature short range ordering of Si – O atoms and specific weak reflections which are typical for low temperature β– quartz (β–SiO_2). Slag samples are magnetic.
3. Slag samples in case of bottom-ash removal mainly consist of high-silica sand in form of low temperature β– quartz (β–SiO_2), and also in form of hydrosilicic acid complex salts – calcium and magnesium silicates – type $CaO{\cdot}SiO_2$, $CaO{\cdot}MgO{\cdot}2SiO_2$, $2CaO{\cdot}MgO{\cdot}2SiO_2$. Quantity of β–$SiO_2$ in samples is from 27 to 66 weight %. Calciferous magnesia siliciferous components (calcium and magnesium silicates) are from 18 to 560 weight %.
4. Average ash dump sample analysis showed that it contains 70 weight % of silicon dioxide, 27 weight % of cementitious component in form of calcium silicates, magnesium silicates and plaster, and about 3 weight % of free iron oxides.

Therefore, produced waste has heterogeneous chemistry, but it is identical in a large part with natural mineral raw materials.

Construction industry is the main line of power stations fuel combustion waste utilization [4, 5].

First of all the waste could be used as a bulk material for automobile road border levees regulated by industry document [6]. Nowadays border levees are formed with sand and gravel mixture, which is quarried with open-pit method. This method in most cases leads to ecological damage of a region with lakes and wetlands formation in site of opencast mines. As shown above ASW is identical with natural silicate compounds which are generally used in this industry.

Secondly, ASW is used for heavy weight concrete, light weight concrete and mortars production for the purpose of concrete and fillers cost-effectiveness, concrete and mortar mixture properties quality level improvement. Ashes which contain no less than 30% of CaO are used as a matrix for lime and concrete partial replacement in autoclaved and non-autoclaved lightweight concrete.

In accord with [7] calcium oxide content by weight should be no more than 10% – in the researched samples it is higher. CaO_{frree} content no more than 5% for cell concrete is not normative – in the samples it is no more than 9.7%. Sulphur and Sulphur oxide compounds content in terms of the equivalent amount of SO_3 is no more than 6% – in the samples it is 2.21%. MgO content is no more than 5%, in the samples – 7.09%, but it is acceptable for MgO content to be higher in some cases: if samples soundness is given during auto-clave test; if these ashes application is based on concrete lifetime advanced studies with consideration of specific application conditions. Na_2O con-tent is no more than 3.5%, in the samples – 2.12%. Ignition weight loss is no more than 5%. The same parameter for ash and slag mixture is exceeded. It is important to mention that marginal excess of tested parameters is possible to level during production by adding a neutral material.

Today resource efficient technology engineering is a topic of immediate interest in all industries where traditional imported raw material is being replaced by local natural resources and industrial waste.

The conducted research shows that according to elementary, granulometric and mineral composition Kansk-Achinsk coal ASM is mostly identical to natural raw material. It is possible to use ash as a raw material for rare elements recovery [8]. It is found that there are over 50 such elements in ash and slag, what let to reduce expenses for crude ore geological survey. ASM contains aluminium, silicon, calcium and ferrum compounds in such quantities which are of some real interest. Besides, there is an important problem of the following toxic elements control in ash and slag: Li, Ba, Ga, B, Cu, Ag, Cd, Hg, Pb [9].

Non-ferrous metals of value might be re-claimed from ASM as a part of a complex process, when main part of ASM is utilized to give construction materials, stonecast products and other.

We undertook the study in order to evaluate possibility to quarry some particular elements from Kansk-Achinsk coal ash. Table 2 represents the results of the study.

Table 2. Content of some chemical microelements in Kansk-Achinsk coal ash.

Descriptor	Weight percent, %
Potassium	0.23
Titanium	0.22
Barium	0.016
Chromium	0.0028
Cobalt	0.0029
Nickel	0.0063
Stannum	0.011
Molybdenum	0.0041
Stibium	0.0008
Arsenium	0.0046
Argentum	0.00023
Zinetum	0.0027
Strontium	0.0037
Lead	0.00019
Cadmium	0.000077
Tungsten	0.034
Lanthanum	0.0012
Lithium	0.0035
Indium	<0.005

4 Conclusions

Analysis of table 1, table 2 and the con-ducted research results shows that silica is ash main component, and ash has the form of ore concentrate analogy in a number of elements. We think that ASM main properties which make it useful for utilization are: high content of calcium oxide; high ash sorption properties; ASM radiation safety class which makes them safe to be used in materials, goods, and structures production for residence and public buildings construction and reconstruction; sufficiently high content of individual elements.

References

1. A.I. Kalachev, Power station ash and slag: disposal, transportation, utilization, storage, 5, Russian ASW market through trader's eyes. Phenix consortium (2014)
2. J. Hot, M. Sow, C. Tribout, M. Cyr, Constr. Build. Mater. **110**, 218 (2016)
3. S.-W. Yoo, Y.K. Cho, S.-H. Jung, K.-M. Lee, S.-J. Kwon, J. Mater. Cycles Waste, 1 (2016)
4. N. Nordin, M.M.A.B. Abdullah, M.F.M. Tahir, A.V. Sandu, K. Hussin, Int. J Conserv. Sci. **7**, 161 (2016)
5. M. Pasetto, N. Baldo, Constr. Build. Mater. **108**, 112 (2016)
6. ODM 218.2.031.-2013, Methodological recommendations for power stations coal combustion fly ash, ash and slag mixtures use in road construction (2015)
7. GOST 25818-91, Thermal power station fly ashes for concrete. Specifications (2015)
8. Y.-X. Wang, L. Peng, Y. Wang, T. Zhang,H.-L. Liu, L. Mu, Huanjing Kexue/Environmental Science **37**, 60 (2016).
9. L.C. Ram, R.E. Masto, N.K. Srivastava, J. George, V.A. Selvi, T.B. Das, S.K. Pal, S. Maity, D. Mohanty, Environ. Monit. Assess. **187**, 14 (2015)

Energy budgets' formation based on incomplete data

Valery Litvak[1] *, *Galina* Klimova[1], and *Veronika* Morozova[1]

[1]National Research Tomsk Polytechnic University, 634050 Tomsk, Russia

Abstract. The universal technique of the current energy budget calculation is recommended for several periods of time. Basing on the data of annual energy budget and considering the plant's features, social, climatic and other conditions, correcting factors are calculated to estimate energy resources consumption for the day, week, month and year. Short-term current energy budget supports important financial, trade, logistic, organizational and other control directions. The developed technique is applicable for subnational entities and industrial enterprises in order to increase level of energy resources consumption planning and forecasting as well as bills' optimization.

Govermental statistic witnessing provides filling statistic data to form region energy budget based on several census report form [1-4].

According to facts it takes about half-year from information acquisition of energy resources using to energy budget formation. Of course, it stops effective sampling and energy efficiency management.

Provided approach makes possible to form and analyze energy budget during a process, it gives opportunities to take energy-saving and efficiency upgrading measures.

Principle of such method is to develop an energy budget model based on using detected internal regularities of historic budgets with incomplete or even irreconcilable data of current period. Statistic budget formation remains to be main machine to improve accuracy and adequacy.

Such approach is clear because of following consideration:

- ✓ Immeasurable elements' calculations are made by detected internal budget relations;
- ✓ Elements' numeric value tolerance definition is defined by errors of energy flow changes and mutual coupling factor calculation;
- ✓ If desired successive refinement of energy budget elements' numeric values design ratios opportunity can be saved.

For approach simplication of budget internal relations formation nonlinear relations are simulated as linear, and multifactorial restrictions are simulated doubled restrictions.

Correlation based on linear equation is formed with direct relationship presence, where individual result intimately increase or decrease with factor changes. It is defined in the shape of following equation [7]:

$$y_x = a_0 + a_1 x, \quad (1)$$

* Corresponding author: litvak@tpu.ru

where y – resulting characteristic individual value, x – factor characteristic individual value, a_0, a_1 – equation parameters, y_x – – resulting characteristic theoretical value.

Equation parameters a_0 and a_1 are defined by the way of least square method:

$$\begin{cases} \sum y = na_0 + a_1 \sum x, \\ \sum yx = a_0 \sum x + a_1 \sum x^2. \end{cases} \quad (2)$$

Parameter a_0 is defined by the following way:

$$a_0 = \frac{\sum y \sum x^2 - \sum yx \sum x}{n \sum x^2 - \sum x \sum x}. \quad (3)$$

Consequently a_1

$$a_1 = \frac{n \sum yx - \sum x \sum y}{n \sum x^2 - \sum x \sum x}. \quad (4)$$

Closeness correlation changes are performed by calculation of linear factor of correlation, theoretical correlation relation and correlation index.

Adequacy and correctness of closeness correlation measuring operation are estimated by Fisher's ratio test.

Empirical Fisher's ratio test value $F_э$ is compared to critical (tabular $F_т$) with significance level 0.01 or 0.05 and number of degrees of freedom (*m*-1), (*n*-*m*). If $F_э > F_т$, correlation empirical estimator is regarded to be significant.

Linear regression equation factors significancy is estimated with Student's *t*-test ($n<30$).

Empirical Student's *t*-test value is compared to critical (tabular) with significance level 0.01 or 0.05 and number of degrees of freedom (*n*-2). If empirical value is bigger than tabular, correlation empirical estimator is regarded to be significant [7].

Done estimation results afford to predicate that elements have relations is proved on the whole range of energy resources. regression factors significancy is checked by Fisher's ratio test.

Nowadays there has been a resumed interest to the energy budget as a tool or analyzing and planning of power generation, for quality control of energy conservation and energy resources efficiency enhancement measures [5-6].

Thus, it is of vital importance to develop approaches towards energy budget calculation for short period of time. This issue has a great value in case of development potential, plotting graphs of product transportation and fuel supply, assessing possibilities of energy resources demand coverage for severe conditions e.g. winter peak and summer low load, repair and maintenance schedule, specific climatic conditions (waterflood, hurricane, fallout and etc.), anthropogenic and natural disasters.

It is rational to divide current energy resources budgets into following time steps: hour, shift, civil day, operating day, weekends and public holidays, weeks with different number of working days and holidays, month, year quarter, heating and non-heating seasons, repair period, annual accounting period.

It is not feasible to build energy budgets for ultra-short periods of time (less than one hour). Methodically it can be done by formulated approaches, but practically such budgets have not a great value. Table 1 shows energy budget of reinforced concrete plant for one year.

Basis for mentioned below calculating is annual energy budget of reinforced concrete plant.

Volumes of energy resources are presented in own units for each, because it is not feasible to present the resources in reference fuel units. Moreover service and drinkable water consumption is shown as well. Water is not an energy resource, but on the other hand, water-supply system is operated by chief power engineer service crew.

Current energy budget for short period of time supports financial, commercial, logistic, organizational and other branches' basis of controlling [3, 8]. Current energy budget could be used for independent assessment of energy production, efficiency of power saving and other cases.

Average calculations of assigned year energy consumption are not difficult if we do not consider short periodical energy budgets structure perturbations:

$$A_t = \int_0^t P_t \cdot dt \cdot \frac{A_{year}}{N_t} \cdot \Delta t, \quad (5)$$

where P_t – current power, kW, A_{year} – annular consumption, kilowatt hour, T – basis time duration, hour, Δt – current period duration, hour, N – number of equal current time intervals per one year [4]. In case of known plant energy consumption per one year (40000 Ths. kilowatt hour) average short periodical consumption volumes will be calculated as follows.

There is actual difference between energy resources use for short period of time and average for one year. Thus it is necessary to adjust temperature, social, process and other factors. Element value adjustment should be done in the following way:

$$A_k = A_{av} \cdot (1 + k_1 \cdot k_2 \cdot k_3 \cdot k_4), \quad (6)$$

where A_k – current element value, A_{av} – average element value, k_1 – correcting factor considering energy resources volumes of operating days and holidays, k_2 – correcting factor considering demand's increment between start and end of the year; k_3 – correcting factor considering energy resources consumption depending on air temperature, k_4 – correcting factor considering energy resources consumption between heating and non-heating seasons.

Energy consumption ratio of holidays and operating days with correcting factor k_1 approaches 0.5:1. This ratio between holidays and operating days for one week reinforced concrete plant operation equaled 0.55:1.18. In relation to year average plant measurements level showed following k_1 values (Table 1).

Table 1. Correction factor k_1.

Correction factor k_1, relative unit	Electrical energy	Thermal energy	Gas	Water
Operating days	1.15	1.1	1.07	1.12
Weekends and public holiday	0.55	0.75	0.825	0.7

Correction factor k_2 considering demand's increment between start and end of the year is defined as:

$$k_2 = 1 + k_{\text{in}} \cdot \frac{i}{365}, \qquad (7)$$

where $k_{in} = 0{,}02 \div 0{,}04$ – average incremental growth of demand equal 2 – 4% per year, i – number of the date starting from 1st of January.

Air temperature correction factor k_3 describes energy resources incremental growth with air temperature lower than $+10^0$C in relation to average temperature for same period of many years. In some cases, temperature may be close to zero degrees. Temperature reference scale origin is shifted to the point of -100^0C to avoid division by zero. Then:

$$k_3 = \frac{100 - \Theta_d}{100 - \Theta_{av}}, \qquad (8)$$

where Θ_d – desired air temperature of design day, ^{0}C, Θ_{av} – average long-term temperature of this day, ^{0}C.

Correction factor k_4 characterizes energy resources consumption between heating and non-heating seasons. It is important to include this factor into calculations because there is a great difference in heat and fuel consumption for both seasons. As a rule difference water and electrical energy consumption is small.

Centralized annual heat and fuel consumption comprises the following:

$$Q = Q_{heat} + Q_{\text{hws}} + Q_{pr}, \qquad (9)$$

where Q_{heat} – heat consumption for heating, Gcal, Q_{hws} – hot water supply, Gcal, Q_{pr} – heat consumption for process requirements, Gcal.

Thermal energy for heating Q_{heat} is supplied only during heating season, hot water supply Q_{hws} – all year round, and Q_{pr} according to operational program. Unfortunately, there is no any division of these three flows in official statistics. In calculations hot water supply system volume in comparison with heating accounts for 10 – 12%. Throwing away Q_{pr} due to its uncertainty, we may assume that:

$$Q = Q_{heat} + Q_{\text{hws}} = Q_{heat} + 0.12Q_{heat} = 1.12Q_{heat},$$

$$Q_{heat} = \frac{Q_{heat} + Q_{\text{hws}}}{1.12} = 0.89 \cdot (Q_{heat} + Q_{\text{hws}}),$$

$$Q_{heat} = \frac{Q_{heat} + Q_{\text{hws}}}{1.12} = 0.89Q.$$

Daily average heat transmission for heating volume during heating season is:

$$q_{\text{av}}^{heat} = \frac{0.89Q}{235} = 3.79 \cdot 10^{-3} Q,$$

and hot water supply all year round:

$$q_{\text{av}}^{hws} = (1 - 0.89) \cdot \frac{Q}{365} = 0.29 \cdot 10^{-3} Q.$$

Daily heat supply during heating season:

$$q_{day} = 3.79 \cdot 10^{-3} Q + 0.29 \cdot 10^{-3} Q = 4.08 \cdot 10^{-3} Q,$$

and during non-heating season:

$$q_{day} = 0.29 \cdot 10^{-3} Q.$$

Correspondingly, the correcting factor k_4 during heating and non-heating seasons is equal to:

$$k_4 = 4.08 \cdot 10^{-3} \cdot 235 = 0.9588,$$

$$k_4 = 0.29 \cdot 10^{-3} \cdot 130 = 0.0377.$$

This correcting factor for electrical energy and water supply can be taken as 1.

Correction factors for other periods of time are defined the same way. Table 2 shows correcting factors' values for accepted design conditions.

Table 2. Correcting factors' values

Current period	k_1, k_o – operating days, k_h – holidays				k_2	k_3	k_4 heat and fuel	
	Electrical energy	Thermal energy	Gas	Water			Heating season	Non-heating season
Hour k_o k_h	1.15 0.55	1.1 0.75	1.07 0.875	1.12 0.7	$1+k_{in}\cdot\frac{i}{365}$	$\frac{100-\Theta_d}{100-\Theta_{av}}$	0.9588	0.0377
Day k_o k_h	1.15 0.55	1.1 0.75	1.07 0.875	1.12 0.7	$1+k_{in}\cdot\frac{i}{365}$	$\frac{100-\Theta_d}{100-\Theta_{av}}$	0.9588	0.0377
Week	$\frac{k_o \cdot n_o}{7}+\frac{k_h \cdot (7-n_o)}{7}$				$1+k_{in}\cdot\frac{i}{52}$	$\frac{100-\Theta_d}{100-\Theta_{av}}$	0.9588	0.0377
Month	$\frac{k_o \cdot n_o}{n_a}+\frac{k_h \cdot (n_a-n_o)}{n_a}$				$1+k_{in}\frac{j}{12}$	$\frac{100-\Theta_d}{100-\Theta_{av}}$	0.9588	0.0377
Heating season	$\frac{k_o \cdot n_o}{n_a}+\frac{k_h \cdot (n_a-n_o)}{n_a}$				$1+k_{in}\frac{n_a}{365}$	$\frac{100-\Theta_d}{100-\Theta_{av}}$	0.9588	0.0377
Non-heating season	$\frac{k_o \cdot n_o}{n_a}+\frac{k_h \cdot (n_a-n_o)}{n_a}$				$1+k_{in}\frac{n_a}{365}$	$\frac{100-\Theta_d}{100-\Theta_{av}}$	0.9588	0.0377

*Foot-note: k_O – operating load factor, k_h – weekend and public holidays load factor, k_Π – increment load factor of the whole year, Θ_d – desired temperature of period, Θ_{av} – average temperature of period, i – day number of the year, n_o – operating days number of period, j – week number starting from the beginning of the year, n_a – day quantity of the period, e – month number starting from the beginning of the year.

Further example of current day budget for one operating day of heating season is calculated.

Basic data. 13th of February is forty fourth operating day starting from the beginning of the year. Increment rate is $k_{in} = 2\%$. Heating season. Average day temperature –14.5^{0}C, desired temperature –16^{0}C.

DOI: 10.1051/matecconf/20179201052

Table 3. Current daily energy budget of reinforced concrete plant (13th of February)

Balance items' nomenclature	Gas		Electrical energy		Thermal energy		Water	
	Calculated	Real	Calculated	Real	Calculated	Real	Calculated	Real
	Ths.m3	Ths.m3	Ths. kWatt hour	Ths.kWatt hour	Gcal	Gcal	Cubic meter	Cubic meter
Generation	-	-	-	-	73.94	74	30.1	30
Received from other source	11.36	11.4	123.7	123	-	-	105.5	106
Total consumption	11.36	11.4	123.7	123	73.94	74	135.6	136
Administration building	-	-	3.59	3.7	1.02	1.0	4.07	4
Reinforcement shop	0.57	0.6	5.44	5.5	1.71	1.7	6.78	7
Welding shop	0.35	0.3	21.4	20	1.02	1.0	4.07	4
Polygons 1, 2, 3	-	-	8.17	8.2	34.13	35	20.34	20
Formwork shop	-	-	4.58	4.5	1.27	1.3	4.52	6
Forging shop	1.13	1.0	10.89	11	0.68	0.5	2.71	3
Asphalt-concrete shop	0.35	0.3	13.61	14	1.02	1.0	4.07	4
Compressor room	0.23	0.2	7.3	7	2.04	2.0	8.14	8
Crushing shop 1,2	0.46	0.5	18.68	20	3.41	3.5	13.56	13
Limestone shop	0.35	0.3	5.44	5	7.05	7	2.71	2
Electrical service	-	-	4.58	4	0.68	0.7	2.71	2
Boiler house	7.92	8.2	16.33	17	1.27	1.3	5.41	6
Others	-	-	3.71	3	18.1	18	52.89	54

Conclusions

Current energy budget calculation represents simple computational task. Experience of current energy budget use for energy consumption and energy saving control allows to plan energy sources volume for different short periods of time more accurately, to calculate faults, to determine energy performance factors in terms of the year for fixed points based upon development prospect (or other side – production curtailment) and etc.

References

1. GOST 27322-87. Power plant balance. Main principles [in Russian]
2. GOST P 51379-99. Energy performance certificate of fuel and energy resources power consumer [in Russian]

3. V.V. Litvak *Zonal energy saving basis. (scientific, technical and industrial aspects)* (STT, Tomsk, 2002)
4. G.N. Klimova, V.V. Litvak, *Seven problems and seven keys of energy saving / Scientific Publication. Tomsk Polytechnic University* (Red banner, Tomsk, 2013)
5. A.S. Nekrasov, Yu.V. Sinyak *Power plant engineering control* (Energiya, Moscow, 1979)
6. Yu.N.Savenko, E.O. Steinhaus *Energy budget* (Energiya, Moscow, 1971)
7. V.M. Simchera, *Multivariate statistical analysis approaches* (Finances and statistics, Moscow, 2008)
8. G.N. Klimova, V.V. Litvak, Industrial power engineering **7**, 5 (2016)

Optimization of Temperature Distributions in Critical Cross-sections of Load-bearing Structures of Measurement Optical Systems of Autonomous Objects

Michael Yu. Livshits[1,*], *Boris B.* Borodulin[1], *Stepan E.* Korshikov[1]

[1]Samara State Technical University, Heat and Power Engineering Department, Russia

Abstract. Problem of the automatic thermogradient stabilization of the constructive elements sizes is discussed. The actual problem is the determination of the control algorithm controlled heat sources, providing them with the minimum possible number of minimal deviation from the desired temperature in a given cross-section design. The proposed procedure for solving optimal control of temperature distribution in the numerical design of its implementation, allows obtaining results that are suitable for approximate implementation in on-Board computers, Autonomous systems and significantly reduce thermal deformation component of the measurement error.

1 Problem statement

Reduction thermodeformation load-bearing structures, which are placed on heat-producing equipment, greatly reduces the accuracy of optical measurements because of thermal deformation of bearing structures violate the optical configuration of measuring devices [1]. In this case, the current task is the determination of the control algorithm controlled heat sources, providing them with the minimum possible number of minimum deviations from the desired temperature in a given cross section of the structure. The choice of section is determined by a complex solution of thermal deformation and optomechanical tasks for the corresponding layout of equipment on the design.

2 Numerical calculations

In [1] for the supporting structure in the form of a rectangular prism is justified for two nonsmooth boundary value problems of optimal control, which, taking into account the specifics of operation of optical equipment placed on the supporting structure, can be modified to a spatially one-dimensional nonsmooth boundary value problems for semi-

* Corresponding author: entcom@samgtu.ru

infinite optimization, for example on the line in a given section $l_y^* \in l_y$, $l_y^* = const$; $l_z^* \in l_z$, $l_z^* = const$; in relative (l_x, l_y, l_z) Cartesian coordinates

1. Maximum accuracy task - the task of achieving the minimum result of ($\varphi = \varphi_k$) the deviation $J_{\min} = \min\limits_{U^{(1)}(\varphi)} J[\Theta(l_x, l_y^*, l_z^*, \varphi_k)]$ of temperature $T(l_x, l_y^*, l_z^*, \varphi_k)$ from a specified base $T_{des} = const$ to the specified section $L(l_x, l_y^*, l_z^*)$

2. Maximum performance task – goals, which embed the minimum time $\varphi_k^{(\min)} = \min\limits_{U^{(2)}(\varphi)} \varphi_k$ resulting temperature distribution $T(l_x, l_y^*, l_z^*, \varphi_k^{(\min)})$ inside the specified region ε_{des}, $J \le \varepsilon_{des}$.

$J = \left\| \Theta(l_x, l_y^*, l_z^*, \varphi_k) \right\|_{L_\infty[l_x \in L]}$. $\varphi_k = \dfrac{\lambda \tau_k}{C x^2_{\max}}$ – the relative end time of the process, λ, C - thermal conductivity and specific heat of the material of construction.

The solution of these problems is reduced to the determination of the optimum law of change of vector control $U^{(\beta\beta)}(\varphi, l_x, l_y, l_z) = [q_x^{(\alpha\alpha)}(\varphi, l_y, l_z), q_y^{(\alpha\alpha)}(\varphi, l_x, l_z), q_z^{(\alpha\alpha)}(\varphi, l_x, l_y)]$, $\beta = 1,2$, providing $J^{opt}(q_{x(opt)}^{(\alpha\alpha)}, q_{y(opt)}^{(\alpha\alpha)}, q_{z(opt)}^{(\alpha\alpha)}) = \min\limits_{q_x^{(\alpha\alpha)}, q_y^{(\alpha\alpha)}, q_z^{(\alpha\alpha)}} J$ at the specified value $\varphi = \varphi_k$ – maximum accuracy task(β=1) or $\min\limits_{q_x^{(\alpha)}, q_y^{(\alpha)}, q_z^{(\alpha)}} \varphi_k$ at the specified value $J = J_{des}$ – maximum performance task (β=2) in terms of restrictions on the allowable temperature $\Theta_d = const$.

$$\Theta(\rho, \varphi) \le \Theta_d, \varphi \in (0, \infty), \rho \in \overline{\Omega}_p \tag{1}$$

and resource control:

$$\begin{aligned} &\left| q_x^{(\alpha\alpha)}(\varphi, l_y, l_z) \right| \le 1 \;; \alpha = 3.4; l_x = 0; l_x = 1; \\ &\left| q_y^{(\alpha\alpha)}(\varphi, l_x, l_z) \right| \le 1; \alpha = 5.6; l_y = 0; l_y = l_{y_{\max}}; \\ &\left| q_z^{(\alpha\alpha)}(\varphi, l_x, l_y) \right| \le 1\;; \alpha = 1.2; l_z = 0; l_z = l_{z_{\max}}; \end{aligned} \tag{2}$$

when exposed to a control object of the internal and external disturbances. In this example (Fig. 1) the temperature of the inner surface of the outer shell of the supporting structure has a constant time value T_{out_i}. The temperature of the outer shell affects the temperature of the faces of the plate by radiant heat transfer. From opposite sides of the supporting structure has a rectangular area, S1 and S2, which, through radiant heat transfer affects the temperature of the environment. On the lateral surfaces of construction installed heat-emitting devices, occasionally included in the work and controlled heat sources 1,2,3,4,5,6 attached to the surface areas S1 and S2 on opposite faces $\alpha = 3$ and $\alpha = 4$ of the supporting construction.

A valid control $\left| U^{(\beta\beta)}(\varphi, l_x, l_y, l_z) \right| \le 1$ in this example can be represented in a multiplicative form: $q_x^{(\alpha\alpha)}(\varphi, l_y, l_z) = U_x^{(\alpha\alpha)}(\varphi) V_x^{(\alpha\alpha)}(l_y, l_z)$, $\alpha = 3.4$; $\alpha = 1,2,5,6$, $q_x^{(\alpha\alpha)}(\varphi, l_x, l_y) = q^{(\alpha\alpha)}(\varphi, l_x, l_y) \equiv 0$, fixed in advance the spatial distribution $V_x^{(\alpha\alpha)}(l_y, l_z)$ of the controllable heat sources on two opposite faces $\alpha = 3,4$ with a specified coordinate location $V_x^{(\alpha\alpha)}$, $\alpha = 3.4$. Under these conditions, the optimal control of autonomous $\overline{U}^{(\beta\alpha)}(\varphi)$ groups of heat sources each α-th face:

$$\overline{U}^{\beta\alpha}(\varphi) = \left[U_x^{(\beta 3)}(\varphi), U_x^{(\beta 4)}(\varphi), U_y^{(\beta 5)}(\varphi), U_y^{(\beta 6)}(\varphi), U_z^{(\beta 1)}(\varphi), U_z^{(\beta 2)}(\varphi)\right] \qquad (3)$$

$$U_{opt}^{(1\alpha)}(\varphi) = \arg \min_{\overline{U}^{(1\alpha)}(\varphi)} J \; ; \; U_{opt}^{(2\alpha)}(\varphi) = \arg \min_{\overline{U}^{(2\alpha)}(\varphi)} \varphi_k \; ; \qquad (4)$$

in the problem of maximum precision (β=1) and task performance (β=2) using the method of moments [2,3] it is possible to parameterize and to reduce to the problem of finding a finite number of parameters $\Delta_\delta^{(i)}$, δ=1,2, *i; i*=1,2,..*J*; where $\Delta_\delta^{(i)}$ the duration δ–th interval of constancy of the total number i, where each vector $\overline{U}^{(\beta\alpha)}(\varphi)$ component alternately takes its maximum value at the level of the constraints (2).

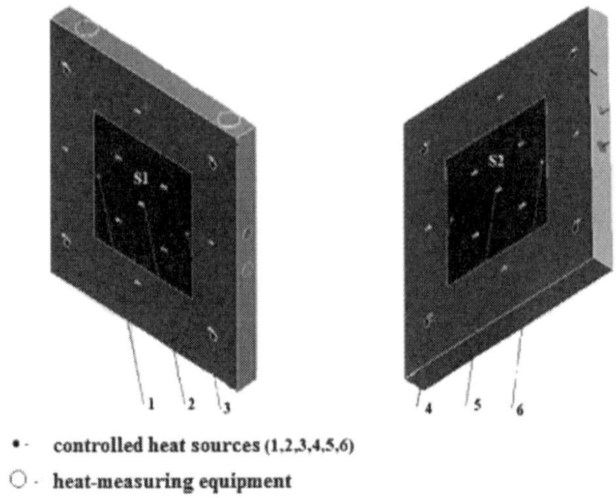

Fig. 1. The location of the heat sources on the supporting structure (example).

The duration $\Delta_\delta^{(i)}$ and the desired number of intervals *i*, is determined by the alternance method [3] from the system of transcendental equations:

$$\Theta(\rho_k, \varphi_k)\big|_{\rho_k \in L_k} = \pm\varepsilon^{(i)}, grad\Theta(\rho_k, \varphi_k)\big|_{\rho_k \in L_k} = 0 \qquad (5)$$

where $\Theta(\rho_k, \varphi_k)\big|_{\rho_k \in L_k}$ - numerical solution of boundary value problems in the final time φ_k points $\rho_k(l_x^{(k)}, l_y^*, l_z^*) \in L_k(l_x^{(k)}, l_y^*, l_z^*, \varphi_k) \subset L(l_x, l_y^*, l_z^*)$ of the considered line section $L(l_x, l_y^*, l_z^*)$, which constitute a countable set of limit points, which is the ratio (5).

The procedure of solving the transcendental system (5) is a specially designed software package [5] based on the numerical solution of boundary value problems in ANSYS. Figure 2 shows the resulting deviations from the set temperature to the responsible of the line-section $L(l_x, l_y^*, l_z^*)$ $l_y^* = 0.5$, $l_z^* = 0.5$ load-bearing structure for optimal accuracy (β=1) 1 and 2-interval control. The maximum on this line $L(l_x, l_y^*, l_z^*)$ the temperature deviation does not exceed 4° when exposed to the most severe variant of the heat load.

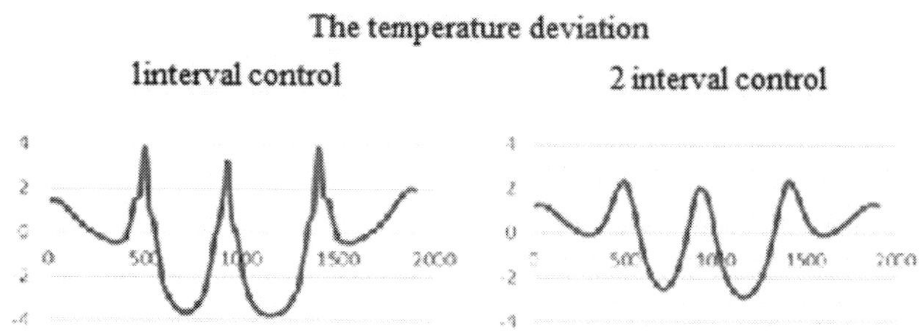

Fig. 2. Chart of deviation from specified temperature on the section line $L(l_x, l_y^*, l_z^*)$, $l_y^* = 0.5$, $l_z^* = 0.5$

3 Conclusions

The proposed procedure for solving optimal control of temperature distribution in the numerical design of its implementation, for example in the ANSYS software environment allows to obtain results that are suitable for approximate implementation in on-board computers, Autonomous systems and significantly reduce thermal deformation component of the measurement error.

References

1. M.Y. Livshits, M.Y. Derevjanov, S.A. Kopytin, Heat Mass Transf. **1**, 719 (2012)
2. A.G. Butkovskii, *Theory of optimal control of distributed parameter systems* (Nauka, Moscow, 1965)
3. E'.Ya, Rapoport, *Optimal'noe upravlenie sistemami s raspredelenny'mi parametrami* (Vy'sshaya shkola, Moscow, 2009)
4. B.B. Borodulin, M.Y. Livshits, EPJ Web Conf. **110** (2016)
5. A.P. Efimov, Conf. – Samara scientific centre of RAS 195 (2006)

DOI: 10.1051/matecconf/20179201054

The effectiveness of absorption heat pumps application for the increase of economic efficiency of CHP operation

Dmitriy S. Luzhkovoy[1*], *Olga* Yu. Romashova[1], *Aleksandr* A. Tubolev[1], *Yuriy* V. Kobenko[1], *Elena* S. Riabova[2]

[1]National Research Tomsk Polytechnic University, 634050 Tomsk, Russia
[2]Samara State Social-Pedagogical University, 443099 Samara, Russia

Abstract. The article deals with a comparative analysis of CHP operational efficiency in various working modes before and after the absorption heat pumps installation. The calculation was performed using a mathematical model of the extraction turbine PT- 80/100-130/13. Absorption heat pumps of LLC "OKB Teplosibmash" were used as AHP models for the analysis. The most effective way of absorption lithium-bromide heat pumps application as a part of the turbine PT-80/100-130/13 turned out to be their usage in a heat-producing mode during a non-heating season with a load of hot water supply. For this mode the dependence of the turbine heat efficiency on the heat load of the external consumer at a given throttle flow was analyzed.

1 Introduction

Efficiency improvement of energy supply application tends to be a worldwide problem as the demand for them, which is increasing every year, inevitably leads to the difficulties of economic, environmental and technical nature. For the Russian energy system, which is dominated by cogeneration, the problem of increasing the efficiency of existing combined heat and power stations is particularly important.

One of the most promising technologies for CHP implementation is to convert low-heat into high-grade heat using absorption heat pumps (AHP).

The positive effect of the heat pumps implementation in CHP operation depends on its working mode. When operating in the condensing mode AHP can be used for fuel-flow condensate heating, and in heat-extraction mode it can be used for heating the main condensate, delivery water, raw water used for the station needs [4, 5] after chemical cleaning (including heat supply makeup).

The effect of the fuel economy while using AHP at CHP becomes evident if at a given throttle flow the operation performed by the whole stem flow of the turbine is increased. It becomes possible in two cases:

* Corresponding author: dsl8@tpu.ru

- if underproduction by vapor system that is selected for the heat pump system will be less than the extra work that is gained as a result of steam consumption reduction for the regenerative, line-operated or other (for example, plantwide) heat exchanges.
- if by means of cooling circulating water it becomes possible to reduce the pressure in the condenser and thus, increase the available heat drop.

The first case corresponds to the turbine operation when the steam of a higher capacity than is required by heat-exchange conditions is supplied to one of the heat exchangers. Most of industrial-heating CHP have plant heat exchangers (the heaters of chemically demineralized, chemically purified, raw water, atmospheric deaerators) that get the steam from the plant collector which is powered through PRDS from the industrial turbine extraction.

The aim of the research is to carry out a comparative analysis of the calculation data of absorption heat pumps inclusion into the scheme of the extraction turbine of PT-type in the context of its operation modes.

2 Features of the analyzed system

As a model, which was used for analyzing the efficiency of heat pumps integration, turbine PT-80/100-130/13 was taken. Its basic diagram is presented in [1]. Initial data for turbine operation modes were taken from [2].The models of absorption heat pumps were AHP of the company of LLC "OKB Teplosibmash" [3].

For the analysis the characteristic turbine operation modes with controlled heating and industrial extraction were chosen: heat-producing mode from a heat graph with a rated capacity load of regulated extraction, condensing mode, heat-producing mode with a plant heat load of DHW that is provided by P-extraction.

In the condensing mode the full-flow condensate before the first LPS was heated by AHP, in the heating mode the same was done with the raw water after chemical cleaning behind the plant heater. Circulating water was cooled in all cases.

Figures 1 and 2 show the calculation model of chemically purified water preparation for the heat supply makeup before and after the introduction of heat pumps.

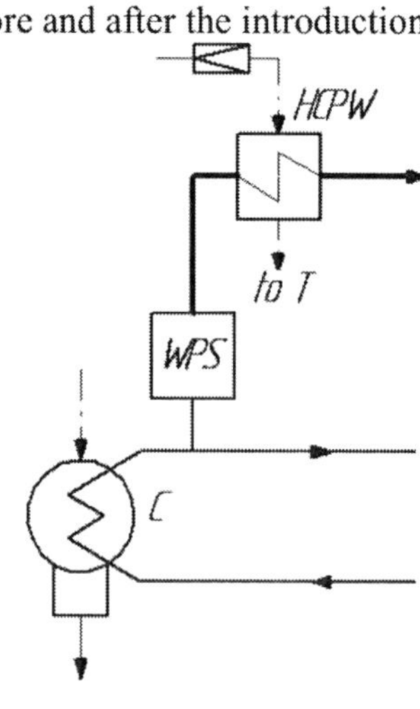

Fig. 1. The model of makeup water preparation before the introduction of heat pumps where C – condenser; WPS – water purification section; HCPW – the heater of chemically purified water; T – turbine

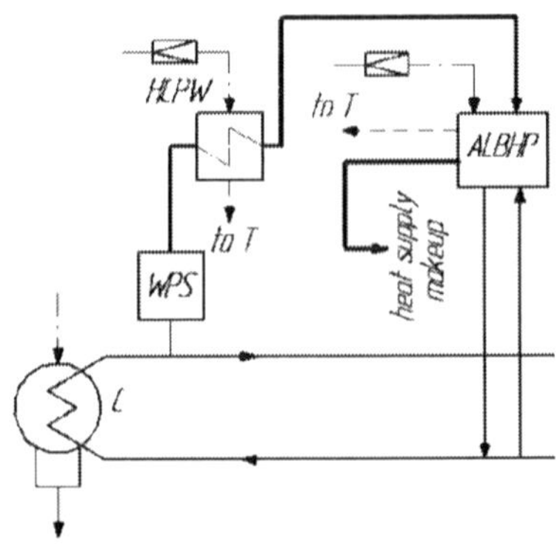

Fig. 2. The model of makeup water preparation after the introduction of heat pumps where ALBHP – absorption lithium-bromide heat pumps; the remaining symbols are the same as in Fig.1

3 Calculation results

Tab. 1 and Fig. 1 and 2 show the calculation data of the turbine in the heat extraction mode with the heat release for the heat supply makeup during the summer period at CHP. The main steam rate is set to (G_0 = 80 kg/sec).

Table 1. The comparison of calculation results of the turbine PT-80/100-130/13 in the heat extraction mode at various heat supply makeup values.

Index	makeup 125 kg/sec		makeup 250 kg/sec		makeup 500 kg/sec	
	without ALBHP	with ALBHP	without ALBHP	with ALBHP	without ALBHP	with ALBHP
Steam consumption for the turbine, kg/sec	80					
Generated electrical power, MW	70.916	72.423	66.752	68.491	53.29	59.794
Steam consumption in the cond., kg/sec	48.61	50.54	40.65	44.76	22.43	30.08
Pressure in the cond., kPa	9.2	9.1	8.4	8.3	6.7	6.5
Temperature of circ. water at the condenser inlet, °C	20	19.3	20	18.9	20	17.7
Temperature of circ. water at the condenser inlet, °C	31.1	30.8	29.3	29.1	25.1	24.5
Circ. water consumption, kg/sec	2222					
Steam consumption from P-extraction at HCPW, kg/sec	7.6	1.72	15.91	4.06	35.1	11.57
Total steam consumption from P-extraction at HCPW and ALBHP, kg/sec	7.6	5.92	15.91	12.31	35.1	28.07
Heat transferred to makeup water, MW	26.19	26.19	52.4	52.4	104.7	104.7
Temperature of heat supply makeup, °C	70					
Heat taken away from the cold supply, MW	-	5.72	-	11.79	-	23.57

Electr. efficiency factor of the gross bucket	0.3956	0.404	0.4296	0.4505	0.5292	0.5936
Specific reference fuel consumption at el. energy efficiency b, gr.ref.fuel / kW*h	388.1	380.1	357.5	340.8	290.2	258.7

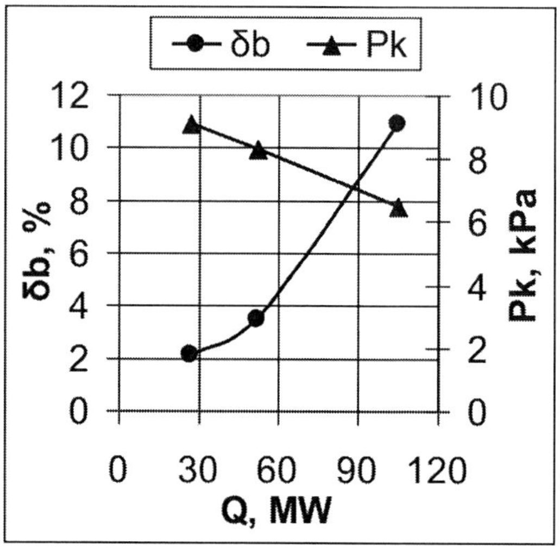

Fig. 3. Dependence of the relative change in the fuel economy on the turbine PT-80/100-130/13 and the pressure in the condenser on the heat load of the external consumer (G_0 = 80 kg/sec)

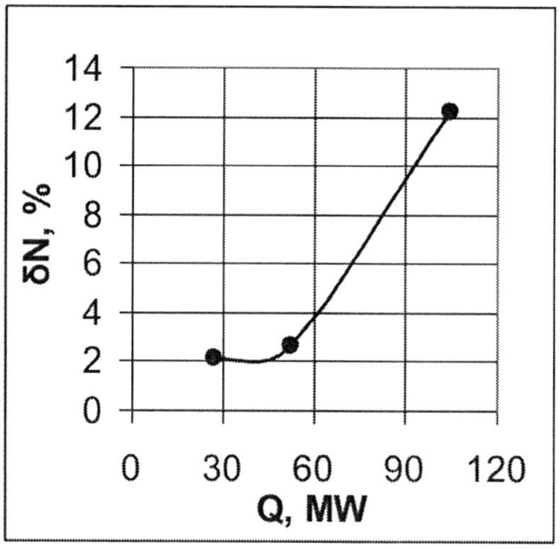

Fig. 4. Dependence of the relative gain of the electric power generated by the turbine PT- 80/100-130/13 on the heat load of the external consumer (G_0 = 80 kg/sec)

4 Conclusions

1. The application of absorption heat pumps in the condensing mode does not provide an appreciable effect of fuel economy because a positive result from the pressure reduction in the condenser and, consequently, an increasing turbine heat drop are balanced by a negative result from the steam replacement that is necessary for regeneration. However, the application of AHP reduces the pressure in the condenser by cooling the circulating water at the condenser inlet in such a case when it is

DOI: 10.1051/matecconf/20179201054

necessary to limit the steam consumption in the turbine to maintain the required vacuum in the condenser (for example, summer limitation of electricity production).

2. The application of AHP in the heating mode on the thermal graphics with rated capacity load of regulated extractions does not lead to the reduction of the specific fuel consumption. In this case, it is not possible to use the pressure reduction effect in the condenser with the help of AHP, as during the cold season the cooling water temperature is low, and there are not any problems with the creation of vacuum in the condenser. Thus, it does not make economic sense to reduce the pressure using AHP.
3. The most effective method of using AHP as part of the turbine PT-80/100-130/13 is its application in the heating mode during a non-heating season with a load of hot-water supply. In this case it is possible to get a positive effect both from the condenser pressure decrease and the steam flow reduction to heat the makeup water.
4. The influence of external heat load Q_m on the efficiency of heat pumps application was analyzed. The increase of Q_m from 26 MW to 105 MW with the steam flow on the turbine G_0= 80 kg/sec results in the reduction of specific reference fuel consumption from 2.1% to 10.9% (fig. 3).
5. The introduction of the absorption heat pumps into the heat-extraction turbine turns out to be an effective means of improving the efficiency of CHP. However, in order to get the maximum benefit from it, it is important to consider the operation modes of the turbine that determine the efficiency of the heat pump system.

References

1. E.A. Bojko, K.V. Bazhenov, P.A. Grachev, *Thermal power stations (steam turbine power stations of CHP)*: (IPC KGTU, Krasnojarsk, 2006)
2. S.V. Tsanev, I.N. Tambieva *Thermal diagram and cogeneration values of CHP and APS*: (MEI, 1987)
3. The official website of LLC “OKB Teplosibmash” [Electronic resource]
4. B.V. Lukutin, E.B. Shandarova, A.F. Makarova, D.L. Matukhin, I.L. Fuks, IOP Conf. Ser. Mater. Sci. Eng. **124**, 012070 (2016)
5. B.V. Lukutin, E.B. Shandarova, I.O. Muravlev, D.L. Matukhin, A.F. Makarova, *International Conference on Mechanical Engineering, Automation and Control Systems (MEACS)*, 1 (2015)

The errors when determining the thermophysical characteristics of liquids by the laser flash method

Mark D. Katz[1,*], and *Ilija* M. Katz[1]

[1]National Research Tomsk Polytechnic University, 634050 Tomsk, Russia

Abstract. The analysis of errors in determining the thermophysical characteristics of typical organic liquids, specified by conductive-radiative heat transfer of ultrathin heated up to high temperature sample layer, under conditions consistent with the implementation of the laser pulse method, when exposed to the surface of the collimated laser pulse of finite duration was made. The influence to the error of the radiation absorption process was discovered.

1 Introduction

Organic liquids are increasingly used in various fields of science and technology. It used as solvents, coolant-cutting fluids and high temperature fluids [1]. Every year, new organic liquids are developing for special purpose of machinery and technology.

A thermal process in the technical fluids and its thermal characteristics plays an important part in reliability control and normal operation of technical systems.

For these reasons, determination of the thermal characteristics with high accuracy becomes more actual task, especially in the development of new products and technical systems.

To investigate the thermal properties of such materials, it is best to use highly efficient procedures, for example, the laser pulse – laser flash (LFA) method [2, 3].

It is based on the absorption of a pulse of radiant energy in a thin layer by the frontal "hot" surface of the sample and the approximate solution of the one-dimensional heat conduction problem for an infinite plate.

From the time-temperature dependence of the reverse ("cold") surface of the sample, one can determine the maximum heating temperature T_{max} and the time τ_{05} for half the maximum heating temperature of the cold surface of the sample to be attained.

The heat capacity c , thermal diffusivity a , thermal conductivity λ of the material being investigated when using LFA method, is calculated as:

* Corresponding author: riki@tpu.ru

$$c = Q / (T_{max} \cdot \rho \cdot L);$$
$$a = 0.1388 \cdot L^2 / \tau_{0.5};, \qquad (1)$$
$$\lambda = a \cdot c \cdot \rho.$$

where Q – is the energy, absorbed by the sample, J/m^2 ; L is the thickness of the sample, m; ρ is the material density, kg/m^3; τ_{05} is the time for half the maximum heating temperature of the cold surface of the sample to be attained, s.

The LFA method is applying for determination of the thermal characteristics of organic liquids and its compositions [4–6]. For example, the laser flash analyzers Netzch are using for determination of liquids thermal diffusivity in a wide range of the thermal characteristics and measuring temperatures.

There is a practical interest for determination of the partially-transparent liquids thermal characteristics. The heating of these liquids is attended with heat energy accumulation. The sources of this heat energy are thermal diffusivity and bulk absorption in certain spectral range, which depend from the heat emission wavelength.

The condition of LFA method applying for determination of the thermal characteristics of partially –transparent solid materials are discussed in [7–8]. For solving of this task the one-dimensional Thermal Conductivitymethod was used. Also, it was discovered that systematic error of applying LFA method for such materials is increasing. The reason for this is that radiative heat transfer is increasing heating process more than the heating process of opaque materials [9].

The solved problem of determination of the thermal characteristics for solid materials[2] is complicated for liquids, because it's necessary to form thin layer of it. Front surface of this sample is heating, and temperature of the back surface is measuring.

The other problem of applying LFA for liquids thermal characteristics determination is providing following conditions: small duration and high heating intensity without boiling of heating surface.

The purpose of this work is analysis of the conditions determining the thermal characteristics of typical organic liquid (Ethanol) by a laser pulse method [2–3], using conductive-radiative heat transfer mathematical model.

2 Statement of the problem

In case of a problem definition it was supposed that the color of the sample is gray. Sample self thermal radiation is far less then external radiative flux. This supposition is not apply restrictions to the problem statement and to the real conditions of the heat-transfer processes of translucent material.

Also, it was supposed, that thermal characteristics of the materials of the measuring cell and organic liquid (Ethanol) doesn't depend on temperature. This assumption is justified for many materials in real-rated range of possible changes in temperature during the experiments even at high values of the heat flux to the on-heated surface of the sample due to the short pulse duration.

In case of a problem definition it was supposed that the transverse size of the measuring cell is many times more liquid layer thickness cannot be taken into account a possible two-dimensional heat-transfer process and heating due to the heat on the side walls of the cylindrical body. The problem is solved in a Cartesian coordinate system. Solutions represent a three layer plate total thickness L3 (fig. 1).

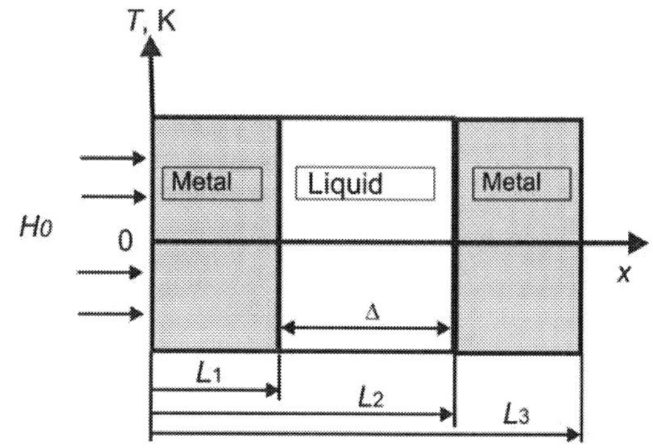

Fig. 1. Scheme of the area of solution.

Surface segment was heated by the collimated heat flux. The heat flux direction is vertical to heating surface. The radiant flux density is H_0 pulse duration is τ_{imp}.

3 Mathematical model and solution method

The mathematical model consists of the three nonstationary heat equation

$$c_1\rho_1 \frac{\partial T_1}{\partial t} = \lambda_1 \left(\frac{\partial^2 T_1}{\partial x^2} \right), \quad 0 < x < L_1 \tag{2}$$

$$c_2\rho_2 \frac{\partial T_2}{\partial t} = \lambda_2 \left(\frac{\partial^2 T_2}{\partial x^2} \right), \quad L_1 < x < L_2 \tag{3}$$

$$c_3\rho_3 \frac{\partial T_3}{\partial t} = \lambda_3 \left(\frac{\partial^2 T_3}{\partial x^2} \right), \quad L_2 < x < L_3 \tag{4}$$

with boundary conditions:

$$x = 0: \quad -\lambda_1 \frac{\partial T_1}{\partial x} = q, \quad \tau_{imp} > t > 0; \tag{5}$$

$$x = 0: \quad -\lambda_1 \frac{\partial T_1}{\partial x} = 0, \quad t > \tau_{imp}; \tag{6}$$

$$x = L_1: \quad T_1 = T_2; \quad -\lambda_1 \left(\frac{\partial T_1}{\partial x} \right) = -\lambda_2 \left(\frac{\partial T_2}{\partial x} \right) \tag{7}$$

$$x = L_2: \quad T_2 = T_3; \quad -\lambda_2 \left(\frac{\partial T_2}{\partial x} \right) = -\lambda_3 \left(\frac{\partial T_3}{\partial x} \right) \tag{8}$$

$$x = L_3: \quad \lambda_3 \frac{\partial T_3}{\partial x} = 0, \quad t > 0; \tag{9}$$

$$t = 0: \quad T_1 = T_2 = T_3 = T_0, \tag{10}$$

where T and T_0 are the actual and initial temperature. Subscripts 1, 3 are referred relate to the material of the measuring cell and 2 the liquid, K; t is the time; t is time, s; τ_{pulse} is laser pulse duration, s.

The radiant flux density of the material was calculated by Bouguer–Lambert–Beer Law[10]:

$$H(x) = H_0 \cdot \exp(-k_\lambda \cdot x), \tag{11}$$

where H(x) – the radiant flux density of the material, J/m^2; kλ - coefficient of power adsorption of collimated laser flux, m^{-1}.

The boundary value problem (2)–(11) was solved by the method of finite differences [11].

Problem (2)–(11) belongs to the category of singularly perturbed problems [12].

This problem is more complicated those problems in [7-8]. The main difficulty of this task is the necessity of solving a system of three heat equations with a significant (hundreds of times) difference of materials heat diffusion of each layer.

Therefore, the numerical solution of heat diffusion problem for three-layer infinite plate with significantly different coefficients of thermal conductivity in a high-intensity pulsed heating is independent problem. Its solution is largely due to the choice of grid parameters. The procedure of the grid parameters choosing is the same as for thermal diffusivity phase [13] and chemical conversions [14].

However, the procedures [13, 14] are used for values of the heat flux less then 10^5 W/m2 and large heating time.

Described above thermal diffusivity was solved for more complicated [13-14] conditions of thermal action. Therefore, the parameters of the grid to solve the problem (1) - (11) were chosen such that further their solution had no effect on results of numerical modeling of temperature pattern.

The values of heat transfer rate and pulse duration taken in accordance with the actual possible characteristics of the laser source of identification of thermal properties of materials research apparatus.

For further simulation following parameters of laser heating source were used: laser pulse energy is less then 25 J/pulse; pulse duration is τimp=0.3…1 ms.

4 Results and discussion

The errors in determining the thermal characteristics of material can be defined [7–9] by comparison of the actual values of thermal diffusivity (a) and heat capacity (c) with the material obtained by numerical simulation of the implementation conditions of experiments on determination of thermal characteristics materials by this method. The errors of liquids thermal conductivity are determined by the method specified in [2–3].

Verification of mathematical models and numerical simulation of the reliability of the results of their study included a comparison with known experimental results as well as verification of conservativeness of difference scheme used. The numerical simulation of thermal diffusivity of the organic liquid Ethanol (thermal characteristics: $\lambda = 0.172$ W/(m·K), $c = 1909$ J/(kg·K), $\rho = 812.1$ кg/m^3) [16] was made in this paper. It was supposed, that the measuring cell was made from iron with wall thickness 0.5 mm ($\lambda = 20$ W/(m·K), $c = 551$ J/(kg·K), $\rho = 7894$ kg/m^3) [17], and flux density is $H_0 = 2.4 \cdot 10^7$ W/m^2, pulse duration 1 ms, T_0 = 273 K. Grid time steps was not above

$1 \cdot 10^{-6}$ s. The initial step in the coordinate $x = 5 \cdot 10^{-7}$ m, and in the areas near the borders of the materials chosen irregular (the algorithm of geometric series) [18 $x_{i+1} = zx_i$, where z – coefficient of variation. At the boundary surface the step of simulation was decreased, and after crossing boundary surface it was increased. The coefficient of variation has been assumed constant and equal 0.02.The wavelength characteristic of Ethanol has four absorption bands in frequency bands 850–870 cm^{-1}, 940–1050 cm^{-1}, 2010–2900 cm^{-1}, 3070-3300 cm^{-1}, where absorption coefficient changes from 5 to 120 m^{-1}.

The bounds range of pulsed heat flux to the heating surface was discovered in [16]. These bounds is conform to conditions of significant determination of thermal characteristics of typical liquids: water, methylbenzene, Ethanol without changing of its physical state during conductive heating of the sample. The maximum value of laser pulse energy density is 18 J/pulse.

At the fig. 2 is shown the errors calculation of heat capacity and thermal diffusivity of Etalon distribution depending on the absorption coefficient k_λ .

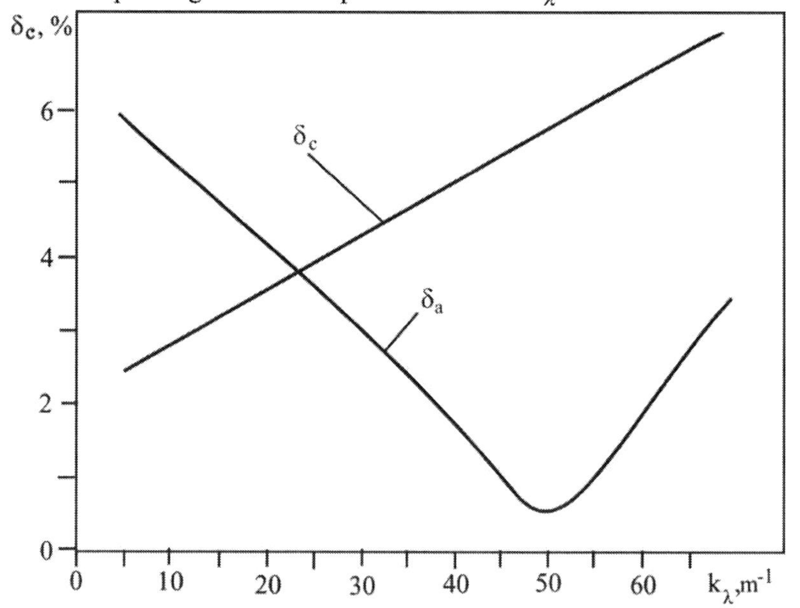

Fig. 2. The errors calculation of heat capacity and thermal diffusivity of Ethanol distribution depending on the absorption coefficient.

As can be seen from fig.2, the values δ_c and δ_a greatly depend of k_λ . With increasing values increase (Fig. 2) and reach 7%, when $k_\lambda = 70$ m^{-1}. The values change monotonically with increasing and reaches a minimum value of about 0.4% at $k\lambda = 50$ m^{-1}.

Behavior of changing can be explained by the nature of radiative heat transfer in conditions of low conductivity of the sample liquid.

Accumulated during the pulse duration energy of ultrathin heated up to high temperature sample layer energy is redistributed by thermal conductivity and radiation on the x coordinate direction. The heat conduction by this direction is used for calculation of liquid specific heat capacitance and thermal diffusivity by the procedure from [2].

For small values of k_λ the heat energy caused by radiative heat transfer is consumed on increasing liquid temperature. When k_λ you increase the radiation component of the heat flow in the area, close to the near-surface layer of the sample is small, and the heat transfer is carried out largely by conduction.

4 Conclusion

Analysis of the results (Figure 2) allows you to draw a conclusion about the existence of the conditions of the experiment providing the minimum methodological error in determining thermal characteristic partially -transparent liquids at other appropriate conditions.

References

1. A. Weisberger, A. Proskauer, J. Riddick, *The organic solvents. Physical properties and cleaning methods* (Foreign literature, Moscow, 1958)
2. W.J. Parker, R.J. Jenkins, C.P Butler, J. Appl. Phys. **32**, 1679 (1961)
3. O. Altun, B. Erhan, A. Kalemtas, J. of Achievements in Materials and Manufacturing Engineering **30**, 115 (2008)
4. Y.Tada, M. Harada, M. Tarigani, W. Equchi, J. Rev. Sci. Instrum. **49**, 1305 (1978)
5. N.S. Srinivasan., X.G. Xiao, S. Seetharaman, J. Appl. Phys. **75**, 2325 (1994)
6. J. Blumm., A. Lindemann, S.Min, Thermochim. Acta. **455**, 26 (2007)
7. G.V. Kuznetsov G., M.D. Katz, Meas. tehn. **55**, 1057 (2012)
8. G.V. Kuznetsov, M.D. Katz, Meas. tehn. **53**, 674 (2010)
9. G.V. Kuznetsov, M.D. Katz, Meas. tehn. **52**, 384 (2009)
10. B. Yavorskii *Handbook of physics* (Nauka, Moscow, 1979)
11. A.A. Samarskiy *Theory of difference schemes* (Nauka, Moscow, 1983)
12. K. Chang, F. House, *Nonlinear singularly perturbed boundary value problems* (Mir, Moscow, 1988)
13. G.V. Kuznetsov , P.A. Stizhak, Russ. J. Phys. Chem. **29**, 60 (2010)
14. O.V. Vysokomornaya, G.V.Kuznetsov, P.A. Stizhak, Russ. J. Phys. Chem. B **5**, 4 (2011)
15. N.B.Vargaftik V.P.Filippov, A.A.Tarizmanov, E.E. Totskiy, *Handbook of heat-conductor fluids and gases* (Energoatomisdat, Moscow, 1990)
16. M.D. Kats, I.M. Kats, MATEC Web Conf. **72**, 01041 (2016)
17. V.V. Chirkin, *Thermal properties of materials of nuclear engineering: Manual* (Atomizdat, Moscow, 1968)
18. A.A. Samarskiy, Y.I. Popov, *Difference methods for solving problems of gas dynamics* (Nauka, Moscow, 1992)

An experimental study of flow over flat and axisymmetric bodies

Evgeny Maslov[1, 2,*], *Valery* Faraponov[2], *Nikolay* Zolotorev[2], *Andrey* Chupashev[2], *Vladislav* Matskevich[2], and *Sergey* Chizhov[1]

[1]National Research Tomsk Polytechnic University, Tomsk, Russia
[2]National Research Tomsk State University, Tomsk, Russia

Abstract. The technique and results of an experimental study of the flow structure and critical parameters of the flow over flat and axisymmetric bodies in a range of the Mach numbers (M = 2 ÷ 5) are presented.

1 Introduction

Nowadays, there has been a tendency to increasing in speeds of aircraft to hypersonic ones. Hence, a problem of development of body frame geometries, optimized for large Mach numbers of incoming flow is becoming urgent. The technique and results of experimental studies of a supersonic air flow over flat (a wedge) and axisymmetric (a cone) models are given in this paper.

2 A technique of conducting experimental studies

Experimental studies were conducted in the aerodynamic unit [1]. The main function of the aerodynamic unit is creating a short-term supersonic flow of gas for carrying out aerodynamic and aerophysical testing. The range of its realized modes of behavior is the following: the Mach numbers M = (2 ÷ 7), the pressure in the stagnation chamber P_f = (1.5 ÷ 8) MPa, the braking pressure at the nozzle section P_0= (0.15 ÷ 0.3) MPa, the static pressure P = (0.03 ÷ 0.07) MPa, the braking temperature of the running air flow T_0 = (17.5 ÷ 250) °C. The operating time of the aerodynamic unit reaches (1.0 ÷ 3.0) s. For the operating time of t_k = 3 s, the values of parameters of the realized modes remain constant during $t_c \approx 2$ s. While testing the models in the aerodynamic unit, there is an opportunity to vary the temperature up to ~ 400 K and the wind velocity in the range of M =(2 ÷ 7). Preliminary heating of the working gas is realized by utilizing a pebble heater. To create a supersonic flow, the axisymmetric bell-shaped nozzles were used. For obtaining the maximum volume of experimental data on critical parameters and the structure of the air flow in the course of testing the models the aerodynamic unit is supplied with a pressure recording system, a visualization system and a three-component aerodynamic (tensometric) balance.

* Corresponding author: maslov_eugene@mail.ru

DOI: 10.1051/matecconf/20179201056

The definition of the pressure fields is based on application of a pneumometric method which consists in measuring the total, static flow pressures and the pressure in the stagnation chamber. In the wind-tunnel experimental chamber for measurement of the total and static flow pressure the Pitot-Prandtl tube was installed. Registration of measurement results was carried out by means of TDM2-A strain gage sensors. The sensor allows to measure pressure up to 0.5 MPa. Strain gage sensors are grouped in blocks. Each block contains five receivers. In the stagnation chamber, the pressure was measured by means of the sensor DM 5007A – DI U2. The sensor allows to measure pressure up to 10 MPa.

In carrying out the tests in the aerodynamic unit, pressure sensors are exposed to short-term pulse impacts. This leads to mechanical deformation of the sensors and, therefore, can lead to inadmissible error measurements. Because of this, there is a need for graduation of the sensors before each series of experiments. To ensure necessary accuracy of measurements, the calibration stand was developed and built up. The scheme (a) and external view (b) of the calibration stand are provided in Fig. 1.

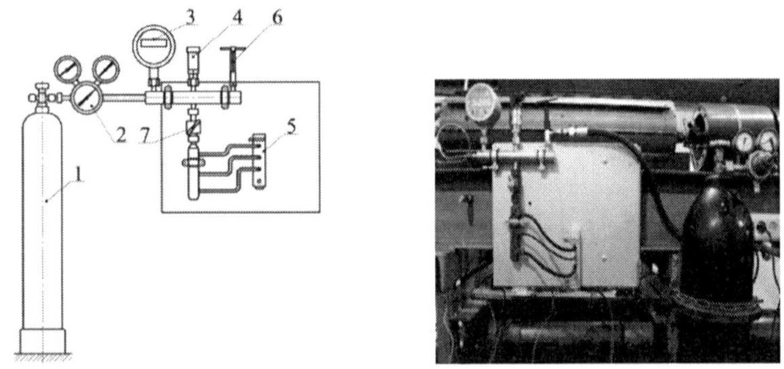

a) b)

Fig. 1. A scheme (a) and a photo (b) of the calibration stand for pressure sensors: 1 – a cylinder with compressed air, 2 – a reducer, 3 – the digital manometer, 4 – the pressure sensor DM 5007 A – DI U2, 5 – the block of strain gage sensors; 6 – the pressure release valve; 7 – the regulating valve.

A distinctive feature of the developed stand is the possibility to graduate pressure sensors of both types used in the aerodynamic tests. Technically, it is implemented by means of the regulating valve 7. The graduation of the TDM2-A pressure sensor (Fig. 2, a) and the pressure sensor DM 5007 A – DI U2 are given as an example in Fig. 2 (Fig. 2, b).

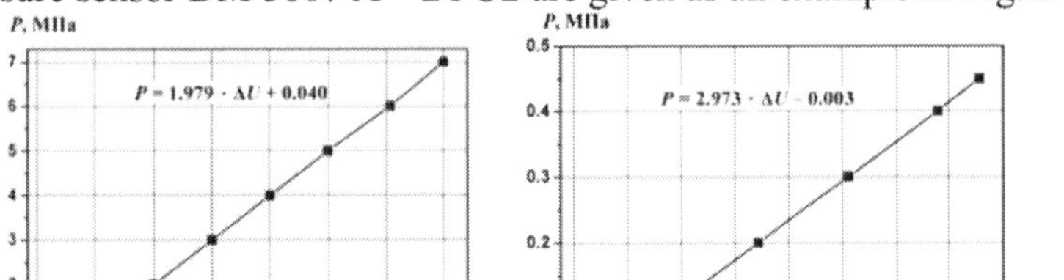

a) b)

Fig. 2. Calibration dependences for TDM2-A (a) and DM 5007 A – DI U2 (b) sensors.

The average relative error of graduation for the TDM2-A pressure sensor amounts to 0.2%, for the pressure sensor DM 5007 A – DI U2 – 0.15%. The visualization system is intended to provide high-quality observations of the airflow over a model. Visualization of the processes which accompany a gas flow in a wind-tunnel experimental chamber is based

 DOI: 10.1051/mateccconf/20179201056

on application of the schlieren method in a parallel beam. The complex of devices realizing visualization consists of a shadow device and a video recorder. The shadow device includes a pointlight, two transmitters and receivers installed coaxially relative to the wind-tunnel experimental chamber and a Foucault knife. Telescopes of Tal-100SR are used as the transmitters and receivers, the high-speed video camera "Phantom 711" is used as the video recorder.

In the experiments to determine the drag coefficient of the bodies at a supersonic flow-past the three-component aerodynamic (tensometric) balance was used. The balance allows measuring the drag force, the lift force and the disturbing moment.

3 An experimental study of supersonic flow over symmetric models

On the basis of the presented technique an experimental study of critical parameters and the flow structure for symmetric models was conducted. The models are made in the form of a cone (axial symmetry) and a wedge (plane symmetry) [2] with diversion channels for pressure measurement upon surfaces. The corner of semi-solution made is 15 °. Drainage slots of diversion channels were carried out in the characteristic fixed points on a surface in the vertical plane. The schemes of arrangement of drainage slots and the photo of the models are provided in Fig. 3.

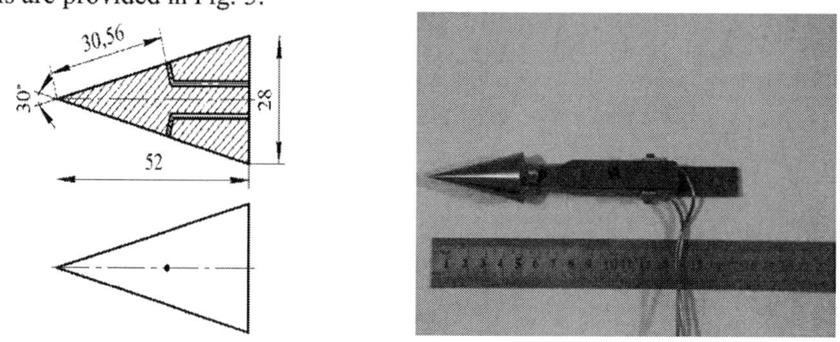

a)

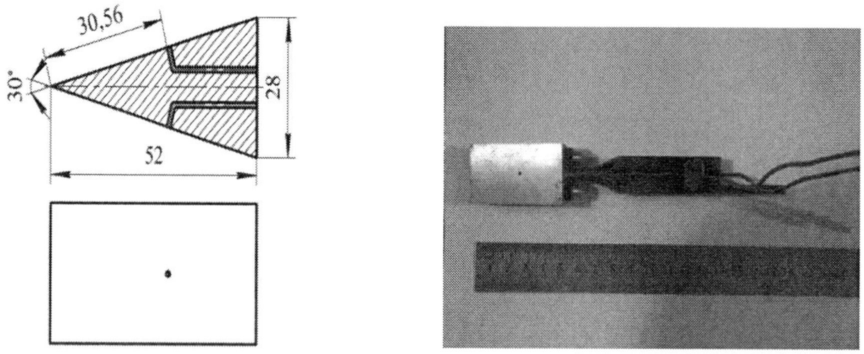

b)

Fig. 3. Schemes and photos of the models in the form of a cone (a) and a wedge (b).

To measure pressure in the experiments (the drainage tests), the models were fixed by means of a holder on the support allowing to vary the pitch angle.

To determine the drag coefficient (the weight tests), the models were installed on the tensometric balance and then placed into the wind-tunnel experimental chamber (Fig. 4).

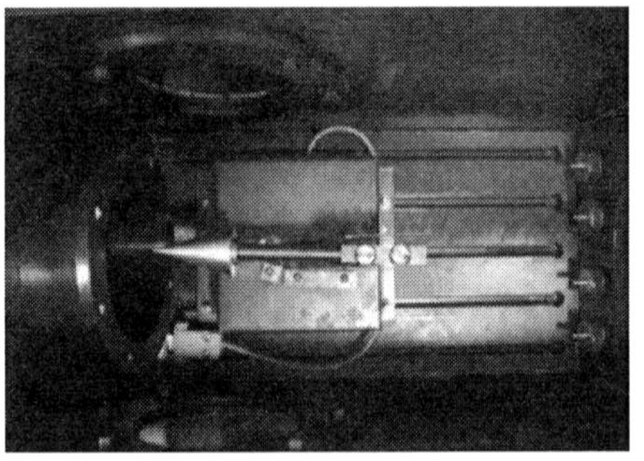

Fig. 4. The model in the form of a cone installed on the tensometric balance in the wind-tunnel experimental chamber.

4 Analysis of the results of the experimental studies

The amplitude-frequency characteristics in the antechamber and in two drainage slot, and visualisation frames of the air flow structure around a cone at M = 3 are presented in Fig.5.

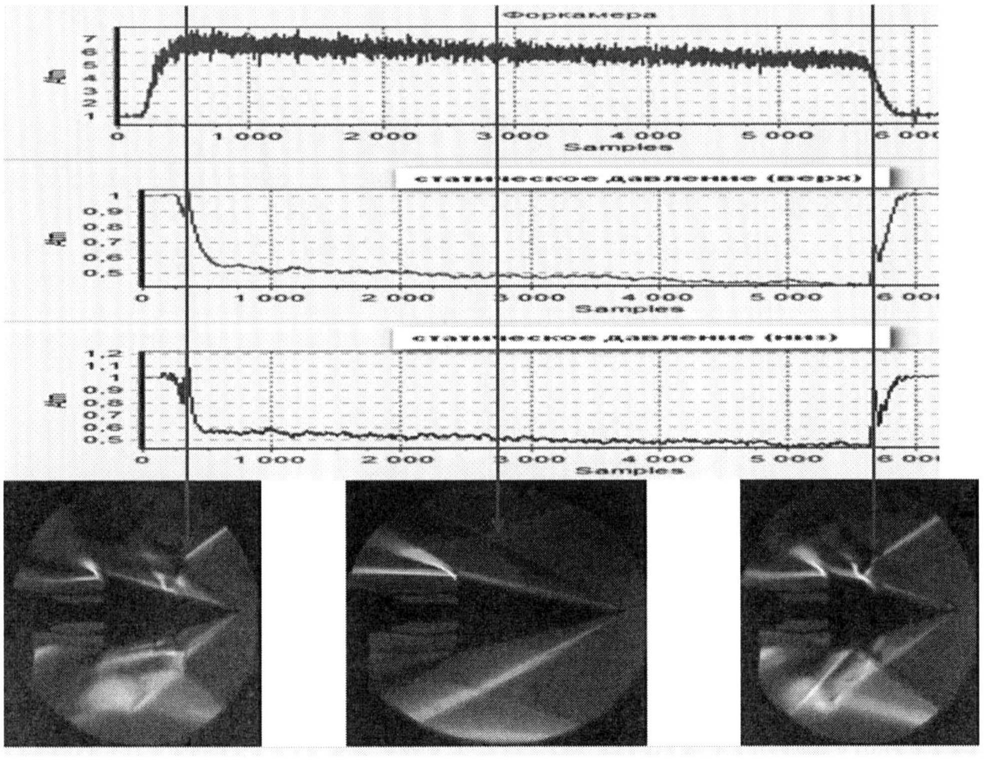

a) b) c)

Fig. 5. Amplitude-frequency characteristics in the antechamber and in two drainage slots and visualisation frames of the air flow structure around a cone, M = 3.

These frames of visualisation describe three significant moments of flow around the model by a supersonic airflow.

The first significant moment is the onset of forming the Mach rhombus (Fig. 5, a). At the same time the shock wave passes through the drainage slots. At this moment there is a pressure jump which is fixed on the oscillograms.

The second image (Fig.5, b) features the stationary flow – the pressure in the antechamber and on the surface of the model is almost constant. At the same time the formed Mach rhombus is well visible on the video footage.

The third picture (Fig. 5, c) shows the moment of "collapse" of the Mach rhombus. At this time the pressure in the antechamber sharply falls, the shock wave again passes through the drainage slots. The signals are registered on the oscillograms as peaks.

The results of the drainage tests of the flow around a wedge and a cone in the range of the Mach numbers M = (2 ÷ 5) are shown in Fig.6, the pitch angle is equal to zero [2].

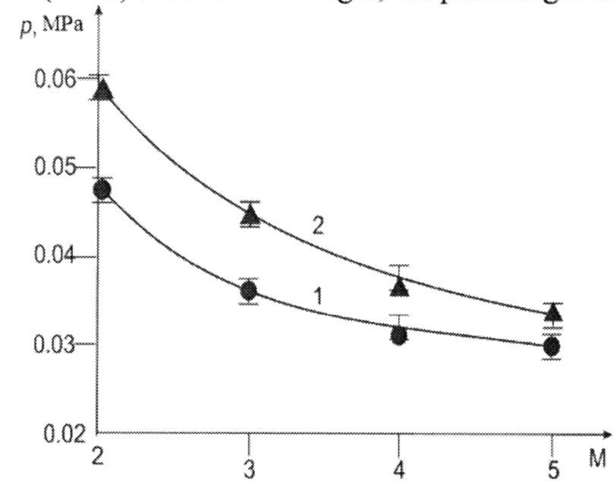

Fig. 6. Static pressure upon surfaces of the models: 1 – a cone; 2 – a wedge.

Analysis of the results obtained showed the difference in values of the static pressure upon the wedge and cone surfaces. The noted effect is explained by the fact that the flow over a cone in the considered speed range has a spatial nature. It promotes smooth change in direction of the gas flow in comparison with the flow over a wedge [2, 3]. At the same time it should be noted that the difference in values of the static pressure the flow over flat and axisymmetric models decreases with increasing the Mach number of the incoming flow.

The drag force and pressure in the antechamber were measured during the weight tests. The aerodynamic coefficient of the drag force C_x was defined by the results of the measurement. Fig. 7 shows the comparison between the obtained values of C_x for the cone and the data from [4, 5].

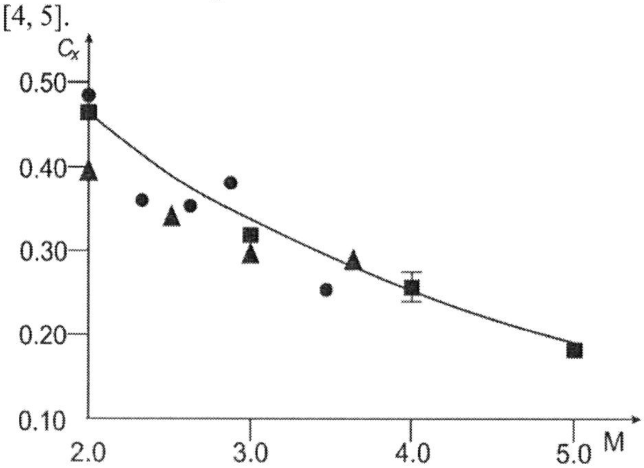

Fig. 7. Coefficient of the drag force for a cone: ■ – results of the experiments; ● – data [4]; ▲ – data [5].

Comparison of the results obtained with the findings of other authors showed their satisfactory agreement. In the range of the Mach numbers M = (2 ÷ 4) the maximum

DOI: 10.1051/matecconf/20179201056

difference between the data of the authors [4] is ~ 8%, between the data of the authors [5] is ~ 15%.

5 Conclusion

An experimental study of supersonic flow over models in the wide range of the Mach numbers has been conducted. The comparison of experimental results with the data of other authors is given and their satisfactory agreement is shown. The analysis of the experimental study results showed that in mathematical modeling of gas-dynamic processes in propulsion systems in the range of M = (2 ÷ 5) it is necessary to consider the spatial nature of the flow over bodies.

Acknowledgements

This research carried out in 2015 was supported by «The Tomsk State University Academic D.I. Mendeleev Fund Program», grant № 8.2.46.2015.

References

1. V.I. Zvegintsev, *Gasdynamic installations of short-term action* (Parallel, Novosibirsk, 2014)
2. E.A. Maslov, V.V. Faraponov, A.A. Chupashev, V.V. Matskevich, MATEC Web Conf. **72**,7201065 (2016)
3. N.F. Krasnov, V.N. Kosheva, A.N. Danilov, *Aerodynamics in questions and tasks* 759 (1985).
4. V.G. Artonkin, P.G. Leutin, K.P. Petrov, Tr. TsAGI. **92** 1413 (1972).
5. Yu.V. Sheludko, *Physical and gas-dynamic ballistic studies* (Science, Leningrad, 1980)

Experimental study of the processes of aerosol production for technical oil cleanup

Gennadiy Medvedev[1,*]

[1]National Research Tomsk Polytechnic University, 634050 Tomsk, Russia

Abstract. From the experimental study of mechanism of fine mist formation in oil-flooded screw compressor it has been drawn the conclusion on evaporation of fine aerosol of mineral oil that has been obtained in two ways: on the basis of oil spray with mechanical burner and on the basis of volume condensation of oil vapour in air. The given paper presents experimental facility for the production of mineral oil aerosol and also comparison of the experimentation results with calculations.

1 Introduction

Mist formation in steam gas flow in most cases turns up to be an undesirable phenomena (sulfuric mist, chimney spouted smoke, etc.), but in some cases it is essential to create mist with special characteristics [1, 2], for example, in development of facilities for testing gas cleaning apparatus or for some other scientific or practical purposes.

2 Experimental facility scheme and its specification

In order to produce mineral oil aerosols the experimental facility has been developed; its scheme is represented in Fig.1. The facility comprises such basic elements as a vapour source, an oil separator and an air control unit.

In the evaporator vessel (1) which is hermetically sealed with a cover (2) the used mineral oil (7) heating with the help of the controllable electric heater (8) is filled. The oil temperature is controlled with the thermoelectric transducer (6) canned with the thermocouple cover (5). In the evaporator the mineral oil vapours are mixed with the air stream coming from the receiver (14) per the air supply tube (4), thus forming here air mist. Out of the evaporator the oil aerosol is delivered into the oil separator where the mineral oil (12) tramped with the special filtering element (11) congregates in the lower part of the filter shell (9). Thereafter the air is let out of the oil separator throughout the pouring hole of the cover (10) into the atmosphere. The output oil concentration is varied on account of air consumption changes and the load upon the evaporator heater. The air consumption in the facility is controlled with the help of the gearbox (15) and the air valve (16). At high concentrations the growth of mineral oil particles before the oil separator may occur and the

* Corresponding author: ggm@tpu.ru

removal efficiency will increase. The air control unit includes the air compressor (13) which pressurizes air to the receiver (14) to the necessary level of pressure.

The operating result of the separation facility is refined from heavy fractions and disperse impurities of the mineral oil (12).

The temperature in the facility elements has been measured with the copper-to-constantan thermocouples. As a secondary instrument millivoltmeter has been used.

To determine the air consumption the pre-gradable float variable-area flowmeter has been used.

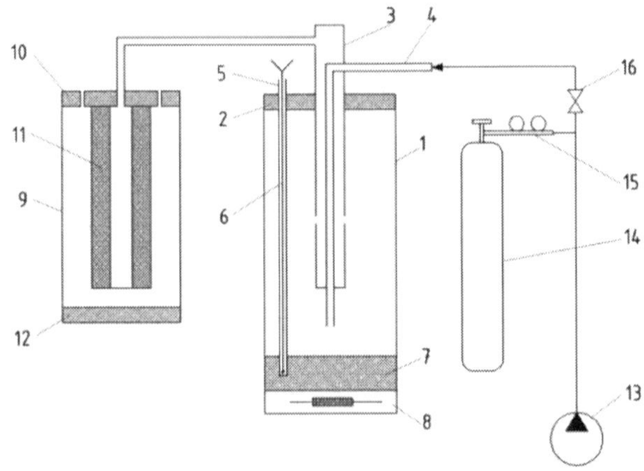

Fig.1. The scheme of the separation facility for used mineral oil cleanup: 1 - evaporator vessel, 2 – cover, 3 – aerosol outlet nozzle, 4 – air supply tube, 5 – thermocouple cover, 6 – thermocouple, 7 – used oil, 8 – heater, 9 – filter shell, 10 – cover, 11 – filtering element, 12 – refined oil, 13 – compressor, 14 – receiver, 15 – gearbox, 16 – control air valve.

3 The study of heat transfer in the oil evaporator and aerosol production

The study of heat transfer included measurements of oil temperature, air temperature on entry of the evaporator and air temperature on exit from it. During experimentation the moments of oil mist occurrence were also recorded.

Data on heat transfer over the air were obtained with the purpose of evaluation of particular parameters of heat balance of air – vacuum oil hot surface interaction. During the first stage there were conducted evaluations of such parameters: oil surface radiation heat transfer Q_p, evaporation ingredient Q_i which was estimated on the quantity of formed aerosol, surface-to-air convection component Q_k. With the values of oil temperature T_3=220°C, air consumption G=0.0001 m^3/s, heat transfer coefficient α =10 W/(m^2·K) the following data were received: Q_p=23 W, Q_i=8.9 W, Q_k=15.7 W and the sum Q_s=47.6 W. Considering that the released by heater heat quantity is equal to 500 W, the developed experimental facility has low efficiency.

During the second stage there was an attempt to clear up the question of mechanism of convection heat transfer between the oil surface and air stream. For this reason we considered the model that is schematically shown in Fig.2.

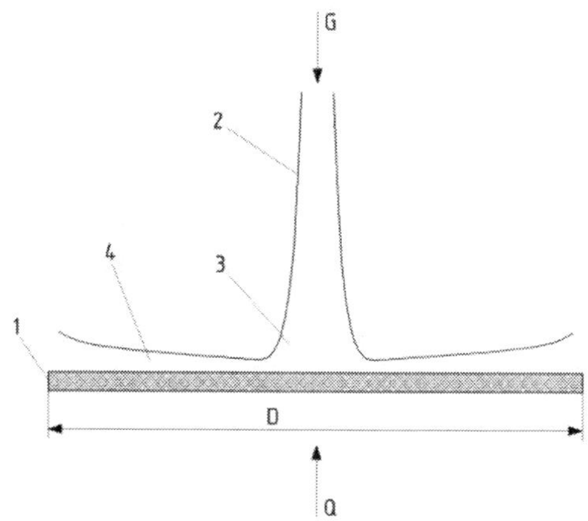

Fig.2. The scheme to the calculation of heat transfer between the air stream and vacuum oil layer.

The oil layer 1 on the surface in diameter D is blown over the nozzle 2 by the air flow with consumption G. In the plan of heat interaction of flow with the oil surface it is required to divide the flow region into region 3 where the turn of air flow takes place and its direct straight leakage on surface occurs (as in flow-around of «stagnation point») and region 4 where surface flow with a kind of a wall jet takes place which velocity can be explained by a well-known «Schlichting velocity profile».

In this particular case the region 3 is much less than the region 4 in magnitude of the surface. Thus, the observed diameter of a dimple of the oil surface amounts to approximately 8 mm in diameter of muzzle equal to 5 mm whereas the diameter of bottom totals 100 mm. Therefore at a first approximation when analyzing convective heat transfer we assumed the use of the so called channeling flow relations. As it was impossible to directly determine the heat transfer coefficient it was suggested to use model laws of heat transfer for calculation of total heat flux adding to air.

To calculate local heat transfer in channeling flows the relationship [1] was taken

$$\mathrm{St_m} = 0.0287\,\mathrm{Re}_x^{-0,2}\,\mathrm{Pr}^{-0.6}(\cos\varphi_0)^{-0.6}, \quad (1)$$

where $\mathrm{St_m} = \alpha/(\rho c_p u_{mx})$ – Stanton number; u_{mx} – maximum longitudinal velocity in a wall jet; ρ, c_p – gas density and heat capacitance; α – heat transfer coefficient; $\mathrm{Re}_x = u_{mx}x/\nu$ – Reynold's number; ν – air kinematic viscosity; Pr – air Prandtl number; φ_0 – angle between axial and и tangential components of velocities скоростей on wall (swirl angle φ_0 =0).

It is essential to take into account that the law of velocity variation u_m in swirl flows is different. For our particular case the law [3] was taken

$$u_m = u_0 c \frac{d_c}{r} \quad (2)$$

Here u_0 – velocity of gas outflow; d_c – nozzle diameter; r – radial coordinate counting off from nozzle axe and oil surface; c – constant.

In accordance with the formula (1) we calculated the local coefficient of heat transfer radially the oil surface, then we calculated the local heat transfer and by means of surface integration – full heat flux.

The Fig.3 presents the results of calculations with the use of different values of constants in the formula (1) in comparison with the experimental data.

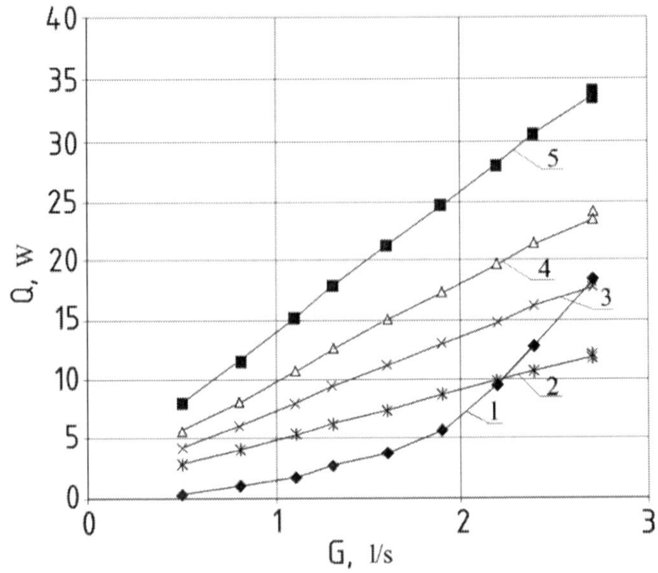

Fig.3. The dependence of convective heat flow on air consumption: curved lines: 1 – experimentation; 2 – A=0.01; 3 – A=0.015; 4 – A=0.02; 5 – A=0.0287.

As it can be seen in the general case the calculation differs from experimentation, but closer values can be received with reduction of constant coefficient in the formula (1).

References

1. S.A. Shutov, A.R. Dorokhov, K.P. Kutzenogi, S.E. Paschenko, *Mekhanizm obrazovaniya melkodispersnogo tumana v maslozapolnennym vintovogo kompressora* (1983) [in Russian]
2. S.A. Shutov, A.R. Dorokhov, K.P. Kutzenogi, S.E. Paschenko, *Eksperimental'nyye issledovaniya ob"yemnoy kondensatsii parov nefteproduktov. Issledovaniye dispersnykh sistem v energeticheskikh khimicheskikh protsessov* (1982) [in Russian]
3. E.P. Dybin, A.I. Mazur, *Konvektivnyy teploobmen pri strunnoy obtekaniya tela* (1982) [in Russian]
4. S.S. Kutateladze, E.P. Volchkov, V.I. Terekhov, *Aerodinamika i teplomassoobmen v ogranichennykh vikhrevykh potokakh* (1987) [in Russian]

Numerical analyses of the effect of a biphasic thermosyphon vapor channel sizes on the heat transfer intensity when heat removing from a power transformer of combined heat and power station

Atlant Nurpeiis[1*], *Gennady* Mamontov[1], and *Lylya* Valieva[1]

[1]National Research Tomsk Polytechnic University, 634050 Tomsk, Russia

Abstract. Numerical analyses of the effect of a biphasic thermosyphon vapor channel sizes on the heat transfer intensity was conducted when heat removing from an oil tank of a power transformer of combined heat and power station (CHP). The power transformer cooling system by the closed biphasic thermosyphon was proposed. The mathematical modeling of heat transfer and phase transitions of coolant in the thermosyphon was performed. The problem of heat transfer is formulated in dimensionless variables "velocity vorticity vector – current function – temperature" and solved by finite difference method. As a result of numerical simulation it is found that an increase in the vapor channel length from 0.15m to 1m leads to increasing the temperature difference by 3.5 K.

1 Introduction

One of the most important units of CHP is power transformer. The stability of the power generation depends on the reliability of power transformer operation [1-3]. Failure of the power transformer leads to huge accidents and further economic repair costs. The temperature of winding is one of the main factors determining the reliability of power transformer operation [4,5]. Currently, there are some cooling systems using in energy engineering: systems with natural oil cooling of winding, oil cooling with blasting and natural circulation of oil, oil cooling with blasting and forced circulation of oil through the air coolers, oil-water cooling with blasting and natural circulation of oil. The economic effect of this heat removing systems is based on saving the insulation resource of the active part of the transformer. Deterioration of the insulation is determined by the evaluation of the changes of the transformer winding temperature. To reduce the temperature of the windings, the forced cooling of oil is used through the application of additional equipment (circulation pumps and external fans). Failure of such additional equipment leads to emergency mode of power transformer. For this reason, perspective direction is the use of autonomous heat transfer devices for heat removing from the transformers. The closed

* Corresponding author: nurpeiis_atlant@mail.ru

biphasic thermosyphons [6, 7] have proven themselves as reliable heat exchangers due to ease of manufacture and design, lack of need for electricity, pumps, etc. Therefore it is necessary to perform the analyses of the possibility of using thermosyphons for cooling power transformers of CHP.

Technical condition diagnostics of the power transformers are often held by thermal imaging method [8]. This approach allows to record the temperature changes inside the power transformers caused by small defects in insulation without the use of additional equipment in order to detect areas of temperature increase. Maximum temperature areas were identified near the top cover of the oil tank of the transformer [8]. For this reason, in this paper we consider the mounting of the closed biphasic thermosyphon to the top cover of the oil tank of the transformer. The principle of operation of such system is as follows. Heat generates under the electric current in the windings. As a result the temperature increases. The heat is transferred from the winding by conduction and natural convection to the coolant (oil). The heated oil moves up due to the natural convection, then it cools due to heat removal to the "top cover of the transformer – closed biphasic thermosyphon" system and falls down.

The purpose of this study is the numerical analyses of the effect of a biphasic thermosyphon vapor channel sizes on the heat transfer intensity when heat removing from a power transformer of CHP.

2 Setting of the problem

There are solutions of similar heat transfer problems in heat pipes [9] in the framework of a boundary layer model. However the use of such heat transfer devices is limited due to the lack of a general theory of heat and mass transfer processes in the cooling system of power transformers of CHP. The basic laws of the complex interrelated physical processes [10–12] occurring in thermosyphon systems are not enough developed

The heat transfer problem in the thermosyphon of rectangular cross section was solved (Fig. 1). The heat supply from the power transformer was carried out through the bottom surface of the thermosyphon ($y=0$, $0<x<(l_1+l_2+L+l_3+l_4)$). Due to intensive evaporation vapor moves in the direction of y-axis (due to the pressure gradient) and is condensed on the top cover of the heat exchanger. Under the influence of gravity the condensate flows down over the side walls of the thermosyphon and spreads over the bottom cover. The outer walls were considered as insulated; heat exchange with the environment was taken into account at the upper boundary. The condition of temperatures and heat fluxes equality was accepted at the "wall/liquid" and "liquid/vapor" boundaries. The following assumptions were accepted: laminar flow regime; heat transfer in the coolant layer occurs in the conduction mode because the effect of convection in a liquid layer is negligible; vapors forming due to evaporation of coolant are considered as incompressible viscous heat-conducting medium satisfying the Boussinesq approximation [14]; thermal properties of material of casing, coolant and vapor do not depend on the temperature; condensate film thickness on a vertical wall does not change over time. Made assumptions allow to simplify the solution and they does not contribute the significant limitations to the problem formulation.

 DOI: 10.1051/matecconf/20179201058

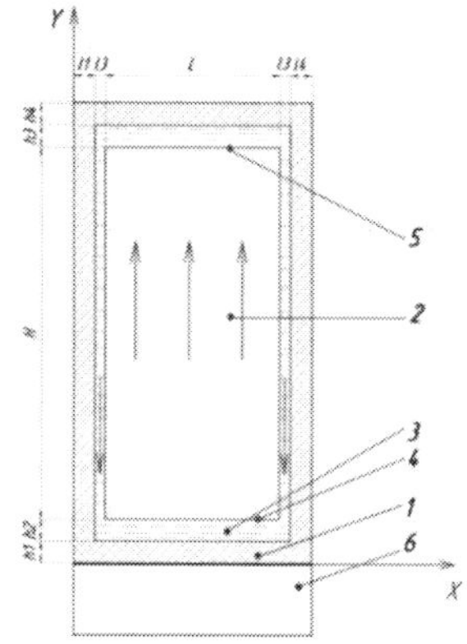

Fig. 1. Solution region: 1 – metal casing; 2 – vapor channel; 3 – liquid film; 4 – evaporation surface; 5 – condensation surface; 6 – top cover of the transformer tank. The vertical arrows indicate the direction of movement of the vapor and liquid.

Study of the heat transfer and fluid mechanics within the stated physical model is described by the two-dimensional Navier-Stokes and energy equations in dimensionless variables as in [13]. The formulated system of equations with nonlinear boundary conditions was solved by finite difference method as in [14,15], with the use of locally one-dimensional scheme of A.A. Samararsky [16]. We used an iterative algorithm [17] developed for solving the problem of conjugate heat transfer in a multiply-connected regions with intensive local absorption and release of heat.

The used algorithm and method of solution was verified on the model problems of free and forced convection [18,19]. Also a comparison of the numerical results with experimental data was conducted [20,21].

3 Results and discussion

Computational study of fluid mechanics and heat transfer were carried out for the closed biphasic thermosyphon of rectangular cross-section.

For clarity, the main results are presented in the form of dimensional fields of temperature and velocities of vapor motion for different geometrical parameters. A heat flow and heat transfer coefficient from the top cover of the thermosyphon was chosen based on typical operating conditions of the power transformers of CHP. The coordinate axes are set in meters, temperature – in degrees Kelvin.

According to the analysis of the results it can be concluded that the change in the thermosyphon height has little influence on the mode of vapor flow. An increase of the vapor channel length is found to effect slightly to the form of isotherms. Isotherms stretch uniformly in the direction of y-axis at a constant speed of vapor outflow and the heat transfer coefficient from the upper cover of the thermosyphon.

Analysis of the obtained distributions showed that the increase in the thermosyphon vertical size from 0.15m to 1m leads to the increase in the temperature difference by 3.5 K in the vapor channel. The temperature of the thermosyphon bottom cover is also reduced slightly (by 2.2°K). Based on these results, we can conclude that the closed biphasic thermosyphon with 1 meter in height can be used for cooling the power transformers of CHP. In this case, it is advisable to carry the upper part of the thermosyphon in the lower temperatures (outside the operating room of the shop), as the ambient temperature is often lower than in the room, which in turn intensifies the heat transfer. This intensification in winter, spring and autumn periods of time can be significant.

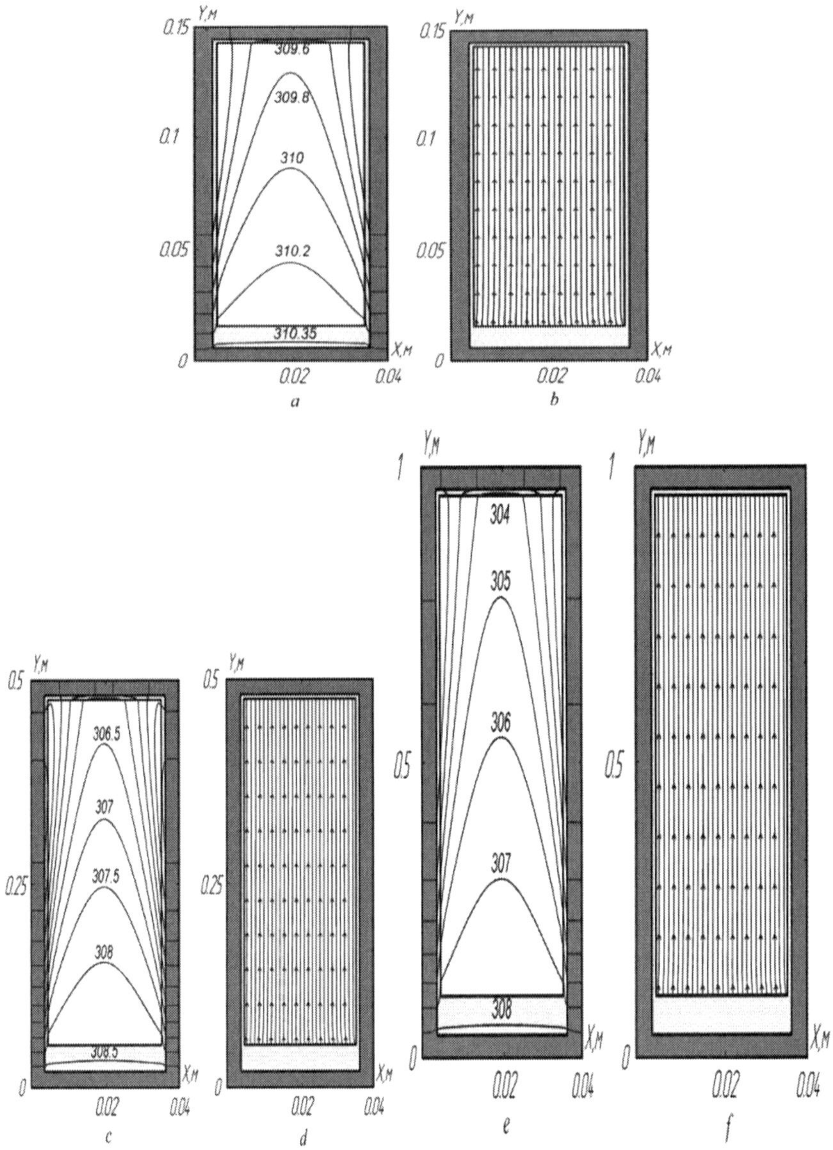

Fig. 2. Isotherms (a, c, e) and velocity fields of vapor motion (b, d, f) at q=1500W/m^2, α=18W/(m^2·K), (a, b) – H=0.15m; (c, d) – H=0.5m; (e, f) – H=1m. q is the heat flow supplied to the lower boundary of the upper wall of the transformer tank.

4 Conclusions

A new approach to modeling of thermal modes of thermosyphon systems linking the main hydrodynamic and thermal processes (forced convection, heat conduction, evaporation, condensation, etc.) occurring during operation of the heat transfer devices was proposed. We presented the temperature distribution and isolines of vapor motion velocities reflecting the impact of changes in the thermosyphon vapor channel height on the heat transfer modes in terms of cooling power transformers of thermal power plants.

The further development of the presented approach will contribute to the creation of a theoretical framework for the efficient design of heat removal systems of power transformers of CHP.

The reported research was supported by Russian Federation President Grant for state support of the Russian Federation leading scientific schools SS-7538.2016.8.

References

1. V.V. Zhukov, *Electrical part of electrical power plants with gas turbine and steam gas units: handbook for high schools,* (Izdatelskij dom MEI, Moscow, 2015) [in Russian]
2. *Power transformers: Hand book*, ed. by S.D. Lizunova, A.K. Lokhanina (Energoizdat, Moscow, 2004) [in Russian]
3. A.M. Golynov, *Cooling devices of oil transformers*, (Energiya, Moscow, 1964) [in Russian]
4. V.A. Starshinov, M.V. Piratov, M.A. Kozinova, *Electrical part of electrical power plants and substations: handbook*, ed. by V.A. Starshinov (Izdatelskij dom MEI, Moscow, 2015) [in Russian]
5. L. Kish, *Heating and cooling of the transformers,* ed. by G.E. Tarle (Energiya, Moscow, 1980) [in Russian]
6. M.K. Bezrodnyi, S.S. Volkov, V.F. Moklyak, *Biphasic thermosyphons in industrial thermal engineering* (Vishcha-shkola, Kiev, 1991) [in Russian]
7. M.K. Bezrodnyi, I.L. Pioro, N.O. Kostyuk, *Transfer processes in the biphasic thermal siphon systems* (Fact, Kiev, 2005) [in Russian]
8. V.I. Zavidej, V.I. Pechenkin, S.V. Kalanchin, Energoekspert. **6,** 64 (2011) [in Russian]
9. G.V. Kuznetsov, A.E. Sitnikov, High Temp. **40**, 964 (2002)
10. G.V. Kuznetsov, D.V. Feoktistov, E.G. Orlova, K.A. Batishcheva, Colloid J. **78**, 335 (2016)
11. G. Kuznetsov, D. Feoktistov, E. Orlova, Thermophys. Aeromech. **23**, 17 (2016)
12. A.E. Nurpeiis, EPJ Web Conf. **76**, 01016 (2014)
13. Y. Dzhaluriya, *Natural Convection: Heat and Mass Transfer* (Mir, Moscow, 1983) [in Russian]
14. G.V. Kuznetsov, M.A. Sheremet, Russ. Microelectronics **37**, 150 (2008)
15. A.E. Nurpeiis, MATEC Web Conf. **72**, 01077 (2016)
16. A.A. Samarsky, *The theory of difference schemes* (Nauka, Moscow, 1977) [in Russian]
17. G.V. Kuznetsov, P.A. Strizhak, J. Eng. Thermophys. **18**, 162 (2009)
18. H. Shabgard, B. Xiao, A. Faghri, R. Gupta, W. Weissman, Int. J. Heat Mass Tran. **70,** 91(2014)
19. S.M. Saedi, J.M. Khodadadi, Int. J. Heat Mass Tran. **49**, 1896 (2006)
20. Yu.A. Dolganov, A. Epifanov, European Researcher **43**, 539 (2013)
21. B. Fadhl, L.C. Wrobel, H. Jouhara. Appl. Therm. Eng. **78**, 482 (2015)

The analysis of flow stability in a vortex furnace model

Igor S. Anufriev[1], *Dmitrii* O. Glushkov[2,*], *Anatoly* P. Papulov[1,3] and *Evgeniy* Yu. Shadrin[1]

[1]Kutateladze Institute of Thermophysics, 630090 Novosibirsk, Russia
[2]National Research Tomsk Polytechnic University, 634050 Tomsk, Russia
[3]Novosibirsk State University, 630090 Novosibirsk, Russia

Abstract. Results of experimental study of the pulsation characteristics of a flow in isothermal model of vortex furnace with vertically oriented nozzles of secondary blast are obtained. With use of laser Doppler measuring system and pressure pulsations analyzer the data about the pressure and velocity pulsations has been received. Spectra of pressure and velocity pulsations at various regime parameters are presented. Absence of non-stationary structures, such as precessing vortex core of a flow, is shown.

1 Introduction

The targeted wide use of low-grade coal in the energy sector requires creating new efficient and ecologically safe combustion technologies for energy generation. One of the promising technologies in this field is pulverized coal combustion in a vortex flow. The key stage in the development of furnaces using the vortex combustion technology is the detailed study of their internal aerodynamics in the laboratory models. It is known that the high swirling flows under certain conditions are characterized by the loss in stability of the stationary regime, which may be expressed in the appearance of the precessing vortex core (PVC). Intensive pressure pulsations associated with the PVC cause wear of power plants and deterioration of the performance of vortex devices and have a negative impact on the furnace processes. In previous works [1, 2] authors visualized the vortex structure of the flow in the model of the improved vortex furnace with distributed fuel-air mixture supply and vertically oriented nozzles of the secondary blast. The aim of this work is to experimentally study the pulsation characteristics.

2 Experimental setup and measurement techniques

The scheme of the experimental setup with the mounted three-component (3D-) LDA system is presented at figure 1. Dimensions of the model are XYZ 320×1200×256 mm, and the diameter of the vortex combustion chamber is 320 mm. A distinctive feature of this

* Corresponding author: dmitriyog@tpu.ru

design compared to the previously studied one [3] is the vertical arrangement of the nozzles of the secondary blast, and the presence of the "visor" inside the vortex chamber to prevent entrainment of fuel particles from the combustion chamber.

Measurements of pressure fluctuations were carried out using noise analyzer Bruel&Kjaer (the limit of pressure measurement up to 103.5 kPa, measurement frequency of 4.2 Hz÷20 kHz, sensitivity of 54.9 mV/Pa). The sensor was placed inside the vortex combustion chamber with a metal sampler – thin-walled tube with a diameter of 2.2 mm and a length of 160 mm. The transfer function of the sampler is presented in [4]. The measurements were carried out close to the conventional center of the vortex chamber ($x=y=160$mm, $z=126$ mm) at various operating parameters ($\gamma=1\div4$, where γ is the ratio of the flow rates through the main and additional nozzles). To diagnose the flow velocity pulsations the two-component (2D-) laser Doppler anemometer LAD-05 (frequency up to 3 kHz) was used. The measurements were performed in the range of Reynolds number $3\cdot10^5<\mathrm{Re}<6\cdot10^5$, calculated on the diameter of the vortex chamber (320 mm) and the velocity module in the top burners ($V_0=5\div25$ m/s).

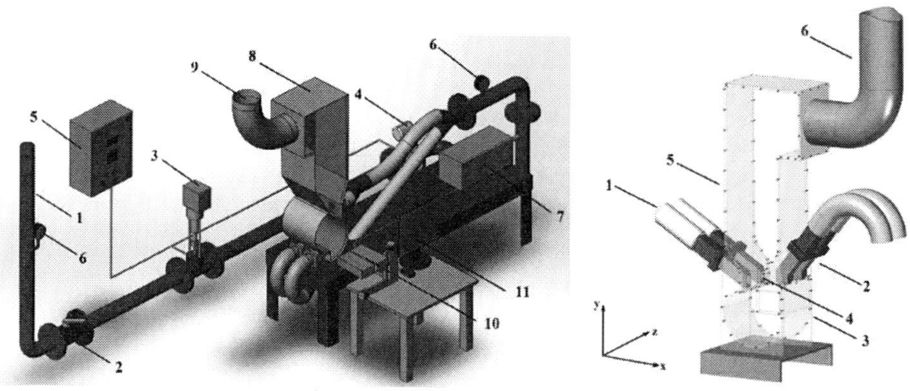

a) b)

Fig. 1. (a) Scheme of experimental stand: 1 – supply line of compressed air, 2 – stop valve, 3 – shut-off control electrically operated valve, 4 – flow converter, 5 – control cabinet, 6 – manometer, 7 – fog generator, 8 – model of the vortex furnace, 9 – ventilation, 10 –measuring system, 11 – computer. (b) Scheme of a laboratory model of the improved vortex furnace: 1 – main nozzles, 2 – secondary nozzle, 3 – combustion chamber, 4 – diffuser, 5 – cooling chamber, 6 – flue.

3 Measurement results and their analysis

Earlier, based on independent measurement techniques, using 3D-LDA [1] and Stereo PIV (Particle Image Velocimetry) systems [2], the structure of the vortex core has been visualized using different criteria [5] and compared with numerical results [6]. To analyze stability of the revealed vortex structure the pressure fluctuations were measured in the central part of combustion chamber. Spectra for the microphone signal obtained for different ratios γ are presented in figure 2. They have a complex form with multiple peaks. However, the position of these peaks neither depends on the flow rate ratio, nor on the magnitude of the total flow, which indicates the absence of the unsteady vortex structures (such as PVC [7]) and the presented peaks characterize the model as an acoustic resonator.

The noise analyzer used in the measurements is limited in frequency range. Therefore, the analysis of pulsations at low frequencies (<5 Hz) was performed with the use of the laser Doppler anemometer LAD-05. The measurements were carried out at different points of the model in the vicinity of the conventional axis of the curved vortex core and near the entrance nozzles, for different values of initial velocity. Figure 3 shows typical spectra of pulsations of the horizontal velocity component in the point (x=150 mm; y=100 mm, z=64 mm) in a plane passing through the centers of the nozzles. The complete absence of peaks confirms the earlier conclusion about the PVC absence. These spectra are characteristic for all points of measurements.

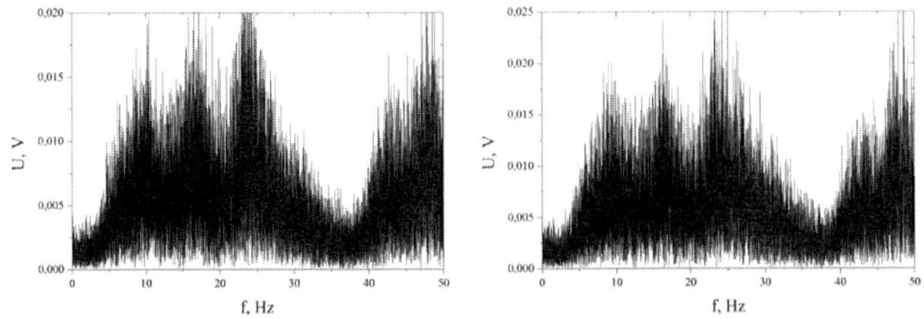

γ=1 γ=4

Fig. 2. Spectra of pressure pulsations in the center of the combustion chamber at different ratios γ.

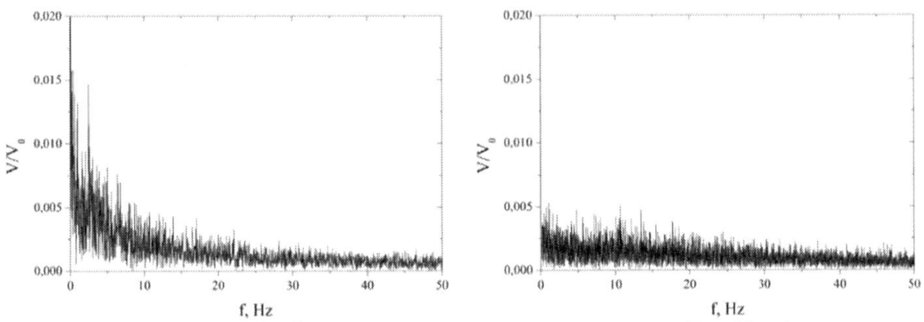

V_0 = 5 m/s V_0 = 20 m/s

Fig. 3. Spectra of the x-velocity component pulsations for different values of initial velocity V_0 (γ=1).

4 Conclusions

Using the modern methods of measurements the pulsation characteristics of the swirling flow have been investigated in the improved model of the vortex furnace with distributed fuel-air supply and vertically oriented nozzles of the secondary blast. The results of the spectral analysis of pulsations of pressure and velocity of the turbulent swirling flow with different prove the stability of the stationary structure of the vortex core flow in the studied improved model of the vortex furnace. The absence of negative effects associated with the vortex core precession is one of the important practical advantages of the studied furnace design.

Research was supported by the Russian Science Foundation (Project No. 14-19-00137).

References

1. I. S. Anufriev, P.A. Strizhak, M.Yu. et al, Tech. Phys. Letters. **41**, 727 (2015)
2. I.S. Anufriev, P.A. Kuibin, E.Yu. et al, Thermophys. Aeromech. **6**, 621 (2016)
3. D.V. Krasinsky, V.V. Salomatov et al, Therm. Eng. **62**, 117 (2015)
4. I.V. Litvinov, S.I. Shtork et al, Int. J. Heat Fluid Fl. **42**, 251 (2013)
5. Y. Dubief, F. Delcayre, J. Turbulence. **1**, 22 (2000)

6. D.V. Krasinsky, O.V. Sharypov, J. Eng. Thermophys. **24**, 348 (2015)
7. V.L. Okulov, I.V. Naumov, J.N. Sorensen, Tech. Phys. **52**, 583 (2007)

Measurement Procedure for Surface Emissivity of Heat-Shielding Materials

Vladimir Arkhipov[1, 2], *Irina* Zharova[1,a], *Viktor* Gol'din[1], and *Natalya* Korovina[2]

[1]National Research Tomsk State University, 634050 Tomsk, Russia
[2]Institute for Problems of Chemical and Energetic Technologies, 659322 Biysk, Russia

Abstract. A procedure is suggested for the measurement of the integral emissivity coefficient of heat-shielding materials in the temperature range close to the thermal destruction temperature.

1 Introduction

The relevance of experimental studies into characteristics of heat-barrier materials is governed by the necessity of using the same in mechanical engineering, construction industry, power engineering, etc. Of most importance is the knowledge about behaviors of heat-shielding materials applied in thermal protection systems of rocket and space hardware while examining thermophysical properties of the materials used as heat barriers and insulators in flow passages of rocket propulsions and the like [1].

One of the basic features of heat exchange is the emissivity coefficient of solid bodies. In the general case, the emissivity coefficient of heat-shielding materials is dependent on many parameters such as properties of the substance itself, surface state and shape of the body under question, temperature, wavelength or wavelength range, direction of emission propagation, etc.

To measure the integral emissivity coefficient of the surface of various materials over a wide temperature range, a number of methods and corresponding designs of experimental instruments are currently in use. A quite complete review on this issue can be found in the monograph [2]. The radiation, calorimetric, and unsteady-state methods have become widely used in the practice of laboratory studies.

The radiation method determines emissivity coefficient using a dedicated thermal detector to comparatively measure the radiant energy emitted by a body under examination or by absolutely black body or by a body whose emissivity coefficient is unknown. In the range between room temperature and melting point or destruction temperature of a material under study, the calorimetric method has the most common use. In contrast to the radiation method, the calorimetric one relies on directly measuring the amount of energy emitted by a body. Unsteady-state methods are applied to measuring the integral emissivity coefficient over a wide temperature range, from helium temperatures to about 1200 K. The unsteady-state measurement methods are premised on using the heat balance equation when a specimen under study is cooled by radiation in vacuum; the equation is written under the assumption that the temperature is uniformly distributed in the specimen volume. The

[a] Corresponding author: zharova@niipmm.tsu.ru

advantage of the unsteady-state methods is that studies can be performed in different gaseous media, including the open air. The merits also include possible adaptation of the relevant devices to nonstandard examination conditions of heat exchange without loss of measurement accuracy [3].

The present work proposes a modified unsteady-state procedure for measuring the integral emissivity coefficient of the surface of heat-shielding materials in the vicinity of elevated temperatures, which can improve the measurement accuracy by taking into account the temperature field nonuniformity in a specimen [4].

2 Modified unsteady-state procedure for emissivity measurement

The modified unsteady-state method for measuring the integral emissivity coefficient is based on using a system of inverse problems of conductivity [5, 6]. In this procedure, a cylindrical specimen of the heat-shielding material is preliminary heated to high temperature (no less than 500 K), T_0, and the temperature is measured during cooling-down in a vacuumed chamber by a thermocouple installed inside the specimen. Two thermocouple measurement options are possible [4]:

– side surface and one of the end faces of the cylindrical specimen are heat insulated, and radiation emanates from the one of the end face surfaces (Fig. 1a);

– both of the end faces of the cylindrical specimen are heat insulated, and radiation emanates from the side surface (Fig. 2b).

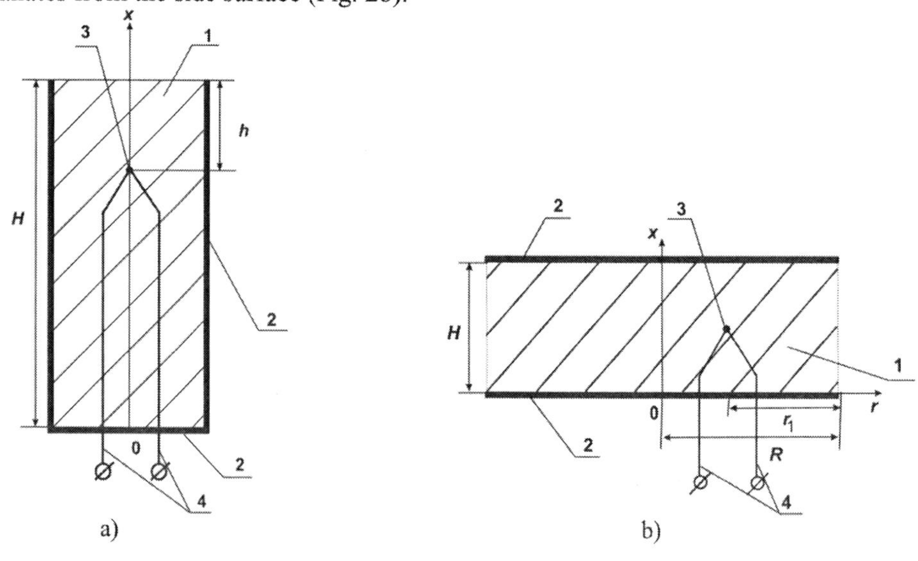

Fig. 1. Emissivity coefficient measurement options: 1 – specimen, 2 – heat insulator, 3 – thermocouple junction, 4 – thermocouple leads.

In case the radiation emanates from one of the end face surfaces (Fig. 1a) when the specimen preheated to high temperature (at least 500 K) is cooled down at cryogenic temperature (or at $T_0 >> T_c$, T_0 – specimen temperature, T_c – ambient temperature), the depth temperature is measured by the thermocouple installed at the symmetry axis of the specimen. With the assumption that the heat cross-flow along the radius of the cylindrical specimen is negligibly small and there is no convective heat exchange with the environment, the integral emissivity coefficient is determined from solving the inverse conductivity problem:

$$\rho c(T)\frac{\partial T(x,t)}{\partial t}=\frac{\partial}{\partial x}\left(\lambda(T(x,t))\frac{\partial T(x,t)}{\partial x}\right), \tag{1}$$

$$0<x<H;\quad 0<t\le t_k;$$

$$T(x,0)=T_0\equiv const, \tag{2}$$

$$0\le x\le H;$$

$$\frac{\partial T(0,t)}{\partial x}=0, \tag{3}$$

$$0\le t\le t_k;$$

$$\lambda(T(H,t))\frac{\partial T(H,t)}{\partial x}=-\varepsilon\sigma T^4(H,t), \tag{4}$$

$$0\le t\le t_k.$$

When the radiation is from the side surface of the cylindrical specimen (Fig. 1b), the integral emissivity is estimated from solving the inverse boundary problem for a two-dimensional equation of heat conductivity:

$$\rho\cdot c(T(r,t))\cdot\frac{\partial T(r,t)}{\partial t}=\frac{1}{r}\cdot\frac{\partial}{\partial r}\left(r\cdot\lambda(T(r,t))\cdot\frac{\partial T(r,t)}{\partial r}\right), \tag{5}$$

$$0<r<R;\quad 0<t\le t_k;$$

$$T(r,0)=T_0\equiv const, \tag{6}$$

$$0\le r\le R;$$

$$\frac{\partial T(0,t)}{\partial r}=0, \tag{7}$$

$$0\le t\le t_k;$$

$$\lambda(T(R,t))\cdot\frac{\partial T(R,t)}{\partial r}=-\varepsilon\sigma T^4(R,t), \tag{8}$$

$$0\le t\le t_k.$$

The following notations are used in Figures 1-2 and problems (1)-(4) and (5)-(8): ρ, c, λ – density, specific heat capacity, heat conductivity coefficient of the specimen; ε – integral emissivity coefficient; T – specimen temperature; t – time; x, r – cylindrical coordinates; T_0 – initial temperature of the emitting surface of the specimen; H, R – specimen height and radius; h, r_1 – distance from the emitting surface to the thermocouple; $\sigma = 5.6687\cdot 10^{-8}$ W·m^{-2}·K^{-4} – Stefan-Boltzmann constant.

The value of integral emissivity coefficient ε in problems (1)–(4) and (5)–(8) is determined by minimizing the functional that includes a numerical solution to $T_1^*(t)$ and to the measured value of temperature $T_1(t)$ at the point where the thermocouple is installed:

$$J(\varepsilon)=\sqrt{\frac{1}{t_k}\int_0^{t_k}\left[T_1^*(t)-T_1(t)\right]^2 dt}\ .$$

Problems (1)–(4) and (5)–(8) were solved numerically by using a variant of the Petukhov method [7] for the second-order boundary equation. The method has a fourth-order approximation in terms of spatial coordinate and a first order in terms of time. The computational algorithm allows simultaneously finding the sought-for function and its derivative. The difference equation system is solved by the matrix sweep method.

The application of the unsteady-state methods requires robust experimental data on specific heat capacity and heat conductivity of heat-shielding materials under examination. Instruments IT–λ–400 (Fig. 2, left) and IT–*c*–400 (Fig. 2, right) were employed to measure the heat conductivity coefficient and specific heat capacity as a function of temperature.

Fig. 2. Instruments for the measurement of thermophysical behaviors of materials.

3 Results of computational experiments

The two options of the suggested measurement procedure have been evaluated by solving a model problem. The computational experiment estimated the integral emissivity coefficient of the cylindrical specimen made of polished pyrolytic graphite during its cooling-down. The pyrolytic graphite had the following characteristics: density $\rho = 2200$ kg/m^3; heat conductivity coefficient $\lambda = 2.8$ W/(m · K); specific heat capacity $c = 1340$ J/(kg · K). The value of integral emissivity coefficient ε of the polished pyrographite surface at $T = 1200$K was 0.83 [8].

In the computational experiment with the radiation emanating from one of the end face surfaces, the cylindrical specimens had the following parameters: height $H = 50$ mm, base radius $R = 5$ mm, initial temperature $T_0 = 1000\ \text{K} \equiv const$. In the numerical experiments, the height h from the emitting surface and the data spread of the thermocouple measurement of the depth temperature, $\Delta T = (\pm 2 \div \pm 10)$ K, were varied. Figure 3 illustrates the relative measurement error for integral emissivity coefficient δε plotted against ΔT and h.

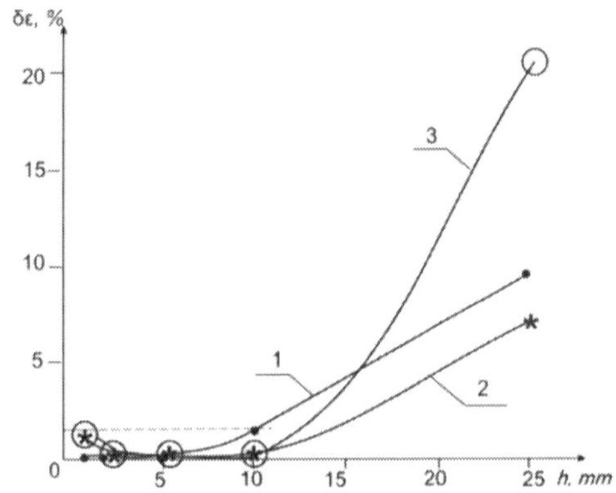

Fig. 3. Relative measurement error for integral emissivity coefficient δε plotted against the temperature measurement error: 1 – $\Delta T = \pm 2$ K; 2 – $\Delta T = \pm 5$ K; 3 – $\Delta T = \pm 10$ K.

The integral emissivity coefficient values subject to random perturbance of the measured temperature ΔT at fixed points and at different distance h from the emitting surface are summarized in Table 1.

Table 1. Integral emissivity coefficient of pyrographite, T_0 = 1000 K.

	h = 1 mm		h = 2 mm		h = 5 mm		h = 10 mm		h = 25 mm	
ΔT, K	ε	δε, %	ε	δε, %	ε	δε, %	ε	δε, %	ε	δε, %
± 2	0.834	< 1	0.835	< 1	0.832	< 1	0.84	1.2	0.91	9.6
± 5	0.82	1.2	0.835	< 1	0.832	< 1	0.835	< 1	0.89	7.2
± 10	0.82	1.2	0.832	< 1	0.833	< 1	0.81	1.2	1.00	20.5

It can be seen from the results presented that when the temperature was measured over the distance range of $h \leq 10$ mm from the emitting surface, the relative error was $\delta\varepsilon < 1.5$ % (in Fig. 3, the error corridor is designated as a dashed line).

The computational experiment with the radiation coming from the side surface had the following conditions: cylindrical specimen height H = 10 mm, base radius R = 10 мм, initial specimen temperature T_0 = 1000 K. The value of the thermocouple spatial coordinate varied, r_1 = (1; 2; 5; 10) mm. The data scattering of the thermocouple temperature measurement varied in the range of ΔT (± 2 ÷ ± 10) K.

Figure 4 displays the relative measurement error for integral emissivity coefficient δε plotted against ΔT and h. It is visible from the data in Figure 4 that when the temperature was measured over the distance range of $r_l \leq 5$ mm from the emitting surface, the relative error was $\delta\varepsilon < 1.5$ % (the error corridor is designated as a dashed line). The scatter in data from the thermocouple temperature measurement was ΔT = (± 2 ÷ ± 10) K.

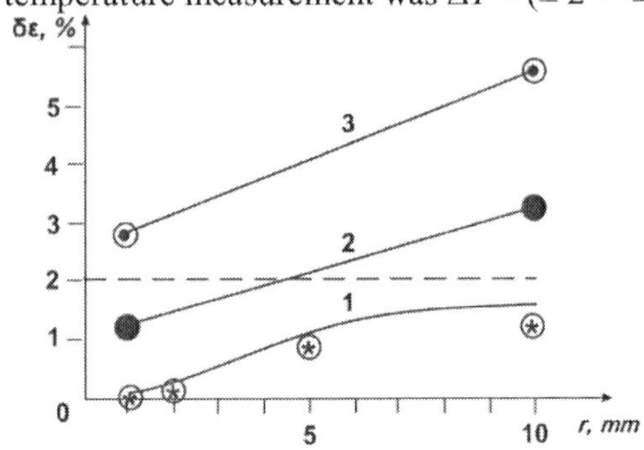

Fig. 4. Relative measurement error for integral emissivity coefficient δε plotted against the temperature measurement error: 1 – $\Delta T = \pm 2$ K; 2 – $\Delta T = \pm 5$ K; 3 –$\Delta T = \pm 10$ K.

The numerical evaluation results of the modified unsteady-state method when the radiation emanates from the side surface of the cylindrical specimen have shown that the measurement accuracy for the integral emissivity coefficient of the surface of heat-shielding materials can be improved if the thermophysical experimental conditions comply with the following requirements:

– data scattering of the thermocouple temperature measurements should not be in excess of ± 5 K;

– it is not advisable placing the informative thermocouple at the symmetry axis of the cylindrical specimen because, in this case, the measurement error for the integral emissivity coefficient may be as high as 3 %.

4 Conclusion

The findings from the computational experiments using the two temperature measurement options demonstrate that the modified unsteady-state procedure can improve the measurement accuracy for the integral emissivity coefficient of materials with low heat conductivity by taking the cognizance of the temperature field nonuniformity in the specimen.

This research carried out in 2015 was supported by "The Tomsk State University Academic D.I. Mendeleev Fund Program", grant No 8.2.46.2015.

References

1. Yu.V. Polezhaev, F.B. Yurevich, *Heat shield* (Moscow, Energiya, 1976) [in Russian]
2. A.E. Sheindlin, *Emitting properties of solid materials* (Energiya, Moscow, 1974) [in Russian]
3. V.A. Arkhipov, V.D. Gol'din, I.K .Zharova, N.I. Kurilenko, G.Ya. Mamontov, Teplofizika i Aeromekhanika **19**, 6 (2012) [in Russian]
4. V.A. Arkhipov, V.D. Gol'din, I.K. Zharova, N.I. Kurilenko, RU Patent 2468360, A method for measuring the integral emissivity coefficient of the surface of heat protection materials, Bulletin No. 33, 27.11.2012. [in Russian]
5. O.M. Alifanov, *Inverse problems of heat exchange*, (Mashinostoyeniye, Moscow 1988) [in Russian]
6. A.N. Tikhonov, V.D. Kal'ner, V.B. Glasko, *Mathematical modeling of manufacturing processes and the inverse problem method in mechanical engineering* (Mashinostoyeniye, Moscow, 1990) [in Russian]
7. V.D. Gol'din,, E.V. Yorkina, I.V., *Issledovaniya po Ballistike i Smezhnym Voprosam Mekhaniki* (Tomsk State University, Tomsk, 2001) [in Russian]
8. Yu.V. Polezhaev, A.A. Shishkov, *Reference Book: Gas dynamic testing of heat shield* (Promedek, Moscow, 1992) [in Russian]

Numerical modelling of combined heat and mass transfer in film absorption

Maria V. Bartashevich[1,*], *Maxim* G. Vlasenko[1], *Nadegda* A. Zyuzina[1], *Andrey* A. Pil'nik[1], and *Andrey* A. Chernov[1]

[1]Kutateladze Institute of Thermophysics, 630090 Novosibirsk, Russia

Abstract. Heat and mass transfer during absorption on a film of lithium bromide water solution flowing by a cooled wall in the steam atmosphere is numerically investigated in this paper. The self-similar solutions are using as the initial conditions for solving the problem beyond the entrance region. The key criteria characterizing heat and mass transfer in the film absorption with uniform velocity profile and with a constant thickness have been determined.

1 Problem statement

Absorption of gases or vapors by liquid takes place in various energy devices [1]. The first theoretical description of combined heat and mass transfer processes at vapor absorption in a laminar falling liquid film was made in [2-4]. The overview of works on film flows with absorption is presented in [5]. Heat and mass transfer in the processes of absorption, desorption, evaporation and condensation in the initial part of semi-infinite axisymmetric film falling under the pressure is studied in [6]. The absorption and desorption processes under the conditions, corresponding to the operating modes of thermal transformers are described in [7]. The analytical solution to the problem of conjugate heat and mass transfer in a laminar falling liquid film with a linear velocity profile is presented in [8]. Data of numerical simulation of absorption are compared with experimental data in [9].

Let us consider the two-dimensional stationary flow of a laminar liquid film (water solution of lithium bromide with constant thickness h) over a plate, inclined at angle θ to the horizon, as in [4]. The liquid surface contacts with stationary steam. Absorption is considered in the framework of usual assumptions [1]. The vapor phase is single-component, and vapor pressure does not change during absorption. Thermal-physical properties of solution are considered constant. We believe that absorption heat is released at the interface and it is spent only for solution heating. For small ranges of concentrations and temperatures, according to [1], we will approximate the dependence of absorbed substance concentration on temperature by linear function $C_i = k_1 - k_2 T_i$, where coefficients k_1 and k_2 are determined by the pressure. The values of $C_i(x), T_i(x)$ are interrelated, unknown and should be determined.

[*] Corresponding author: bartashevichmv@gmail.com

Let us introduce Cartesian coordinate system with axis Ox along the flow and axis Oy normal to the plate. The velocity profile in the film takes a form

$$u = 3\bar{u}\left(2y/h - y^2/h^2\right)/2,$$

where $\bar{u}$ is an average velocity of liquid motion in the film. The process of heat and mass transfer at film absorption is described by the equation of heat conductivity and diffusion:

$$u\,\partial T/\partial x = a\,\partial^2 T/\partial y^2, \quad (1)$$

$$u\,\partial C/\partial x = D\,\partial^2 C/\partial y^2, \quad (2)$$

with the following boundary conditions. At the inlet for $x = 0$:

$$T = T_0, C = C_0.$$

At the interface for $y = h$, similarly to [1,4] there is the equilibrium solution-vapor system:

$$C_i = k_1 - k_2 T_i, \; -\lambda\,\partial T/\partial y = -r_a(\rho D/(1 - C_0))\,\partial C/\partial y.$$

This equilibrium condition connects the equilibrium temperature with concentration. On the wall at $y = 0$, we set:

$$\lambda\,\partial T/\partial y\big|_{y=0} = q_w, \partial C/\partial y\big|_{y=0} = 0,$$

where q_w is heat flux on the wall. The heat transfer coefficient α_w is describing the heat transfer from the liquid film to a wall and defined from the relation $q_w = \alpha_w(T_f - T_w)$, where T_w is a dimensional wall temperature, $T_f = \int_0^1 uTdy \Big/ \int_0^1 udy$ is dimensional bulk temperature of solution. Let us use dimensionless variables $\xi = x/(Peh)$, $\eta = y/h$, $v = u/\bar{u} = 3\left(2\eta - \eta^2\right)/2$, $\theta = (T - T_0)/(T_e - T_0)$, $\gamma = (C - C_0)/(C_e - C_0)$. Here $C_e = k_1 - k_2 T_0$ is equilibrium concentration, corresponding to initial temperature of solution T_0, T_e is equilibrium temperature, corresponding to initial concentration C_0 ($C_0 = k_1 - k_2 T_e$), similarly to [1]. $\mathrm{Pe} = \bar{u}h/a$ is Pecklet number. The system of equations (1) – (2), reduced to the dimensionless form, is written as:

$$v\,\partial\theta/\partial\xi = \partial^2\theta/\partial\eta^2, \quad (3)$$

$$v\,\partial\gamma/\partial\xi = \mathrm{Le}\,\partial^2\gamma/\partial\eta^2, \quad (4)$$

here $\mathrm{Le} = D/a$. The boundary conditions at the inlet for $\xi = 0$:

$$\theta = 0, \; \gamma = 0.$$

At the interface for $\eta = 1$, similarly to [1]:

$$\theta = \theta_i, \gamma = \gamma_i,$$

$$\theta_i + \gamma_i = 1,\ \mathrm{KaLe}\partial\gamma/\partial\eta = \partial\theta/\partial\eta. \quad (5)$$

Here $\mathrm{Ka} = r_a(C_e - C_0)/(C_P(T_e - T_0)(1 - C_0))$. On the wall at $\eta = 0$, we obtain:

$$\partial\theta/\partial\eta\big|_{\eta=0} = \bar{q}_w, \partial\gamma/\partial\eta\big|_{\eta=0} = 0,$$

where $\bar{q}_w$ is dimensionless heat flux.

2 Numerical calculations

Numerical calculations were performed by the finite-difference method. The relationship of temperature and concentration along the interface was calculated via conjugated calculation of the modified coefficients, taking into account the conditions at the interface.

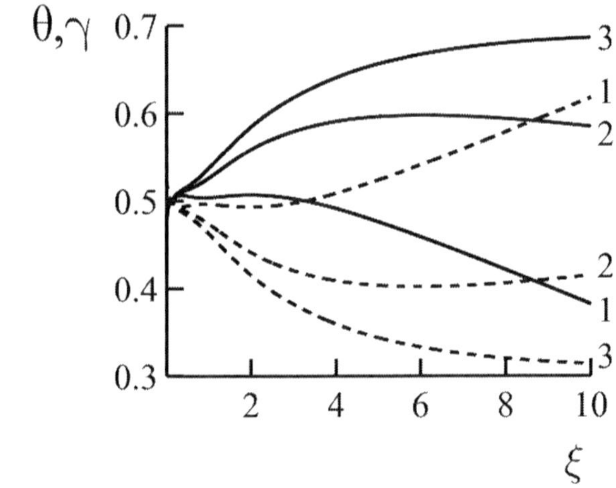

Fig. 1. Dimensionless temperature (solid lines) and concentration (dashed lines) on the film surface. 1 – $\bar{q}_w$ = 0.2; 2 – $\bar{q}_w$ = 0.1; 3 – $\bar{q}_w$ = 0.05.

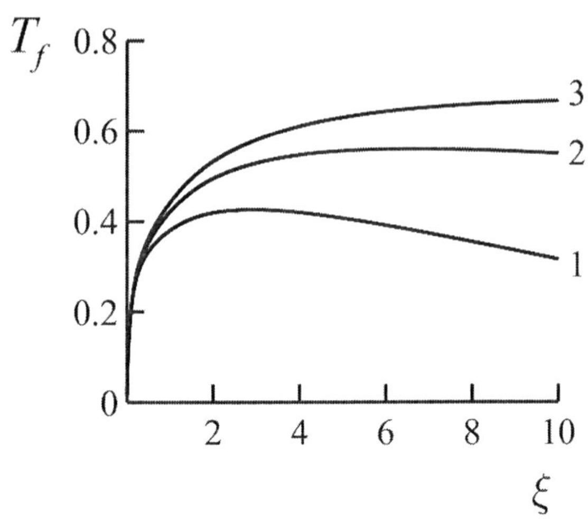

Fig. 2. Dimensionless bulk temperature of solution. 1 – $\bar{q}_w$ = 0.2; 2 – $\bar{q}_w$ = 0.1; 3 – $\bar{q}_w$ = 0.05.

Calculations were carried out at Ka = 10, Le = 0.01. Calculated distributions of dimensionless temperature and concentration along the film surface are shown in Fig. 1 for different heat fluxes on the wall $\bar{q}_w$. Dimensionless bulk temperature of solution is shown in Fig.2. Dependences of Nusselt number on longitudinal coordinate are presented in Fig. 3 for different heat fluxes along the initial region of the flow.

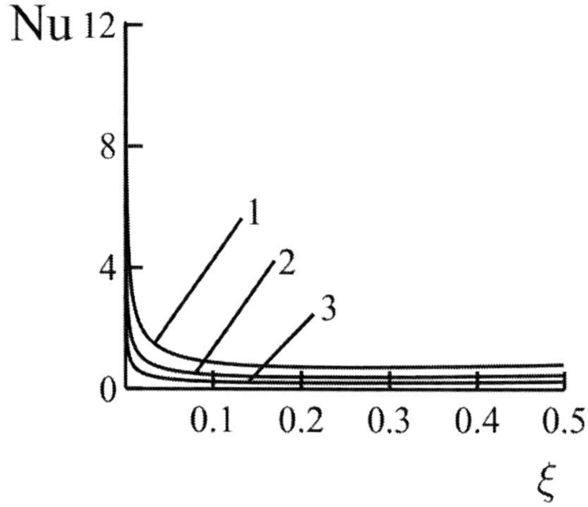

Fig. 3. Nusselt number along the film-flow direction at the initial region. 1 – $\bar{q}_w$ = 0.2; 2 – $\bar{q}_w$ = 0.1; 3 – $\bar{q}_w$ = 0.05.

3 Conclusions

The problem of conjugated heat and mass transfer at film absorption on a cooled wall with a given constant heat flux is studied numerically. The analytical solution is used as an inlet condition for further numerical solution by the finite-difference method. The key dimensionless criteria characterizing heat and mass transfer at absorption on the falling liquid film have been defined.

This work was carried out at the Kutateladze Institute of Thermophysics SB RAS and financially supported by the Russian Science Foundation (project number 15-19-10025).

References

1. V.E. Nakoryakov, N.I.Grigorieva, *Non-Isothermal Absorption in Thermal Transformers* (Nauka, Novosibirsk, 2010)
2. S. H.Chiang, H. L. Toor, AIChE J. **10**, 398 (1964)
3. N. I.Grigorieva, V. E. Nakoryakov, J. Engin. Phy. Thermophys. **33,** 1349 (1977)
4. G.Grossman, International J. Heat Mass Transf. **26,** 357 (1983)
5. D. Killion, S. Garimella, Int. J. Refrigeration. **24,** 755 (2001)
6. V.E.Nakoryakov, N.I.Grigoryeva, M.V. Bartashevich, Int. J. Heat Mass Transf. **54,** 4485 (2011)
7. M. Mittermaier , F. Ziegler, Int. J. Refrigeration. **59,** 91 (2015)
8. M. Mortazavi, S. Moghaddam, Int. J. Refrigeration. **66,** 93 (2016)
9. E. Garcia-Rivera, J. Castro, J. Farnos, A. Oliva, Int. J. Thermal Sci. **109,** 342 (2016)

Static contact angle versus volume of distilled water drop on micro patterned surfaces

Kseniya Batichsheva [1,*] , *Dmitriy* Feoktistov [2], *Vladimir* Ovchinikov[2], and *Sergey* Misyura[2]

[1]National Research Tomsk Polytechnic University, 634050 Tomsk, Russia
[2]Kutateladze Institute of Thermophysics, 630090 Novosibirsk, Russia

Abstract. Static contact angle was determined experimentally in the condition of wetting of polished and laser patterned surfaces of stainless steel substrates by distilled water drops with different volumes. In contrast with polished surface, the contact angle was found to depend on drop volume on micro patterned surfaces. In addition, the enhancement of both hydrophilic and hydrophobic properties was observed on laser patterned surfaces.

1 Introduction

Nowadays there is a search of ways to create surfaces with controlled wetting. Studies in this field [1-5] are motivated by developing the scientific bases of creation of self-cleaning, anti-microbial, anti-corrosion, waterproof materials. It is known [6-8] that many factors affect the intermolecular interaction of condensed liquid and solid phases. The main factors are the chemical composition of the liquid and the surface, and the roughness. To create special micro or nanostructures modifying wetting properties, the following methods are used: sublimation, templating methods, chemical deposition of structures from the gas phase, lithography. All of them require significant financial costs; their technical implementation takes much time. The laser light is promising as a method of surface treatment to create the necessary wetting properties. It is relatively low-cost and technologically simple. However, the scientific bases of using laser are not developed at a level to create the metal surfaces with hydrophobic or hydrophilic properties [9-23].

The purpose of this work is to determine experimentally the dependence of the static contact angle (θ) on the distilled water drop volume (V) on the treated surfaces by the laser radiation.

2 Experimental procedure

Experimental studies were carried out on the setup; the principle of its operation is shown in [24]. We used eleven laser patterned and one polished surfaces made of stainless steel. Technical characteristics used for each surface in "MiniMarker 2 M 20" laser system are listed in Table 1.

* Corresponding author: bka1801@mail.ru

The drop was placed on the surface by the precision dispenser. The following drop volumes were used: 10, 20, 30, 40, 50, 60μL.

Table 1. Technical characteristics of "MiniMarker 2 M 20" laser system used during surface treatment.

No of samples made of stainless steel	**1**	**2**	**3**	**4**	**5**	**6**	**7**	**8**	**9**	**10**	**11**
Frequency of laser beam, Hz	99	99	99	50	50	20	20	99	99	99	99
Traverse speed of laser beam, 10^{-3} m/s	200	400	800	800	400	200	800	800	800	800	800
Power of laser radiation, Wt	20	20	20	20	20	20	20	10	14	16	18

The average arithmetic roughness (Ra) was defined by using "HOMMEL TESTER T1000" profilometer.

3 Results and discussion

Systematic component of the error was neglected. Parasitic errors were rejecting during processing stage. Each experiment was repeated at least five times. All experimental results at fixed parameters were average. The relative measurement error did not exceed 5%.

According to the experimental results, dependences of static contact angle on liquid volume were obtained (Fig.1).

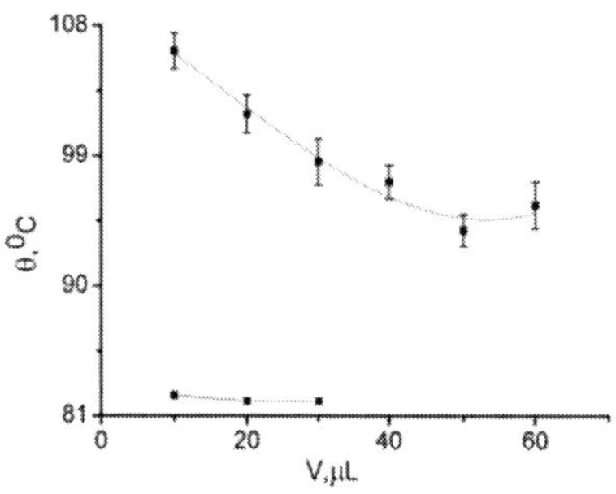

a

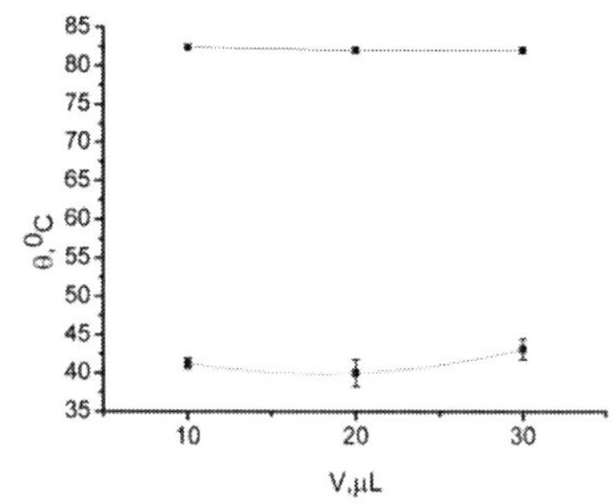

b

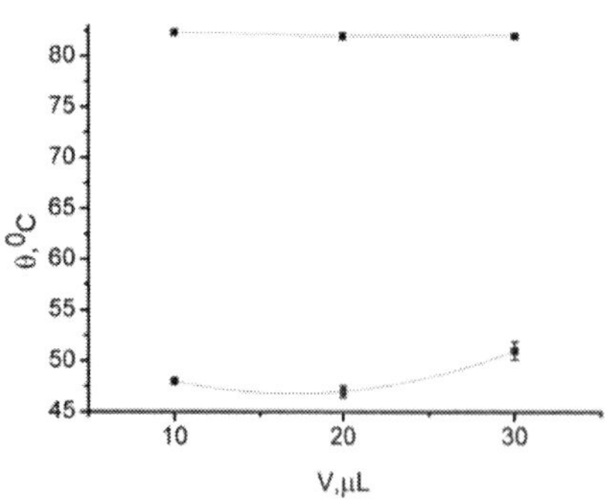

c

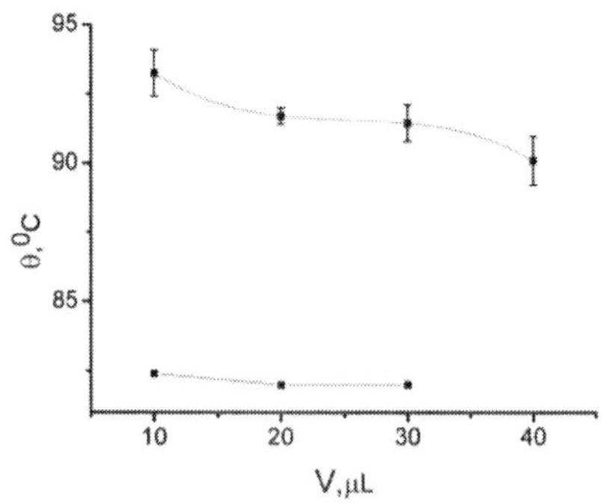

d

DOI: 10.1051/matecconf/20179201062

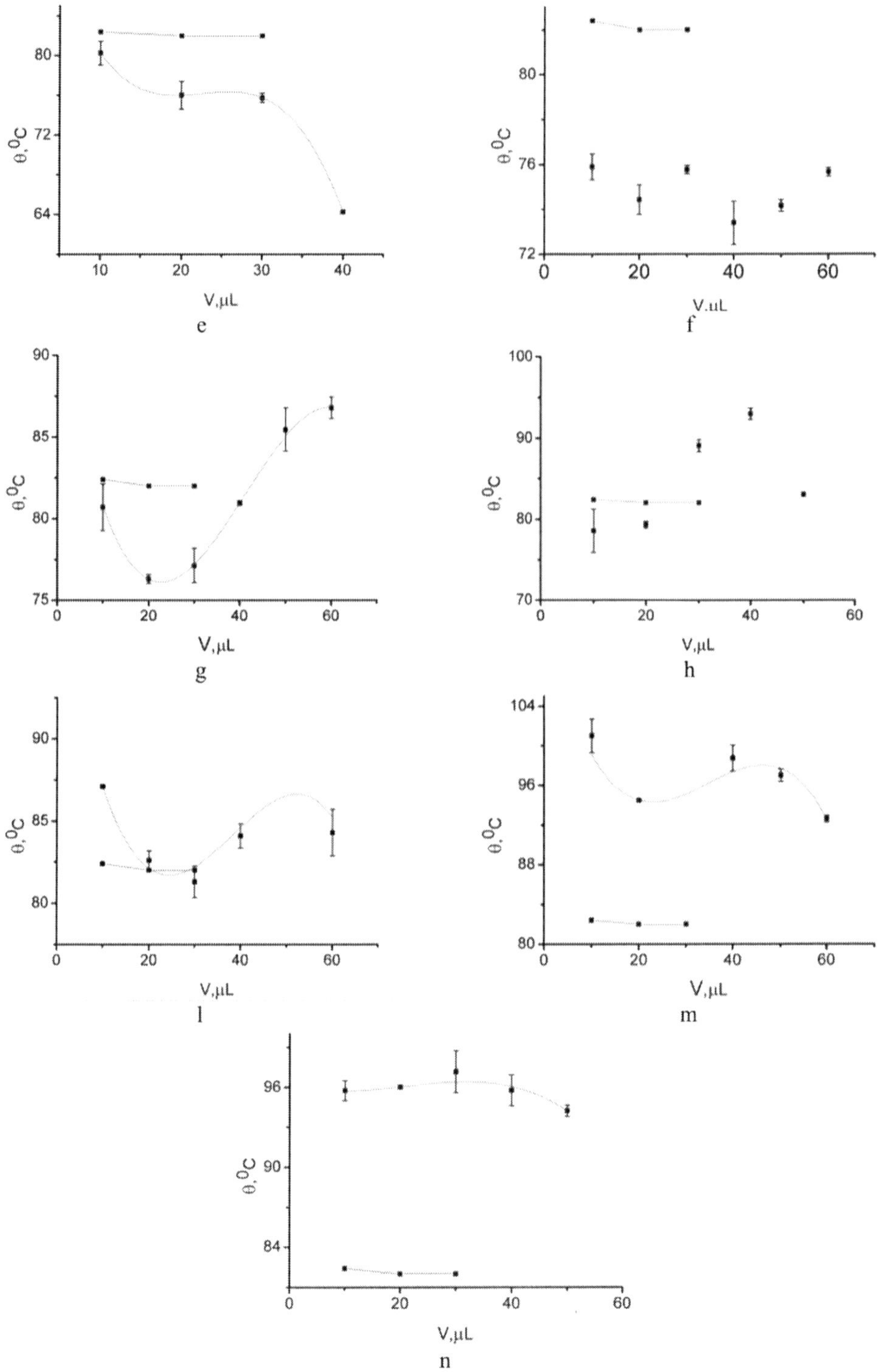

Fig. 1. Dependences of static contact angle on distilled water drop on stainless steel surfaces with numbers: a) 1; b) 2; c) 3; d) 4; e) 5; f) 6; g) 7; h) 8; l) 9; m) 10; n) 11. Contact angle on: 1 – polished surface; 2 – patterned surface.

The static contact angle is found not to depend on the volume on stainless steel polished surface (Ra=0.079 μm). However, on laser patterned surfaces the static contact angle depends on the liquid volume and microstructure (Fig.3). The change of the contact angle θ can be explained by the following way. The behaviour of drop on the surface is known to be described by to models, Wenzel and Cassie-Baxter states. The first model characterizes heterogeneous wetting when the liquid does not penetrate into the micro cavities on the surface. The second one is homogeneous; the liquid penetrates into the micro cavities. In the Cassie-Baxter state (1) the proportion of projection of wetting area on substrate surface changes at increasing liquid volume:

$$\cos\theta = f(\cos\theta_0 + 1) - 1 \qquad (1)$$

Decrease in this proportion leads to increase in static contact angle. It can be concluded that the portion of curves (Fig.3), where θ decreases with increase in volume, correspond to the Cassie-Baxter state. A further increase in liquid volume results in complete filling of the cavities on the surface. There is a transition from heterogeneous to homogeneous wetting mode. It can be assumed that the portion, where θ changes sharply, corresponds to the transition from the state of Cassie-Baxter to the Wenzel state.

It is found that the static contact angle increased on surfaces No 1, 4, 9, 10, 11 (the hydrophobic properties improved). However, θ decreased on the surfaces No 2, 3, 5, 6 (the hydrophilic properties improved). The curves of static contact angle on surfaces No 7, 8 have a sinusoid nature.

4 Conclusion

The influence of the volume of distilled water on the value of the static contact angle of a drop placed on laser patterned surface made of stainless steel was studied. The possibility of controlling the wetting by laser treatment of surfaces was approved. The micro asperities and cavities are found not to impact the static contact angle on relatively smooth surfaces (Ra = 0.079 μm). The static contact angle on patterned surfaces compared to polished depends on the liquid volume. Change of laser radiation power, impulse frequency and traverse speed of laser beam allows to improve hydrophilic, as well as hydrophobic properties.

This work was carried out at the Kutateladze Institute of Thermophysics SB RAS and financially supported by the Russian Science Foundation (project number 15-19-10025).

References

1. O.A. Kabov, D.V. Zaitsev, Multiphase Sci. Technol. **2**, 249 (2009)
2. D.V. Zaitsev, O.A. Kabov, V.V. Cheverda, N.S. Bufetov, High. Temp. **42**, 450 (2004)
3. V.E. Nakoryakov, S.Y. Misyura, S.L. Elistratov, J. Eng. Thermophys. **22**, 1 (2013)
4. S.Y. Misyura, V.E. Nakoryakov, S.L. Elistratov, Int. J. Energy Technol. **4**, 1 (2012)
5. G.V. Kuznetsov, A.V. Zakharevich, N.S. Bel'kov, Chem. Pet. Eng. **50**, 424 (2014)
6. A.V. Zakharevich, N.S. Bel'kov, EPJ Web Conf. **76**, 01081 (2014)
7. D.O. Glushkov, G.V. Kuznetsov, P.A. Strizhak, Therm. Sci. **19**, 1541 (2015)
8. D.O. Glushkov, G.V. Kuznetsov, P.A. Strizhak, R.S. Volkov, Therm. Sci. **20**, 131 (2016)
9. V.P. Lebedev V.V. Lemanov, S.Y. Misyura, V.I. Terekhov, Fluid Dyn. **28**, 624 (1993)
10. S.Y. Misyura, Exp. Therm Fluid Sci.**75**, 43 (2016)
11. D.O. Glushkov, J.-C. Legros, P.A. Strizhak, A.V. Zakharevich, Fuel. **175**, 105 (2016)
12. V.I. Maksimov, T.A. Nagornova, I.A. Shestakov, EPJ Web Conf. **82**, 01048 (2015)
13. D.V. Zaitsev, D.P. Kirichenko, O.A. Kabov, Tech. Phys. Lett. **41**, 551 (2015)

14. A. Sivkov, A. Ivashutenko, Y. Shanenkova, I. Shanenkov, Adv. Powder Technol. **27**, 1506 (2016)
15. A. Sivkov, A., I. Shanenkov, A. Pak, D. Gerasimov, Y. Shanenkova, Surf. Coat. Technol. **291**, 1 (2016)
16. G.V. Kuznetsov, D.V. Feoktistov, E.G. Orlova, Thermophys. Aeromech. **23**, 17 (2016)
17. Y.V. Lyulin, D.V. Feoktistov, I.A. Afanas'ev, E.S. Chachilo, O.V. Kabov, G.V. Kuznetsov, Tech. Phys. Lett. **41**, 665 (2015)
18. D.V. Feoktistov, E.G. Orlova, A.G. Islamova, Eur. Phys. J. Conf. **110**, 98 (2016)
19. K.O. Ponomarev, E.G. Orlova, D.V. Feoktistov, EPJ Web Conf. **110**, 01060 (2016)
20. E.G. Orlova, G.V. Kuznetsov, D.V. Feoktistov, EPJ Web Conf. **76**, 01039 (2014)
21. O.V. Vysokomornaya, G.V. Kuznetsov, P.A. Strizhak, Russ. J. Phys. Chem. B. **5**, 668 (2011)
22. G.V. Kuznetsov, M.A. Sheremet, J. Comput. Thermal Sci. Internat. **1**, 341 (2009)
23. R.S. Volkov, G.V. Kuznetsov, P.A. Strizhak, Int. J. Heat Mass Tran. **85**, 1 (2015)
24. G.V. Kuznetsov, D.V. Feoktistov, E.G. Orlova, Thermophys. Aeromech. **23**, 17 (2016)

Thermal hydraulic efficiency of a single diaphragm established with a clearance in a circular tube

Tatyana Bogatko[1,*], and *Viktor* Terekhov[1]

[1] Kutateladze Institute of Thermophysics, 630090 Novosibirsk, Russia.

Abstract. Results of a numerical study of the flow structure and turbulent heat transfer in a round tube with a mounted single diaphragm with height h and some clearance c between the tube and the diaphragm. The size of the clearance between the diaphragm and the tube wall was varied within range A = c/h = 0 – 0.33. It was determined that an increase in the gap between the diaphragm and the tube wall alters the structure of the recirculation region and eliminate the stagnation zones in the region of secondary vortex. The flow regime, when the values of average heat transfer behind the diaphragm in the presence of a clearance exceed those for the attached rib, was distinguished. When increasing the gap height from A = 0 to 0.33, the thermal enhancement factor increases by 30%.

1 Introduction

Study of the flow structure and heat and mass transfer behind a flat obstacle and the system of such obstacles were relatively extensive since the 60s of the last century. Attention to the study of the features of the flow around detached ribs was attracted much later. The idea of their use for heat transfer improvement is based on formation of the wall jet flow, which destroys the stagnation zone directly behind an obstacle. We can note only a few papers of the experimental and calculation nature devoted to this field [1-6]. All of them were carried out mainly for the rectangular channels and disconnected square obstacles. The exception is a series of works [7-9], where development of the flow with disconnected diaphragms in a rounds tube was studied, and periodic destruction of the flow was performed by the rings of a round wire or twisted ribbon. Alternation of connected and disconnected obstacles along the channel leads to additional possibility of changing the integral heat transfer coefficient [10].

The detailed experimental studies were carried out in [2-3], where they studied the effect of Reynolds number, obstacle sizes and relative height of the clearance on the flow structure and turbulent heat transfer in a rectangular channel with the ribbed wall. Investigation results show that the thermohydraulic efficiency at this way of heat transfer enhancement is significantly higher than the use of attached ribs.

* Corresponding author: tbogatko@gmail.com

As can be seen from the above, the mechanism of heat transfer enhancement in the channels with disconnected obstacles is studied insufficiently. This is explained by the multifactor character of the complex heat transfer process, three-dimensional nature of the flow, and the features of the vortex flow formation in the rectangular channels with different numbers of ribbed walls in comparison with the round tubes. Moreover, additional difficulties in analyzing are caused by interference of separated flows from the alternating obstacles. In this connection it is necessary to study in detail the flow structure under the simpler conditions of the flow around a single two-dimensional obstacles with varying disconnection parameter A = c/h, where c is the size of a clearance between the diaphragm and the tube wall, and h is the height of the diaphragm (Fig. 1). At that, it is important to determine the effect of parameter A on the local and integral characteristics of heat transfer and resistance. These data can serve the basis for the analysis of more complex transfer processes in channels with the system of obstacles.

2 Flow scheme. Calculation method

The scheme of computational domain is shown in Fig. 1. A flat annular diaphragm is placed into the cylindrical channel with the diameter of 100 mm and length of 900 mm at the distance of 100 mm from the inlet. Its height was constant h = 15 mm, and the size of the clearance was varied in range c = 0 – 5 mm. The velocity profile at the tube inlet was uniform, the working medium was air of 10° C. A change in thermal-physical properties was neglected in calculations. The Reynolds number, calculated by the tube diameter and the mean mass velocity, was constant Re = 27500.

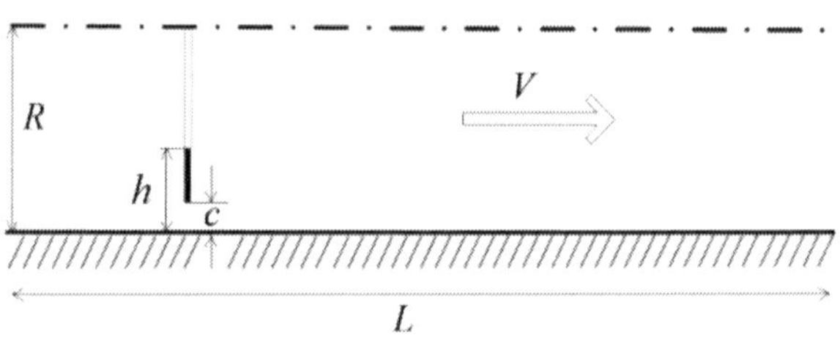

Fig. 1. Scheme of the computational domain.

The calculations were carried out in the framework of the model of incompressible liquid based on the system of stationary Navier-Stokes and energy equations, Reynolds averaged (RANS). The main tool for the study is the universal calculation complex FLUENT. The problem statement is two-dimensional; the flow is stationary and axisymmetrical. The turbulence model k-ω SST is chosen for calculations [11], as the most appropriate for calculation of turbulent separated flows. The previous studies of the authors prove also that for the models, implemented in the given set, the results obtained with application of the mentioned model are the best to match the physics of the given flow type.

3 Calculation results

The flow structure near the disconnected rib is shown in Fig.2. The field of longitudinal velocity component is shown by the colour; the streamlines are shown by the vectors.

DOI: 10.1051/matecconf/20179201063

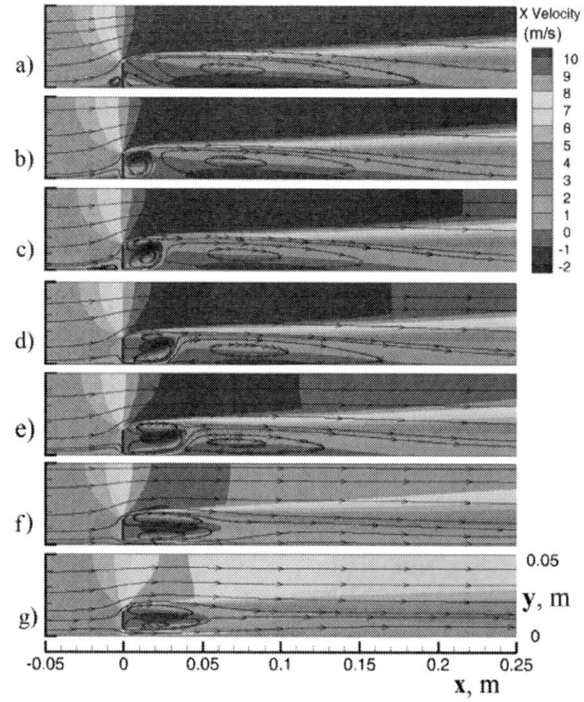

Fig. 2. The stream lines and the longitudinal component of the velocity field in the area of the diaphragm. a) c/h = 0, attached diaphragm; b) c/h = 0.03; c) 0.07; d) 0.1; e) 0.13; f) 0.2; g) 0.33.

The flow pattern for the attached rib, c=0, is shown in Fig. 2a. When liquid moves, near the obstacle the flow is pressed considerably to the tube axis, and this leads to a significant velocity increase in the flow core. An extended recirculation zone and secondary vortex, typical to such a zone, are formed behind the rib. The direction of this secondary vortex is opposite to the main one, and the rotation speed there is relatively small, and all these cause formation of the stagnation zone. In the presence of the heat flux on the tube wall, this zone is characterized by the high temperature and low heat transfer coefficient, respectively. Therefore, from the point of engineering applications, this zone is not the most efficient.

The flow pattern starts changing, when a clearance between the tube wall and diaphragm appears. The wall jet brings significant changes in the structure of the recirculation region. The area of reverse flows, representing a pair of counter-rotating vortices, is formed behind a rib like behind a poorly streamlined body. At that, the velocity in the flow core decreases significantly.

It is interesting to consider the transitional regime, when a pair of vortices just starts to be formed, and a separation bubble reduces in size and moves downstream. One can observe this flow regime in Fig. 2-c-e, at the size of a clearance between the tube wall and diaphragm of 1-2 mm. Already at c = 3 mm the above picture is observed, when the pair of equal vortices replaces the primary and secondary vortices in the recirculation zone.

Distribution of the surface friction coefficients on the tube wall (Fig.4) also show a considerable reduction of the recirculation zone. At that, the reverse flow area along the tube wall degenerates significantly for the large clearance values c/h = 0.2 - 0.33. Actually, two vortices directed oppositely to the center of the recirculation zone remain behind the rib.

These features of the effect of parameter c/h also affect the behaviour of the surface friction coefficient, whose calculation results are presented in Fig. 4. The largest size of the zone of reverse flows, where the friction is negative, is observed at flow separation behind the attached obstacle (c/h=0). Respectively, the value of the friction coefficient for the attached obstacle recovers slower.

Distributions of the friction coefficient (Fig. 4) demonstrate that behind the separated rib with large gaps c/h = 0.2 and 0.33 the separation zone has the different structure, and the reverse flows are not formed along the tube wall. That is why in Fig. 5, which shows the relative length of the recirculation region, the value in these two cases is zero ($X_R/h=0$). The attachment point here was determined as the coordinate, where the surface friction on the wall is zero. Calculation results shown in Fig. 5 clearly demonstrate the influence of the clearance height on the size of recirculation zone. It can be seen that the length of a separation bubble is inversely proportional to the height of the clearance between the diaphragm and tube wall.

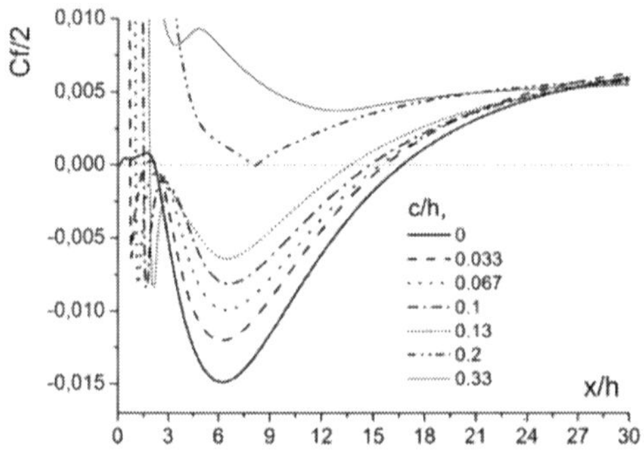

Fig. 4. The distribution of the friction coefficient on the tube wall behind the obstacle.

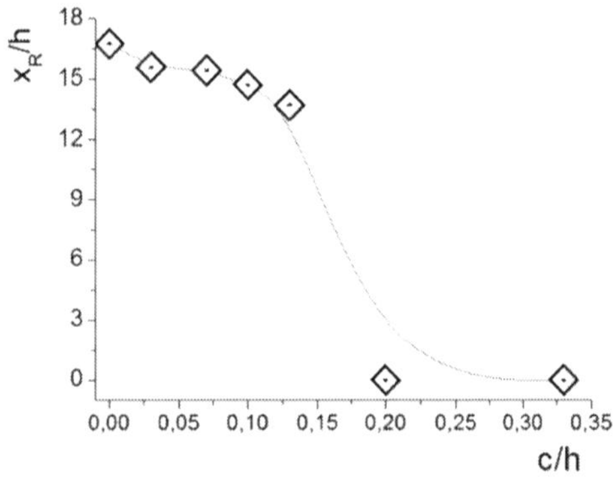

Fig. 5. Relative coordinate of the reattachment point.

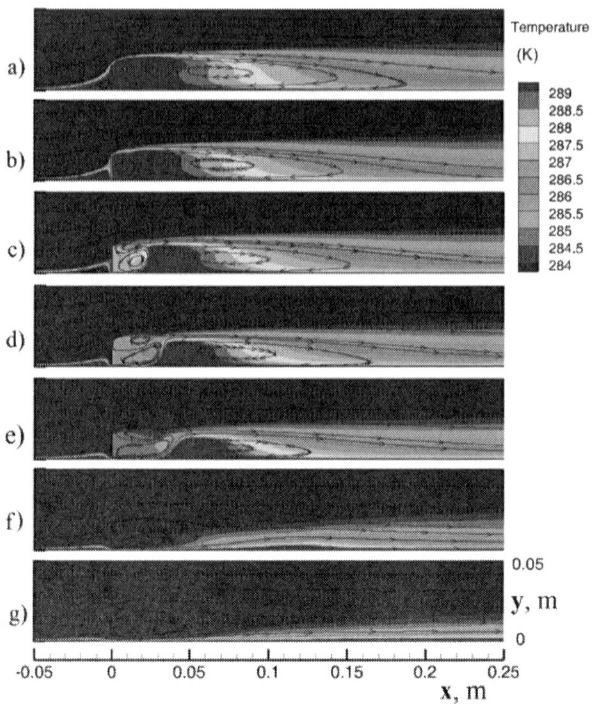

Fig.6. The stream lines and temperature field around the diaphragm. a) A = 0, attached diaphragm; b) 0.03; c) 0.07; d) 0.1; e) 0.13; f) 0.2; g) 0.33.

The temperature fields at the flow around a rib, removed at the different distances from the wall, are shown in Fig.6. Here, we can clearly see the stagnation zone with the increased temperature values (Fig. 6 a). It is obvious that with an increase in the clearance between the diaphragm and tube wall, this zone is removed downstream (Fig. 6 b - e) and degenerates gradually (Fig. 6 f - g).

When choosing the optimal shape of an obstacle for the purposes of heat transfer intensification, the determining parameters are the integral heat transfer coefficient, total hydraulic losses and thermohydraulic efficiency. The last parameter represents the ratio of heat transfer gain to the power increase, spent to overcome the increased hydraulic losses. In this paper, the calculation of the average heat transfer coefficient was determined by integrating the local distributions in the area, which covers the distance of 5 heights to the obstacle, zone directly under the obstacle, and distance of 25 heights after the obstacle ($-5 \leq x/h \leq 25$).

The results of calculation of the integral Nusselt number Nu_{sr} for different values of the studied parameters are shown in Fig. 11. Data are presented in the relative form, where Nu_0 is heat transfer in a smooth tube along the stabilized region.

When considering the average values of heat transfer in the specified region (Fig. 7), the maximum is observed at c/h=0.03. This is caused by the fact that at the clearance height of 0.5 mm, the recirculation zone remains almost of the same size as in the case of

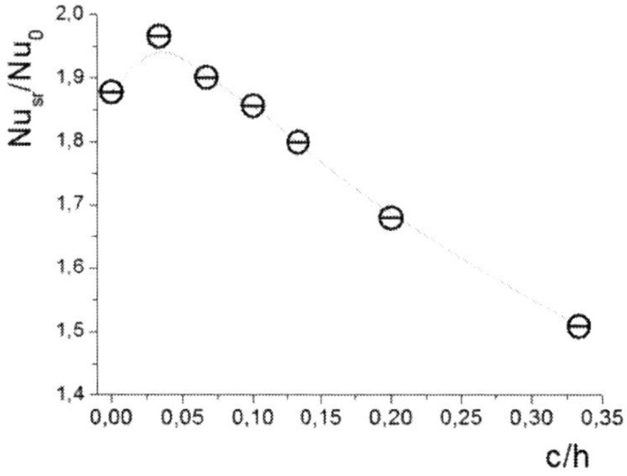

Fig.7. The relative values6 of the average heat transfer coefficient.

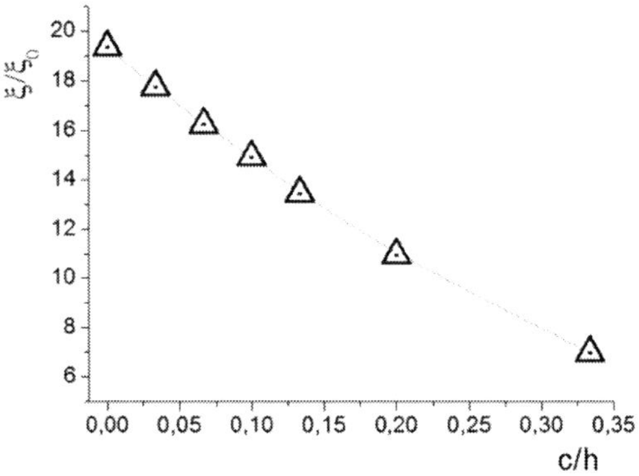

Fig. 8. The relative values of the coefficient of hydraulic losses.

theattached rib, but the formed wall jet gives the increased heat removal in the region under the diaphragm. Figure 8 shows the data for the hydraulic friction coefficient reduced to this value for a smooth tube.

We see that an increase in the scale of wall jet affects significantly investigation results. It is worth noting that with an increase in clearance height c from 0 to 5 mm, the area of the obstacle surface obstruction changes by the factor of 1.6. A decrease in the drag associated with the above fact also contributes significantly to the overall result. Finally, we observe the three-fold decrease in resistance with variation of the tested parameter from 0 to 0.33.

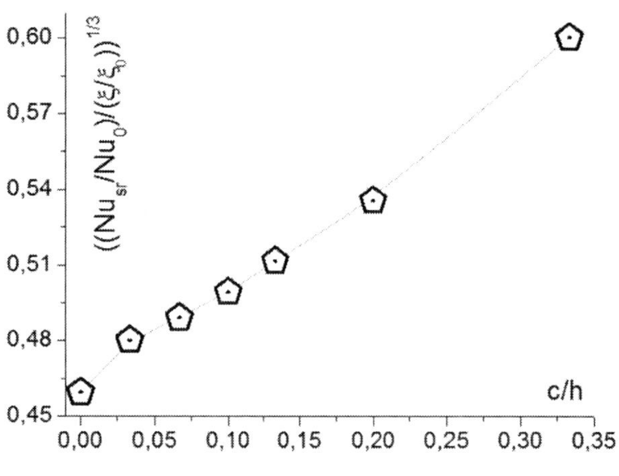

Fig. 9. Thermal enhancement factor depending on the height of the clearance between the diaphragm and the wall of the tube.

Data of thermal enhancement factor for different values of tested parameter c/h are shown in Fig. 9. As it can be seen from these data, with an increase in the clearance between the diaphragm and tube wall from 0 to 5 mm, thermal enhancement factor increases by 30%. Basically, this occurs due to a significant reduction in the coefficient of hydraulic losses (Fig. 8) because the relative value of Nu_{sr}/Nu_0 in this region of tube reduces only by 20%. It should be specially emphasized that an obstacle studied here gives the high values of hydraulic resistance and cannot be used directly as a heat transfer intensifier. However, the comparative analysis and identified flow and heat transfer features for the obstacles with different values of parameter c/h can be useful when selecting and optimizing the surfaces with intensified heat exchange.

4 Conclusion

The turbulent structure and heat transfer behind a flat diaphragm in a round tube were studied numerically with variation of a clearance between the tube wall and diaphragm A =c/h = 0 – 0.33.

It is shown that a wall jet formed in the slot leads to such a strong deformation of separated flow behind the diaphragm that for the large scale of the jet (A > 0.2) there is no repeated attachment of the flow. Due to the breakdown of recirculation zone, the level of kinetic energy decreases and it is taken downstream by the jet.

It is found out that an increase in the distance between the diaphragm and tube wall reduces maximal heat transfer significantly.

The integral characteristics show that that for clearance values c/h = 0.033 and 0.067 heat transfer becomes more intensive due to the increased heat removal on the tube wall under the diaphragm.

The pressure losses decrease by the factor of 3 at a change in the wall jet size from c/h = 0 to 0.33. At that, thermal enhancement factor increases by 30%.

This work was supported by the Russian President's grant NSh-8780.2016.8.

References

1. J.P. Tsia, J.J. Hwang, Int. J. Heat Mass Transfer **42**, 2071 (1999)
2. T.M. Liou, C.P. Yang, H.L. Lee, ASME J. Heat Transfer **119**, 383 (1997)
3. T.M. Liou, S.H. Chen, International Int. J. Heat Mass Transfer **41**, 1795 (1998)
4. A. Valencia, Numer. Heat Transfer, Part A. **37**, 289 (2000)
5. J. Ahn, J.S. Lee, Int. J. Heat Mass Transfer **53**, 445 (2010)
6. S. Eiamsa-ard, W. Changcharoen, Heat transfer - Asian Research **40**, 20357 (2011)
7. C. Thianpong, K. Yongsiri, K. Nanan, S. Eiamsa-ard International Communications in Heat and Mass Transfer **39**, 861 (2012)
8. V. Ozceyhan, S. Gunes, O. Buyukalaca, N. Altuntop, Applied Energy **83**, 280 (2006)
9. K. Nanana, M. Pimsarnb, W. Jedsadaratanachaib, S. Eiamsa-arda, Int. Commun. Heat Mass **48**, 133 (2013)
10. Jenn-Jiang Hwang, Chung-Hsin Chao, J. Thermophys. Heat Transf. **14**, 579, (2000)
11. F.R. Menter, AIAA J. **32**, 269 (1994)

Physical and numerical simulation of two-phase flow in the area of mini-channel

Boris V. Borisov[1*], *Evgeny* S. Chachilo[1]

[1]National Research Tomsk Polytechnic University, 634050 Tomsk, Russia

Abstract. The results of physical and numerical modeling to determine the rate of evaporation of ethanol, at a flow of air flow velocities in the range of 0.0139 - 0.139 m/s and temperature in the range of 10 - 45 °C are placed. The time of the establishment of the diffusion equilibrium in the work area and differences in the results of numerical and physical modeling are have determined.

1 Introduction

Currently, there are an increasing number of studies in the field of heat and mass transfer and hydrodynamics of currents in minichannel. There are growing need for installations able to transfer large thermal flows in the limited space and volume. These equipments have a wide range of applications, from compact evaporators and condensers of power devices to cryogenic cooling systems of computer processors.

Not enough studies on the heat-mass transfer when driving two-phase flow in the area of Mini- channels and so often the experimental date are difficult evaluatived and are explaned. To solve these problems the numerical simulation is applied.

The aim of the work is physical and numerical modeling of heat and mass transfer on the surface of a thin liquid film with air stream.

2 Physical modeling

Sufficiently detailed description of the methodology for determining the mass evaporation rate is contained in [1].

Work area (Figure 1) consists of a liquid chamber-term expanse 40 mm and a height of 3 mm and a rectangular gas duct with height of 3 mm, width 40 mm and length 150 mm. The upper wall of the liquid passage - stainless plate with a square neckline 100 mm2 of area , tin which occurs contact of liquid with the gas and formed a local area of evaporation.

We study the dynamics of evaporation of ethanol fixed film (ГОСТ Р 51723-2001) with a thickness of 3 mm at a flow of air. Work area is horizontal with the atmospheric pressure in the circuit. During the experiment, the gas flow rate ис varied ranged from 100 to 1000 ml / min, which corresponds to a gas velocity of 0.0139 to 0.139 m /s. Gas mixture

* Corresponding author: bvborisov@tpu.ru

flow rate was determined at the outlet of the working area. The experiment was repeated for gas and liquid temperatures ranging from 10 to 45 ° C, respectively.

3 Numerical Simulation

The two-dimensional problem of two-phase flow in the rectangular mini-channel with length L = 120 mm and height H = 3 mm (Figure 1) is investigated. The liquid is in saturation condition, at the saturation pressure determined for a given temperature.

The following assumptions have been taken to solve this problem:

a) Since the gas channel is much greater than its height and width, a flat case is considered.
b) Maximum velocity of the gas in the channel in agreement Re≈100, so it is assumed laminar flow regime.
c) The pressure drop in the channel and the gas velocity are low.
d) Liquid temperature and gas is the same, which excludes the effect of temperature gradient on heat and mass transfer process in the channel.
e) Couples vaporized of fluid not introduce substantial changes to the dynamic of flow, since the density of gas in the mixture is about 9 times higher than the density of the vapor.

From these assumptions, it follows that the gas in the channel profile corresponds to the profile Poiseuille, which is used for modeling the concentration field in the moving medium:

$$\frac{\partial c}{\partial \tau}+u_g(y)\frac{\partial c}{\partial x}=D\left(\frac{\partial^2 c}{\partial x^2}+\frac{\partial^2 c}{\partial y^2}\right), \tag{1}$$

where c - the concentration of the substance, τ - time, c; D - the diffusion coefficient, m^2/s; $u_g(y)$ - distribution of the horizontal component of gas velocity along the channel height, m /s; wherein u_g - averaged over the cross section rate of gas flow.

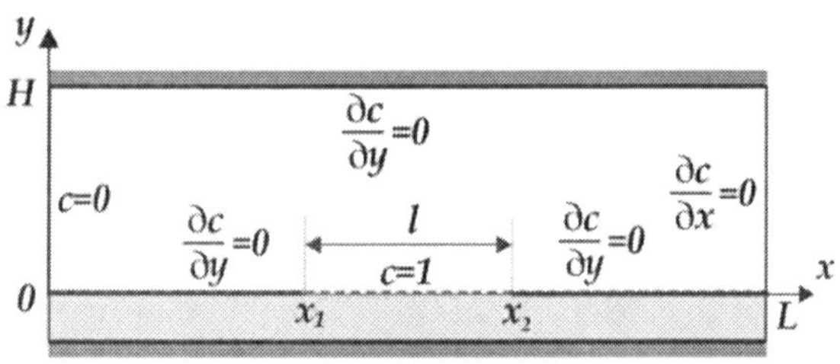

Fig. 1. Schema of simulation area.

At the initial time are believed that the velocity of the gas and liquid vapor concentration is equal to zero:

$$\tau=0,\ 0<x<L, 0<y<H:\ u_g=0,\ c=0. \tag{2}$$

It is believed that the gas entering the work area is a clean air:

$$\tau>0,\ \ x=0,\ \ 0<y<H:\ c=0. \tag{3}$$

On the right edge is believed that the concentration profile and speed profile are installed and there are no further changes of its:

$$\tau > 0,\ \ x = 0,\ \ 0 < y < H: \ \frac{\partial c}{\partial x} = 0. \tag{4}$$

On the solid boundaries of the channel set impermeability conditions:

$$\tau > 0,\ \ (\ 0 < x < x_1;\ x_2 < x < L, y = 0\ ) \cup (0 < x < L, y = H): \ \frac{\partial c}{\partial x} = 0. \tag{5}$$

On the border of the "gas – liquid" section relies presence of saturated vapor in volume:

$$\tau > 0,\ \ y = 0,\ x_1 < x < x_2 : \ c = 1. \tag{5}$$

The differential equation (1) was solved by the sweep method for implicit scheme with second-order approximation in space.

4 Analysis of the results

Figure 2 shows how the rate of evaporation of the liquid is changed over time in numerical modeling. Parish mass due to evaporation of the liquid from the gas-liquid interface per unit time is denoted as J_1, the mass of the evaporated liquid from the outlet channel is denoted as J_2.

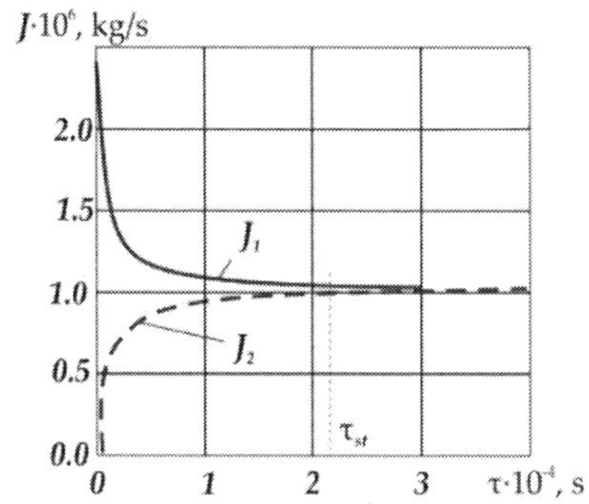

Fig. 2. The dependence of the rate of evaporation of the liquid from the experiment: J_1- flow mass vapor from the liquid surface, kg/s; J_2 - mass flow of vapor through the outlet section of the working part of the channel, kg/s.

From the analysis of the results (Figure 2) shows that with increasing time of the experiment, the difference between the values of J1 and J2 is reduced and becomes almost zero, which corresponds to the steady process of convective diffusion in the work area. The time required to establish equilibrium is referred to as τ_{st}. It is selected so that the difference in the values of J_1 and J_2 does not exceed 4%. It is with such an accuracy was fixed volumetric flow of vapor-gas mixture at the outlet of the channel during the experiment. The result found that τ_{st} is placed in the range of $17.8 \cdot 10^4$ s to $2.3 \cdot 10^4$ s.

Figure 3 shows a comparison of the results of determining the rate of evaporation of the liquid in the numerical and physical modeling, depending on the temperature of the gas and liquid. Analyzing the difference in the results of numerical and physical modeling, it can be concluded that the maximum discrepancy is observed at a temperature of liquid and gas ~ 23.5 °C and further significantly decreases with increasing temperature. In general, the difference was between 1 and 36%.

DOI: 10.1051/matecconf/20179201064

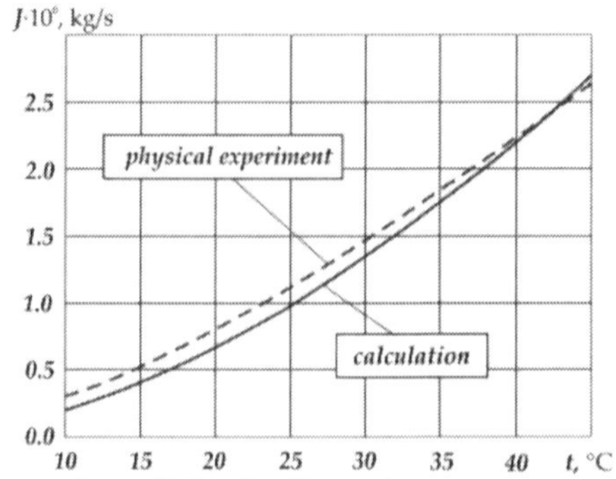

Fig. 3. The velocity of evaporation of the liquid at the temperature of a liquid and a gas and u_g =0.1 m/s.

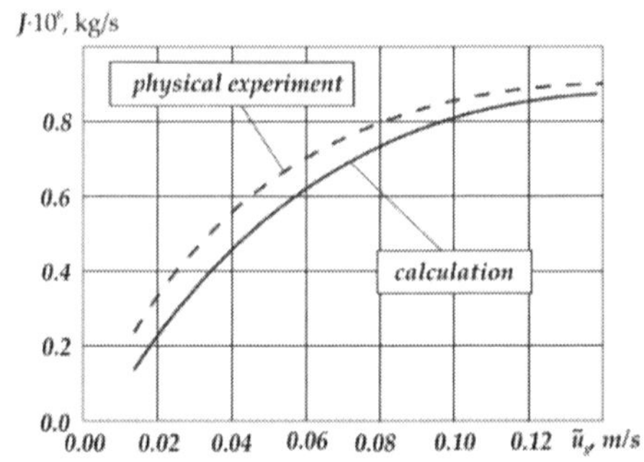

Fig. 4. The dependence of the rate of evaporation of the liquid at gas velocity and at T = 20 °C.

The discrepancies in the results of numerical and physical model-ing, depending on the velocity of the gas, have a similar relationship, the maximum is observed at lower gas velocities of ~ 0,035 m/s and decreases with increasing speed. In general, the difference is from 4 to 38%.

It is assumed that the differences in the results are explained by the difference in the duration of the experiment. Physical experiment was performed at the same setting temperature of the liquid and gas, to eliminate the influence of a temperature gradient in the process mass-transfer liquid film and the gas flow, however, is not taken into account the beginning of a stationary diffusion. Numerical modeling results show that the time required to establish equilibrium of about 5-6 hours, while the experiment was carried out in about 5 -. 60 minutes after the start of the plant work.

4 Conclusion

Thus, have done physical and numerical study of the dynamics of evaporation of ethanol in dependence on the gas velocity, as well as different temperature and gas liquid.

The value of the rate of evaporation. The discrepancy between the results of the numerical and physical modeling is in the range from 1% to 38%. Detected discrepancies maxima at 23.5 °C and a constant gas velocity of 0.1 m / c, and at a rate of 0.035 m/s and a constant temperature of 20 °C.

Reference

1. E.S. Chachilo , M.N. Grekhov, D.V. Feoktistov, MATEC Web Conf. **19**, 1 (2014)

Enhancing efficiency of using water due to explosive breakup of liquid drop

Anastasia G. Borisova[1], *Vladislav* M. Gumerov[1], *Maxim* V. Piskunov[1,*], and *Denis* V. Ushmaev[1]

[1]National Research Tomsk Polytechnic University, 634050 Tomsk, Russia

Abstract. Using high-speed video recording, the experiments were performed to research quantitative characteristics of explosive breakup phenomenon of 5–15 μl water droplets containing 2×2×1 mm and 2×2×2 mm solid inclusions, when heated in a tube furnace at temperatures of 1070–1370 K. Experimental results report number and size of the droplets detached during explosive breakup. We show that the fragmentation of liquid layer covering solid particles facilitates the increase the evaporation surface area 15-fold versus the initial surface area of a drop.

1 Introduction

Experimental studies [1, 2] report the phenomenon of explosive breakup of liquid layers of heterogeneous water drops (liquid volume – 5–15 mcl, size of single solid inclusion is 50–70 % of liquid volume), which is implemented by the intensive evaporation at the interface of solid particle / liquid at high temperatures. Fig. 1 illustrates processes of heating and explosive breakup of heterogeneous drop. Analysis of research results [1, 2] allows us to select conditions facilitating the process of explosive breakup. To provide practical applications discussed in [1, 2] with data on quantitative characteristics of the explosive breakup, it is necessary to estimate such parameters in experiments as number, size and total surface area of the liquid after breakup. Such information will contribute to the enhancing resource efficiency, energy efficiency of high-temperature thermal treatment technologies of liquids [3] and firefighting by liquids with impurities [4].

The aim of this paper is to estimate number, size and surface area increase in the process of explosive breakup of heterogeneous liquid drops at high-temperature heating.

* Corresponding author: piskunovmv@tpu.ru

DOI: 10.1051/matecconf/20179201065

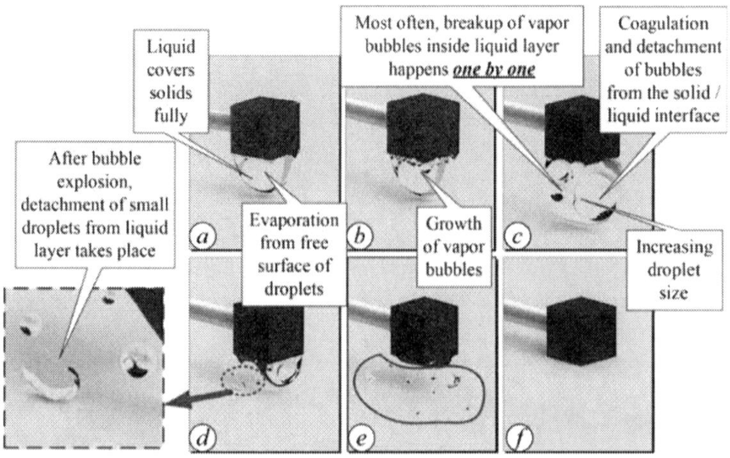

Fig. 1. Scheme of explosive breakup of heterogeneous liquid drop.

2 Experimental setup and research methods

Experimental procedure is generally similar to that used in [1, 2]. The main difference is a using the tube furnace to produce the high-temperature area. The tube furnace has an integral regulator and type *S* thermocouple, which provide the temperature monitoring and the demanded temperature regime. Temperature varies in the range of 1073–1373 K. Importantly, the tube furnace is equipped with the heating tube. Such design enables to install high-speed camera from the one side and to introduce a heterogeneous drop in heating area from the other one. Heating of a drop is performed at the center of the tube.

The heterogeneous drop consists of the solid particle made of natural graphite in shapes of cube and parallelepiped with dimensions 2×2×1 mm and 2×2×2 mm, as well as the water layer (water volume – 5–15 mcl; temperature – 298 K), which covers an inclusion fully. As in [1, 2] the solid particle is mechanically fixed on the thin rod made of ceramics. Then, the dosing device produces drops of the required volumes. Introducing heterogeneous liquid drop – thin rod system into the high temperature air in the tube body is implemented by the automatic mechanism. High-speed video recording allows determining linear dimensions of, for example, the formed small water droplets. The recorded video data are processed by Phantom Camera Control software. Together with drop size, owing to high recording speed the number of the droplets formed during the breakup is determined visually. These obtained data are used to calculate the total area S_{out} of the formed water droplets. This parameter basically characterizes increase of evaporation surface area. Errors of measuring dimensions of the droplets formed during the breakup are not more than 0.001. A series of the experiments at identical conditions includes seven tests, which are performed in the mentioned temperature range with step of 50 K. In each series, size of solid particles and initial water volume vary.

3 Results and discussion

Figs. 2 and 3 illustrate the influence of the ratio between initial volumes of water and inclusion for different dimensions of the latter on the increase of the total evaporation surface area due to the breakup of the liquid layer. Note that the maximum number of small droplets corresponds to the minimum ratio between initial volumes of water and inclusion. This effect is the most obvious for the droplets with the size up to 0.5 mm (Fig. 3).

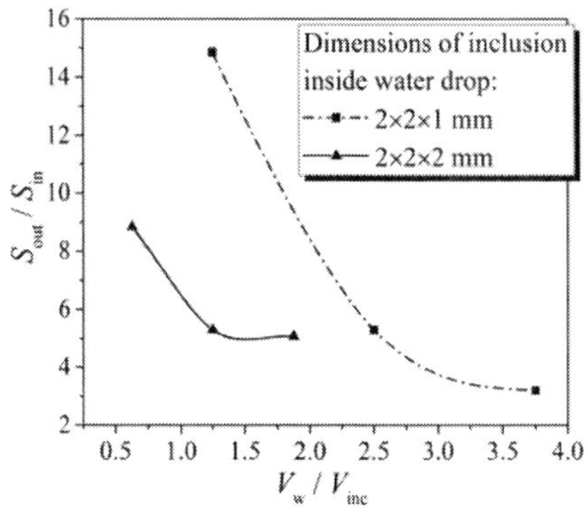

Fig. 2. S_{out} / S_{in} versus V_w / V_{inc} for three initial volumes of water (5 μl, 10 μl and 15 μl) and two dimensions of inclusions (2×2×1 mm and 2×2×2 mm).

Taking into account experimental data in Figs. 2 and 3, the maximum ratio of S_{out} / S_{in} will be under condition of V_w ~ V_{inc} or even V_w < V_{inc}. If the introduction of additional volume of solids is difficult or inappropriate, then the selection of optimum ratio between V_w and V_{inc} can be implemented by using the data reported in Figs. 2 and 3.

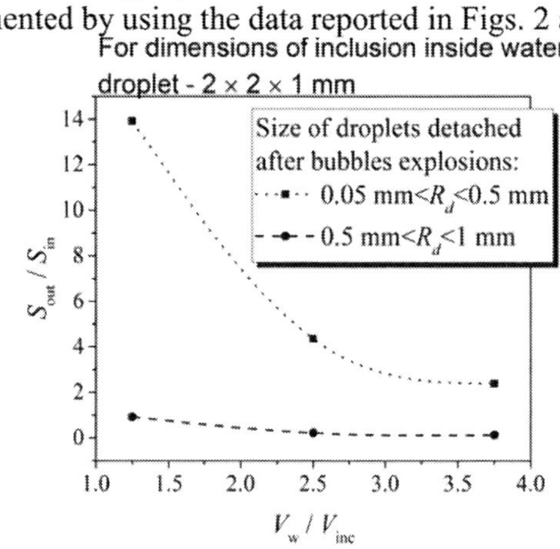

Fig. 3. S_{out} / S_{in} versus V_w / V_{inc} for three initial volumes of water (5 μl, 10 μl and 15 μl) and inclusion dimensions of 2×2×1 mm (taking into account the size distribution of the formed droplets)

4 Conclusion

After explosive breakup of the liquid layer covering solid particles into small droplets the evaporation surface area increases 15-fold versus the initial surface area of a drop.

The investigation was supported by the grant (MD-1221.2017.8) and scholarship (SP-1049.2016.1) of President of the Russian Federation.

References

1. G.V. Kuznetsov, M.V. Piskunov, P.A. Strizhak, Int. J. Heat Mass Tran. **92**, 360 (2016)
2. R.S. Volkov, G.V. Kuznetsov, M.V. Piskunov, P.A. Strizhak, J. Heat Transf. **138**, 014502 (2016)
3. R.S. Volkov, G.V. Kuznetsov, P.A. Strizhak, Int. J. Thermal Sci. **88**, 193 (2015)
4. D.I. Kofman, M.M. Vostrikov, A.V. Antonenko, Chem. Pet. Eng. **45**, 577 (2009)

Engineering computational method of process of fluorination metal tungsten

Roman V. Brendakov[1*], *Alexander* V. Shvab[1], and *Vladimir* N. Brendakov[2]

[1]Tomsk State University, 634030 Tomsk, Russia
[2]Seversk Technological Institute NRNU MEPhl, 636070 Seversk, Russia

Abstract. In this paper, we consider the process of obtaining the tungsten hexafluoride gas. On the basis of a simple one-component diffusion model of the process of fluorination of tungsten metal recorded an equation relating the rate of a chemical reaction and change in concentration of a key component of the length of the reaction zone. The numerical calculations showed good agreement with experimental data. The statistical treatment of experimental data, constructed regression model for the reaction speed parameter fluoridation as a function that depends on the temperature of the supplied gas and tungsten metal particle size. The results can be used to optimize the design of existing devices and new designs for the fluorination process of metal tungsten.

1 Problem statement

Process of fluorination of a metal tungsten is carried out according to the chemical reaction presented by a formula

$$W + 3F_2 = WF_6 \tag{1}$$

To realize fluorination process, gaseous fluorine is passed over a layer of metal tungsten. Passing through the active zone of the reactor, fluorine interacts with tungsten powder to form tungsten hexafluoride (WF_6). Figure 1 shows a schematic diagram of the reaction zone.

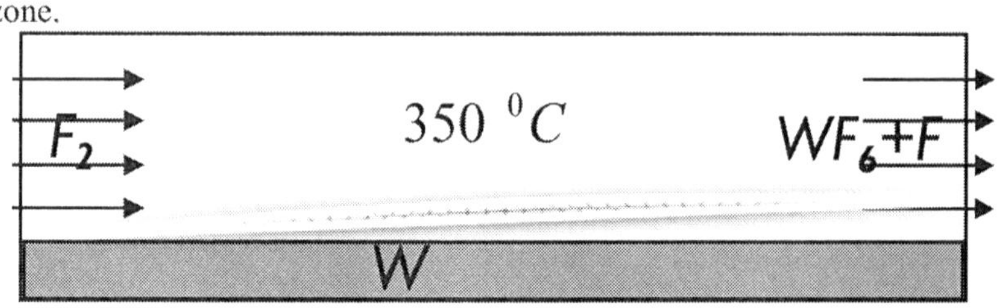

Fig. 1. Scheme of the reaction zone.

As a result of such reaction, concentration of fluorine in a gas stream gradually decreases due to decrease of amount of fluorine and a dilution by its formed WF_6. Decrease

[*] Corresponding author: vnbrendakov@mephi.ru

of concentration of fluorine in bipropellant mix leads reaction rate of fluorination to decrease. At the same time, according to stoichiometric coefficient equal 3, there is a decrease of speed of a gas stream.

In work it is modelled steady-state conditions bipropellant mix of gases of a one-dimensional current. The mathematical description of the considered task can be submitted by the equation of one-component diffusion model and the corresponding boundary conditions:

$$\frac{1}{Pe}\frac{d^2C(x)}{dx^2}=\left(1-\frac{2}{3}C(x)\right)\frac{dC(x)}{dx}-k\cdot(1-C(x)) \quad (2)$$

where $Pe = U_0H/D$ – number of Peclet, $C = C(x)$ - concentration of a key component (tungsten hexafluoride), k – a specific reaction rate.

In work [1] the formula for calculation of reaction rate of fluorination is offered:

$$V = k\cdot C_F, \quad (3)$$

where C_F - concentration of gaseous fluorine on the surface of reaction of fluorination.

2 Numerical calculations

As practice shows, the specific reaction rate of k keeps value only under the given conditions. For example, in work [2] it is directly told "Through a difference between temperature on the reacting surface here and temperature of a gas stream expressed reaction rate of fluorination of a tungsten. Temperature difference of subjects is more, than reaction rate is more". And in work [3] the table on dependence of value of a constant of reaction of fluorination on temperature is provided in a zone of reaction (T) and distribution of sizes (δ) of powder of a metal tungsten. Our regression analysis of the available experimental data allowed to build a functional dependence of the reaction rate constants for fluoridation:

$$k(\delta, T) = 180\cdot\delta^{(-0{,}073)}\cdot e^{(-3970/T)}. \quad (4)$$

For check of adequacy of the developed engineering computational method of process of fluorination of a metal tungsten calculations at value of the constants Pe were carried out = 1.0, by k = 0.14 and comparison of the theoretical graph of concentration of a key component from the longitudinal coordinate and the experimental data from works is executed [4, 5].

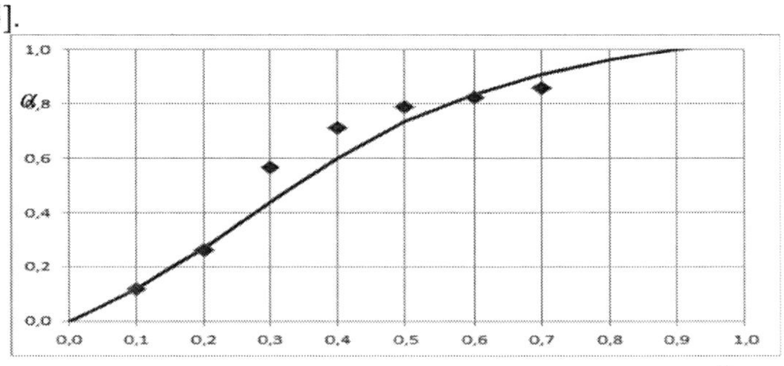

Fig. 2. Dependence of concentration of a key formulation constituent on length reaction zones, solid line – calculation for model, ▲ - data of works [4, 5].

3 Conclusions

Reasonably good agreement between the calculation and experimental data leads to the conclusion of the performance created engineering calculation method. This approach to the calculation of tungsten fluorination process will carry out rapid assessment of options for the optimal conditions of the process of fluorination of refractory metal powders.

References

1. Yu. M. Korolev, V. I. Stolyarov, *Restoration of fluorides of refractory metals hydrogen* (Metallurgy, Moscow, 1981)
2. Patent No. 2310608 – Way of receiving hexafluoride of tungsten. Authors: A. A. Galatian, A. G. Krupin, S. A. Kuzminykh, V. V. Lazarchuk, E. P. Marinenko, A. I. Mines., V. A. Khokhlov (Date of the publication 11/20/2007)
3. Yu. M. Korolev, *International Symposium on Inorganic fluorides: Chemistry and Technology* (TPU Publishing, Tomsk, 2014)
4. T. Sh. Agnokov, Yu. M. Korolev, M. F. Svidersky, Nalchik: KBGU, **4**, 22 (1978)
5. T. Sh. Agnokov, Yu. M. Korolev, M. F. Svidersky, Nalchik: KBGU, 18 (1979)

Numerical simulation of hydrodynamics and heat transfer in the technological reservoir with the heat pump

N.V. Demyanovich[1], *V*.I. Maksimov[1,*], *T*.A. Nagornova[1]

[1]National Research Tomsk Polytechnic University, 634050 Tomsk, Russia

Abstract. The mathematical modeling of mixed convection of viscous incompressible fluid in a rectangular reservoir with the sources of input and output of mass has been carried out. Is studied the influence of heat sink as a result of the work of heat pump and its position on the formation of convective flows and a change in the temperature conditions. Are revealed basic laws governing the heat transfer in the small environment of heat pump. The analysis of the influence of Reynolds numbers on the intensity of the heat exchange is carried out.

1 Introduction

The contemporary tendencies of the production of thermal and electrical energy by eco-friendly and cost-effective methods caused development and application in power engineering of heat-pumping systems and installations [1,2]. But to use them effectively requires an analysis of thermal conditions water reservoirs [3,4]. The solution to this problem can be carried out within the framework of natural convection models for an incompressible fluid in a closed area with a local energy sources (eg, as in [5,6]). But the results of experimental studies show that in same cases of liquid flow in the reservoirs with a heat pumps more greatly correspond to the regime of mixed convection. Therefore with the estimation of effectiveness in the work of heat pumps it is expedient to consider the special features of flow of liquid near this energy converter in the regime of the mixed convection. The results of the numerical simulation of the combined convective-conductive heat transfer in the rectangular gas region with source of heat emission, are known [7-8]. But the simulation of convective flows in the reservoir under the conditions of complex heat exchange with the environment, and also inflows and the heat of sink, so far not been carried out.

Therefore, it is appropriate numerical study of mixed convection of a viscous incompressible liquid in a rectangular area with a local heat sink and sources of input and output mass.

* Corresponding author: elf@tpu.ru

2 Problem statement and the method for the solution

Viewed rectangular cavity with a local heat sink (heat exchanger-evaporator heat pump) having input and output portions and a liquid free surface (Fig. 1).

Thermal properties of the liquid and the wall material is not temperature dependent. Liquid was considered heat-conducting, viscous, newtonian, that satisfies the Boussinesq approximation. Flow conditions – laminar, the boundaries of cavity (with exception of the sections of input and output of water) – are not penetrated. The draining of mass due to the evaporation from the free surface was not considered. It was assumed that the temperature of the entire region in question is constant and identical at the initial moment of time.

These assumptions are not introduced into the formulation of the problem of significant limitations, but they make it possible to simplify its solution. The temperature of heat exchanger- vaporizer was received as constant during entire process.

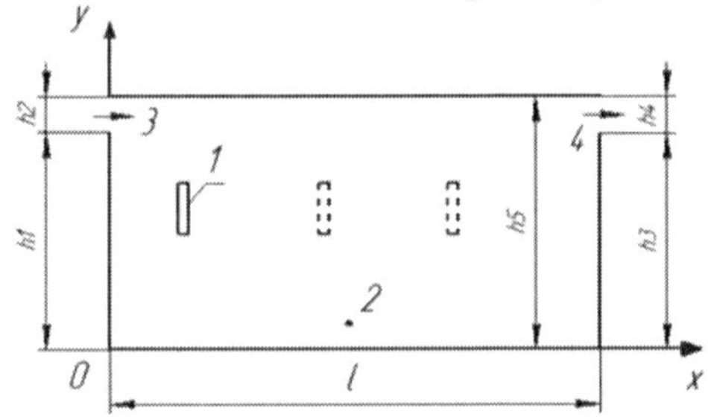

Fig. 1. The area of solving the problem: 1- heat exchanger-evaporator; 2 – fluid; 3, 4 - areas of fluid input and output.

In the section of the mass input (figure 1) speed in the direction *X* is received as constant and the conditions of symmetry are accepted at the output from the cavity. Are analyzed three versions of the position of heat exchanger- vaporizer (Fig. 1).

In the real reservoirs the realization of the three-dimensional distributions of temperatures and components of speeds is possible, but under the conditions in question the velocity component in the direction *Z* there will be much less than other two components, which correspond to the reference plane (*X*,*Y*) of the movement of the water introduced into the reservoir. Therefore it is possible to consider assumption about the two-dimensional formulation of the problem as that substantiated.

Process of heat transfer in the liquid (Fig. 1) described by the system of nonstationary two-dimensional Navier-Stokes equations in the Boussinesq approximation with the nonlinear boundary conditions [8]. The problem was solved in dimensionless formulation similar to [9].

The solution of equations (1–3) with initial and boundary conditions is carried out by the finite-difference method on the uniform grid. Used algorithm [5,6] developed for the numerical solution of problems of natural convection in a closed rectangular areas with local energy sources.

3 Results and discussion

Numerical studies are carried out with the following values of the dimensionless quantities, which correspond to the regime of the mixed convection: Re = 500 ÷ 1000, Gr = $10^5 \div 10^6$. Temperatures: T_0 = 278 K, T_{in} = 293 ÷ 313 K, T_i = 274 K, T_e = 258 ÷ 278 K.

Figure 2 shows the current lines and temperature field, the characterizing regime of mixed convection of a viscous incompressible fluid in the rectangular region for different variants of the position of the heat pump vaporizer in the reservoir. It is seen, that the heat overflow occurs mainly in the direction of movement of the main stream. In the first position of the heat pump evaporator can selected two circulation flows: before and after heat exchanger (Fig. 2 b).

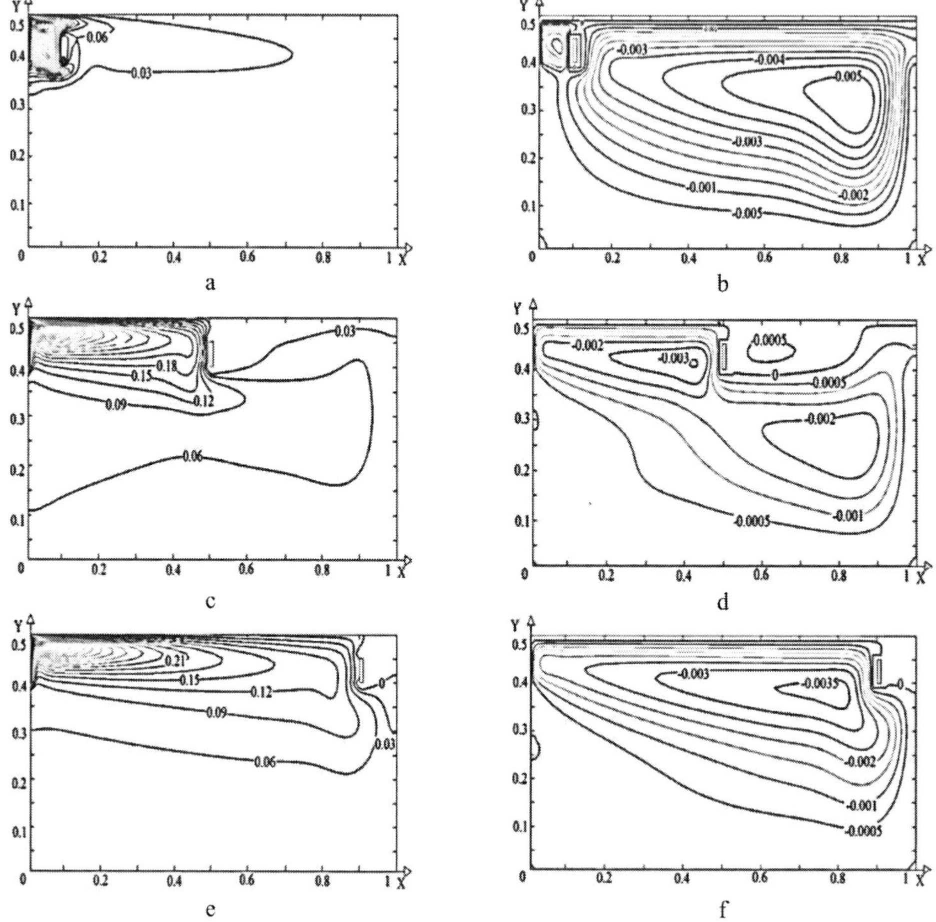

Fig. 2. Field of temperature (*a*, *c*, *e*) and current lines (*b*, *d*, *e*) for various positions of the heat pump.

As a result, immediately after the section of the input of liquid is formed the zone, which prevents the intensive mixing of liquid in the reservoir, which leads to the decrease of the intensity of the heat exchange of vaporizer with the surrounding liquid. In other variants of the position of heat exchanger is formed a main extensive circulation area. This circulation flow appears as a result of the flow around heat exchanger and contributes to the intensive mixing of liquid. In all three versions of the vaporizer location the cold water, which is located in the left lower region angle, does not get mixed with the main flow of liquid.

Displacement of vaporizer to the right wall (Fig. 2 *d*, *e*) lead to the deformation of extensive stagnant low-temperature zones and intensify heat exchange between liquid and vaporizer. The temperature distribution in this case is sufficiently even throughout entire volume of reservoir (Fig. 3).

Figure 3 shows the temperature profiles in the vertical section passing through the middle of the reservoir at $X = 1/2$, at different positions of the heat pump heat exchanger-evaporator. In all cases, besides the arrangement of vaporizer in the position 2 (Fig. 2 *c*, *d*), with which the heat exchanger falls in the section, the maximum temperature is observed in the upper part of the cavity.

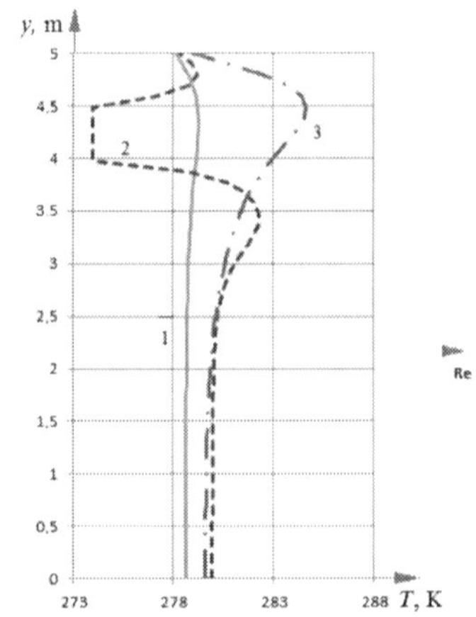

Fig. 3. Temperature distribution in the section, which passes on to center cavities with $X = l/2$, during different arrangements of the heat exchanger of the heat pump vaporizer.

The analysis of the influence of Reynolds number on the values of average Nusselt number [6-8] on three boundaries of reservoir and free surface of a liquid is carried out. It can be seen (Fig. 4), that with increasing number Re at the boundaries $X = 0$ and $X = 1$ (Fig. 4, *b*) a decreases the dimensionless heat transfer coefficient values for all variants of placement of the evaporator heat exchanger. The position of heat pump in the left and middle part of the reservoir prevents to the developed circulation flow in the center zone.

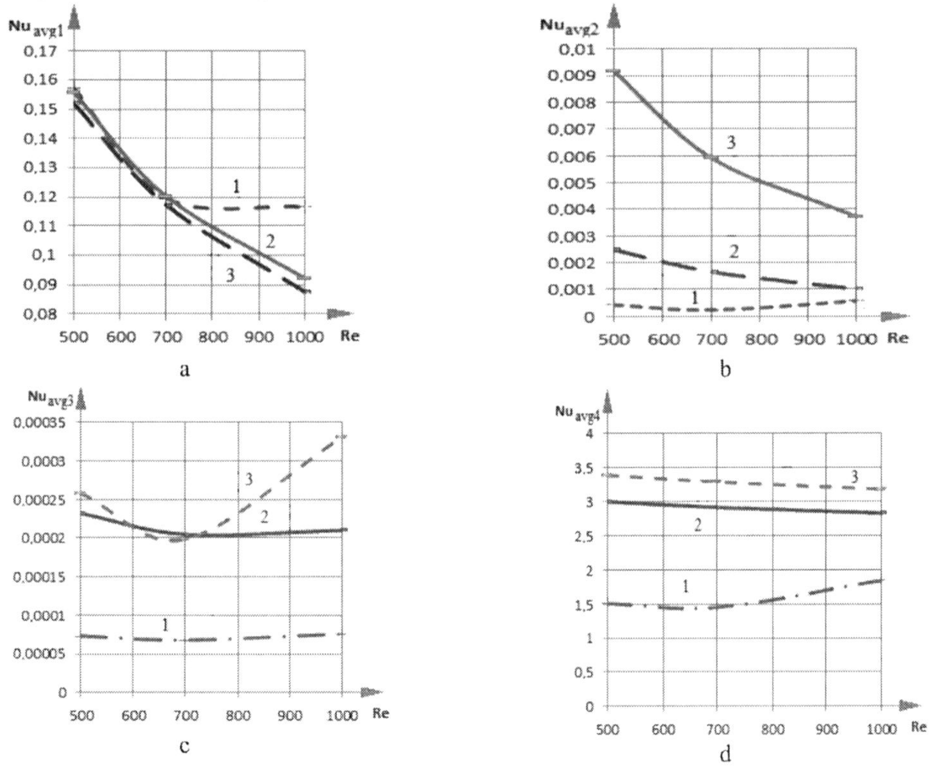

Fig. 4. Dependence of average Nusselt number on Reynolds number with different position of the heat exchanger- heat pump vaporizer.

At the same time, depending $Nu_{avg} = f\,(\mathrm{Re})$ for the boundaries $Y = 0$ and $Y = 1$ of extreme left and right placement of the heat exchanger are different. Are separated well corner points (Fig. 4 *c*, *d*). This position of the heatsink with the low speeds of the water entry prevents the advance of fluid flow with the temperature $\Theta = 1$ to the outlying zones (in the angles) of the considered region and contact with the adjacent walls. But with increasing circulation rates are achieved sufficient conditions for the overflow of the part of fluid energy behind of evaporator located to the left (Fig. 2 *a*, *b*). Position of heat exchanger to the right (Fig. 2 *e*, *f*) prevent the moving fluid flow to reach the lower part of the boundary $X = 1$. This factor has an effect on the average coefficient of heat exchange (Fig. 4 *a*, *d*).

With numerical studies it is established that Nusselt number on the free surface of reservoir (Nu_{avg}) (Fig. 4 *d*) is relatively high. On the solid boundaries Nu_{avg} does not exceed 0.16, on a free same - the average Nusselt number reaches 3.4. On the liquid border $Y = 1$ is achieved the intensive heat outflow, but with an increase of Reynolds number the range of a change in the values of average on this boundary of Nusselt number is insignificant.

4 Conclusion

It was found that the dependences $Nu_{avg} = f(Re)$ change with different schemes of placing the evaporator heat exchanger, which allows to make a conclusion about changing temperature regimes due to displacement of heat sink in the reservoir.

The formulated model of the mixed convection of liquid in the reservoir, used as the low-potential energy source for the heat pump vaporizer, with the local input and output sections of the mass can be used to improve the methodology for calculating the thermal conditions of the reservoirs – low potential energy source.

The work was supported by the Russian President's grant (Scientific School project - 7538.2016.8).

References

1. K.J. Chua, S.K. Chou, W.M. Yang, Appl. Energ. **87**, 3611 (2010)
2. Abdeen Mustafa Omer, Renew. Sust. Energ. Rev. **12**, 344 (2008)
3. J.I. Shun-wen, Z. Yue-ming, Q. Sheng, J. Hydrodyn. **20**, 507 (2008)
4. M.C. Palancara, J.M. Arago´n, F. Sa´nchez, R. Gil, J. Environ. Manage. **81**, 210 (2006)
5. G.V. Kuznetsov, N.I. Kurilenko, V.I. Maksimov, G.Ya. Mamontov, T.A. Nagornova, J. Eng. Phys. Thermophys. **86**, 519 (2013)
6. G.V. Kuznetsov, M.A. Sheremet, Fluid Dyn. **41**, 881 (2006)
7. G.V. Kuznetsov, V.I. Maksimov, M.A. Sheremet, J. App. Mech. Techn. Phys. **54**, 588 (2013)
8. G.V. Kuznetsov, V.I. Maksimov, J. Eng. Thermophys. **16**,270 (2007)
9. A.S. Krasnoshlykov, V.I. Maksimov, EPJ Web Conf. **82**, 01046 (2015)

The analysis of combustion heat of the metallized solid fuel compositions

Yana Dubkova[1,*], and *Alexander* Zhukov[1]

[1]National Research Tomsk State University, 634050 Tomsk, Russia

Abstract. The technique and results of the pilot study of power characteristics of the solid-fuel compositions containing metal additives are shown. Results of the comparative analysis of influence of various additives on the caloric content of fuel compositions are given.

1 Introduction

Fuel for the modern aircraft has to answer a big number of requirements: to have high specific impulse, high density, the required aggregate state of components under operating conditions, has to be stable, safe in the address, nontoxical, compatible to constructional materials, have raw material resources etc. These conditions, in turn, provides reliable and long operation of aircraft. Along with above-mentioned one of the major power characteristics is combustion heat or caloric content of fuel Q. Combustion heat (caloric content) of material is defined as amount of thermal energy which is emitted at the complete combustion of a mass unit of fuel [1], [Q] = J/kg or [Q] = a cal/kg. Now much attention is paid to development of high-energy materials (HEM). For effective use of HEM it is necessary to know their main characteristics. The technique and results of a research of combustion heat of the exemplars made of new compositions of high-energy materials are given in this work.

2 Materials and methods of the experimental research

Exemplars of the mixed solid fuels [2] containing in quality combustible-binding divinyl rubber of the SKDM-80 brand as an oxidizer – bidispersny powder of perchlorate of PHA ammonium / were experimentally investigated by. For increase in power characteristics fuel compositions powders of diboride of a titanium TiB_2, AlB_2 dialuminium boride, amorphous pine forest B, AlB_{12} polyaluminium boride, powder of aluminum of the ASD-4 brand were entered. Component structures of the studied HEM are brought in table 1

Table 1. Component structures of the studied HEMs.

Component of HEM, mass %	No. HEM sample				
	1	2	3	4	5
SKDM-80	14.8	13.6	11.0	11.8	16.0

[*] Corresponding author: y.a.dubkova@niipmm.tsu.ru

Ammonium perchlorate	70.2	71.4	74.0	73.2	69.0
TiB_2	15.0	–	–	–	–
AlB_2	–	15.0	–	–	–
B	–	–	15.0	–	–
AlB_{12}	–	–	–	15.0	–
ASD-4	–	–	–	–	15.0

Measurement of combustion heat of the studied exemplars of high-energy materials is taken on calorimetric installation [3]. A basic element of calorimetric installation is the bomb calorimeter. Schemes of a calorimeter and a bomb calorimeter are provided on figures 1, 2. In the figure 3, a), b) the general view of installation and a bomb calorimeter are given.

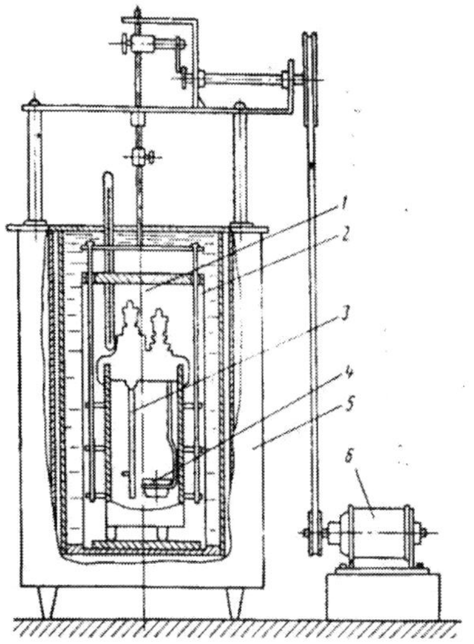

Fig. 1. The scheme of calorimetric installation for definition of combustion heat of fuel: 1 camera, 2 mixer, a 3- flame igniter, 4 crucible with the examinee fuel, 5 housing of a calorimeter, 6 electric motor.

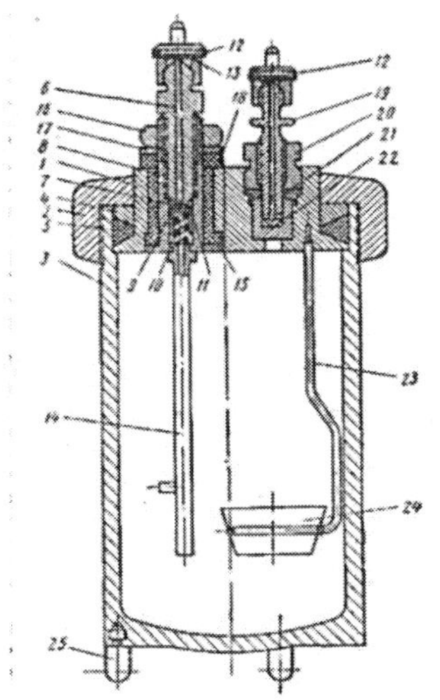

Fig. 2. Scheme of a bomb calorimeter: 1-cover; 2-tightening ring; 3 - glass; 4 - metal ring; 5 - rubber ring; 6, 19 - the union; 7 - ftoroplastovy laying; 8 - housing; 9 - cylindrical valve; 10 - spring; 11 - aluminum laying; 12 - removable caps; 13 - ftoroplastovy laying; 14 - electrocontact tube; 15 - ebonite insulator; 16 - nut; 17 - washer; 18 - ebonite laying; 20 – omental nut; 21 - epiploon; 22 - saddle; 23 - current carrying pin; 24 - cup; 25 - cylinder foot.

a

b

Fig. 3. A general view of a calorimeter (a) and a bomb calorimeter (b) for measurement of combustion heat of HEM.

Solid-fuel elements were carried out in the form of cylinders (fig. 4). In the course of manufacture of exemplars entered a hardener in number of 0.5 masses % into powder system over the mass of mix.

Fig. 4. HEM samples prepared for placement in a calorimeter bomb.

Before carrying out experement exemplars measured, weighed then prepared them for landing in a calorimeter bomb. In a vessel of a calorimeter poured distilled water, lowered a bomb and took water temperature the metastatic Beckmann thermometer (fig. 5 and), b), c) allowing to receive a high precision of measurements. After the executed measurements ignition and step-by-step measurement of water temperature in a calorimeter was carried out. Calibration was carried out by combustion of standard solid fuel of ballistidny typ ("N" blasting powder) with the known combustion heat [4].

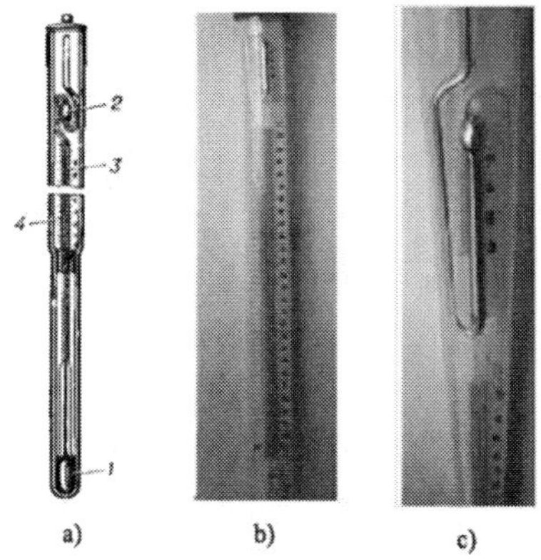

Fig. 5. Beckman thermometer (a): 1 – tank; 2 – padding camera; 3 – capillary; 4 – main scale; (b), (c) – a general view.

According to theoretical bases of the experimental technique, the total of the warmth which is marked out at combustion of the studied HEM exemplar with the mass of m is determined by a formula

$$Q' = c_p(M + W + B)(\Delta T \pm \Delta\theta) \quad (1)$$

where $W = \frac{c_k}{c_p} M_k$ is a thermal capacity (water equivalent) of a calorimeter; $B = \frac{c_в}{c_p} M_k$ is a allowance for availability of air in a calorimeter; $\Delta T = T_k - T_n$.
Combustion heat Q fuel is determined by a formula

$$Q = \frac{Q'}{m} = \frac{c_p(M + W + B)(\Delta T \pm \Delta\theta)}{m}$$

The mass of water in calorimetric installation makes M= 3000 g; water equivalent of a calorimeter of W = 449 g, B = 1.219 g. Assessment of extent of correction of B in a formula (2) on availability of air in a calorimeter when determining caloric content of fuel on a formula (3) showed that at change of temperature and pressure in actual measuring ranges T=(283÷303)K, p=(740÷780) millimetre of mercury the mass of air in a calorimeter changes slightly. The maximal difference of extent of correction of B from its value under normal conditions makes ~ 5%. At the same time the mass of air in a calorimeter makes ~ 0.06% of cooperative mass of water and a water equivalent of the device. Therefore, in the considered conditions the air contribution (extent of correction of B) when determining caloric content of fuel on a formula (2) is negligible, and the formula (2) can be presented in the following form:

$$Q = \frac{Q'}{m} = \frac{c_p(M + W)(\Delta T \pm \Delta\theta)}{m} \quad (3)$$

The sign "+" or "-" in (3) corresponds to heating or cooling of the device owing to heat exchange with a surrounding medium.

Taking into account the aforesaid the formula (2) can be presented in the form

$$Q' = c_p(M + W)\Delta T \pm c_p(M + W)\Delta\theta, \quad (4)$$

where a summand in a right member is the amendment to definition of value of combustion heat on heat exchange of calorimetric installation with a surrounding medium.

The accounting of heat exchange of calorimetric installation with a surrounding medium when determining combustion heat of fuel on a formula (2) is carried out on the basis of the comparative analysis of reference value of combustion heat Q'_{ref} and results of measurement of the quantity of heat Q'_{instr} allocated at combustion of reference HEM. The total of warmth which is allocated at combustion of a standard sample of HEM with the mass of m of g is determined by a formula

$$Q'_{ref} = Q_{ref} \cdot m, \tag{5}$$

where Q_{ref} is the caloric content of reference HEM the known value.

Mean value of allowance for heat exchange of the device with a surrounding medium is determined by results of a series of the experiments put on reference HEM ("N" blasting powder). Results of measurement of combustion heat of reference HEM are given in table 2.

Table 2. Results of measurement of combustion heat of reference HEM ("N" blasting powder).

No	m, g	ΔT, K	Q'_{ref}, cal	Q'_{instr}, cal	$\Delta Q'$, cal	Q_{real}, cal/g
1	2.520	0.76	2142	2621	479	862
2	2.400	0.72	2040	2483	443	847
3	2.416	0.73	2054	2518	463	856
4	2.349	0.70	1997	2414	417	836
Average					450	850

Thus, on the basis of results of measurement of combustion heat Q_{real} of reference HEM value of allowance for heat exchange of calorimetric installation with a surrounding medium $\Delta\overline{Q'}$= - 450 cal feces is received. Substituting in (3) received value and values of sizes corresponding to parameters of the unit, the working formula for calculation of combustion heat of the studied HEM has an appearance:

$$Q = \frac{c_p(M+W)\cdot\Delta T - \Delta\bar{Q}'}{m} = \frac{3449\cdot\Delta T - 450}{m},$$

where $[\Delta T]$ = K; $[M]$ = g; $[Q]$ = cal/g.

3 Results

The results of measurement of combustion heat of the view HEM received on the basis of the offered technique are given in table 3.

Table 3. Combustion heat of exemplars of HEM.

No. sample	m, g	ΔT, K	Q'_{instr}, cal	Q, cal/g	$\bar{Q}$, cal/g
1.	2.250	0.87	3001	1134	1126
	2.065	0.80	2759	1118	
2.	1.720	0.81	2794	1363	1338
	1.917	0.86	2966	1312	
3.	1.410	0.83	2863	1711	1476
	2.250	0.94	3242	1240	
4.	1.880	0.98	3380	1559	1534
	2.079	1.04	3587	1509	

5.	1.797	0.84	2897	1362	1352
	2.004	0.91	3139	1342	

In table 4 the received values of combustion heat of exemplars of HEM are given in the International System of Units.

Table 4. Combustion heat of samples of HEM.

Q, MJ/kg	No. sample					
	1	2	3	4	5	Reference
	4.72	5.61	6.18	6.43	5.66	3.56

4 Conclusion

The method and results of the pilot study of combustion heat of the solid-fuel compositions containing various metal additives are presented (TiB_2, AlB_2, AlB_{12}, Al, B). The tmethod is based on use of a calorimetric way of definition of combustion heat of materials. By results of a series of the experiments put on reference HEM ("N" blasting powder), an allowance for defined value heat exchange $\Delta\overline{Q'}$ of a calorimeter with a surrounding medium.

On the basis of the offered technique combustion heat of high-energy materials with additives of metal powders is measured. The comparative analysis of influence of effectiveness of various additives on fuel caloric content is carried out. It is received that the maximal heat effect is reached at combustion of HEM containing AlB_{12} polyaluminium boride (exemplar No. 4).

The work was financially supported by the Ministry of Education and Science of the Russian Federation within the framework of the Federal Target Program. Agreement No. 14.578.21.0034 (Unique identifier RFMEFI57814X0034).

References

1. Ya. M. Shapiro, G. Yu. Mazing, N. E. Prudnikov, *The theory of the rocket engine on solid fuel* (Voyenizdat, Moscow, 1966) [in Russian]
2. V. A Arkhipov., A. V Pesterev., L. A. Savelyeva, R. A. Yusupov, Higher Education Institutions. Physics. **9**, 8 (2012)
3. K. N. Godovskaya, L. V. Ryabina, G.Yu. Novik, M. M. Gerner, *Commercial analysis* (The higher school, Moscow, 1972) [in Russian]
4. A. P. Babichev, N. A. Babushkin, A. M. Bratkovsky, *Physical quantities: Reference book*, (Energoatom, Moscow, 1991) [in Russian]

MATEC Web of Conferences 92, 01069 (2017) DOI: 10.1051/matecconf/20179201069
Thermophysical Basis of Energy Technologies - 2016

Experimental study of evaporation of distilled water and 10% NaCl and $CaCl_2$ aqueous salt solutions droplets under their free falling on a heated surface

*D.*V. Feoktistov[1,*], *E.*G. Orlova[1], *T.*M. Sharifulin[1], *A.*G. Islamova[1]

[1]National Research Tomsk Polytechnic University, 634050 Tomsk, Russia

Abstract. The paper presents the experimental results of evaporation of distilled water and 10% aqueous salt solutions of NaCl and $CaCl_2$ droplets under their free falling on a heated surface. It is proved that it is more expedient to conduct the experimental research in this field according to classical multifactorial experiment. Laser treatment of surfaces is found to increase the evaporation rate and to biases the point of boiling crisis in the region of lower surface temperatures. In this case, in the conditions of boiling crisis the frequency of contact of a droplet with a heated surface will decrease.

1. Introduction

There are many experimental and numerical studies of falling, spreading and evaporating droplets on a heated surface [1-12]. However, it is worth noting that there are few studies in this field with using laser patterned surfaces. The processing of metal surfaces is known to change their properties; it can improve both hydrophilic and hydrophobic properties [13, 14]. The effect of this method of patterning on the spreading (changes in the geometric shape of the droplet) and evaporation when droplets free-fall on a heated surface remains unknown. Experimental study of such processes will expand the available scientific contributions in the wetting field.

This paper presents the experimental results of a study of distilled water and 10% aqueous salt solutions of NaCl and $CaCl_2$ droplets under their free falling on a heated surface.

2. Experimental procedure

When choosing a research procedure, we compared the plan of classical multifactorial experiment, which is a sequence of single-factor experiment (change of one parameter at constant rest parameters, for example, change of the surface temperature at a fixed height, droplets volume, composition of the liquid), and a full factorial plan, which is to find the

* Corresponding author: fdv@tpu.ru

required number of experiments on the number of factors. As a result, a number of important features were highlighted. Several factors are varied in the full factorial experiment (surface temperature, height of dosing, and droplet volume). This leads to the fact that the dispersion in evaluating the coefficients is typically an order of magnitude less than the experimental error. The nature of the dependences between the main studied factors (evaporation rate of a droplet on the surface temperature, as well as the evaporation rate of liquid droplets in different volumes of dosing) are not known in advance and not found the mechanism of influence of these factors. So the use of a full factorial plan can lead to erroneous results. Consequently, it is more expedient to conduct classical multifactorial experiment for studying the processes of falling, spreading and evaporation of droplets onto a heated metal surface.

According to results of preliminary experiments, the main ranges of variable factors were defined (Table 1).

Table 1. Ranges of variable factors.

Factors	Values of factors	
Surface temperature	60, 80, 100, 120, 130 °C	
Surfaces	polished aluminum surface	laser patterned surface
Working liquid	Distilled water 10% aqueous solution of NaCl 10% aqueous solution of $CaCl_2$	
Volume of distilled water droplet	0.015 mL	
Height of the droplet fall	300 mm	

Experimental studies of falling, spreading and evaporation of droplets onto a heated metal surface were conducted on the setup with using the shadow method (Fig.1).

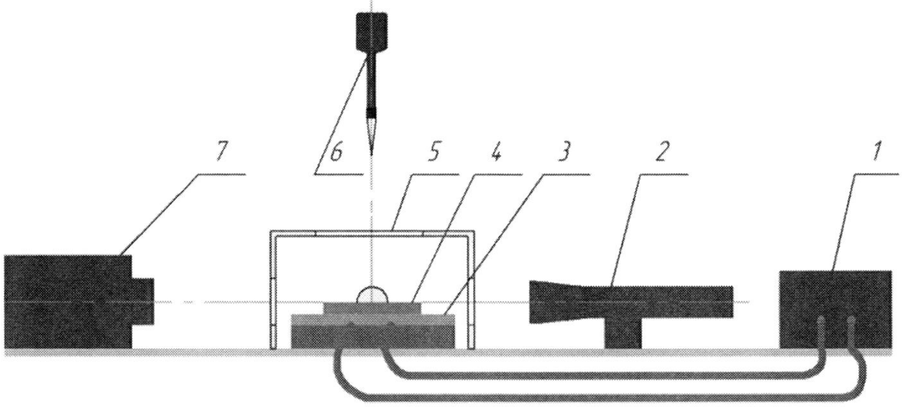

Fig. 1. Sketch of the experimental setup: 1 – power source of electrical heater; 2 – source of plane-parallel light; 3 – heating element; 4 – substrate; 5 – box; 6 – dosing device; 7 – high-speed video camera.

Source of plane-parallel light 2 lighted a droplet on substrate 4. Its shadow was recorded by the lens of high-speed video camera 7. A droplet was squeezed by single-channel electronic dosing device 6 (Lenpipette Stepper ThermoScientific), which was mounted vertically above the substrate at a distance of 300 mm. Substrate temperature was controlled by heating element 3 connected to power source 1. To maintain constant conditions of heat and mass transfer with the environment, the test cell was covered by glass box 5 [15-17].

Experiment was conducted with using two heated surfaces: polished and laser patterned surfaces made of aluminum alloy (AMG-6). Photographic images of surfaces were obtained on the microscope "Hitachi-3000M" (Fig. 2). The parameter of roughness (Ra) was defined on the profilometer.

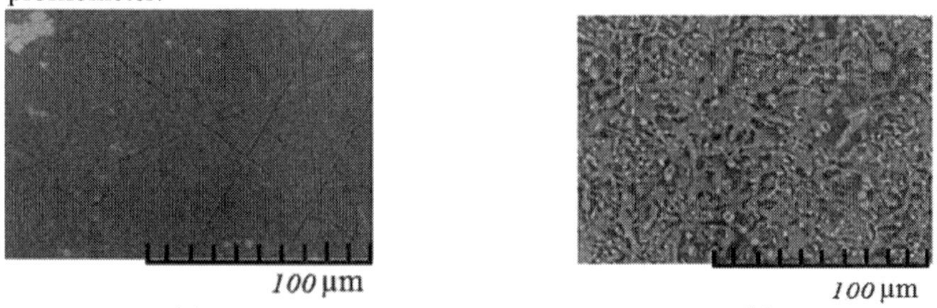

(a) (b)

Fig. 2. Photographic images of aluminum alloy surfaces (AMG-6) (magnification×1000): (a) polished (Ra=0.374 μm); (b) laser patterned (Ra=1.290 μm).

Surface of polished substrate (Fig. 2 (a)) is formed by multi-directional grooves obtained after polishing with diamond paste. The width of the grooves is less than 1 mm, the depth is 0.374 μm. There are impurities of arbitrary shape (white – magnesium; gray and black – manganese, iron, silicon, zinc, copper, titanium, and beryllium). Area of one impurity does not exceed 10 μm^2. It is known that the impurity with such size does not affect the behavior of the liquid droplet on a solid surface [5, 6].

After processing the substrate by laser radiation, chaotically arranged structure of molten metal with micro cavities and asperities was formed (Fig. 2 (b)). Local area of asperities does not exceed 10 μm^2. It should be noted that after the laser treatment of the surface, the roughness is increased by 3, 4 times.

3 Results and discussion

According to experimental results dependences of mass evaporation rate of water droplet and 10% aqueous salt solution droplet (NaCl, $CaCl_2$) on the temperature of polished (a) and patterned (b) surfaces (Fig. 3) were obtained.

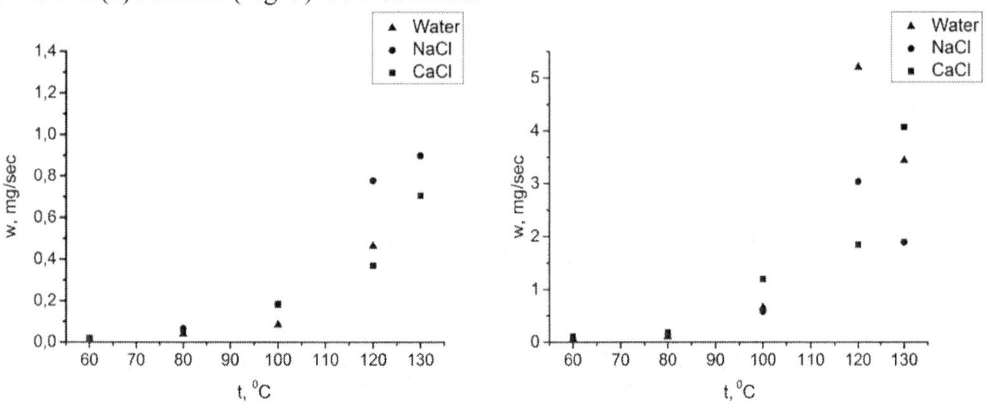

(a) (b)

Fig. 3. Mass evaporation rate of distilled water and aqueous salt solutions droplets versus surface temperature. Substrate: (a) polished; (b) laser patterned.

After breakdown of the droplet from the tip of dosing device, the droplet moves due to gravity. At the moment of impact of the droplet with substrate attraction force transforms to inertial force directed from the center to the periphery – a droplet spreads to the maximum diameter. In this case, the inertial force is balanced by the surface tension of the liquid. Further, the surface tension of the liquid compresses the droplet to an equilibrium state. After establishing the equilibrium mode, the evaporation of water droplets and salt solution droplets has a similar consistency and depends on the temperature and the substrate structure. This is due to low concentration of salt in its aqueous solution. However, the evaporation of salt solution droplets is accompanied by crystallization of salt on the substrate surface.

According to analyses of the obtained dependences (Fig.3) it can be concluded that the microstructure biases the point of boiling crisis in the region of lower surface temperatures. In this case, in the conditions of boiling crisis the frequency of contact of a droplet with a heated surface will decrease. It means reducing the intensity of heat transfer from the heated metal to the spheroid.

The reported study was supported by the grant of President of Russian Federation for the government support of young Russian scientists (MK-6810.2016.8).

References

1. D.O. Glushkov, G.V. Kuznetsov, P.A. Strizhak, R.S. Volkov, Therm. Sci. **20**, 131 (2016)
2. G.V. Kuznetsov, A.V. Zakharevich, N.S. Bel'kov, Chem. Pet. Eng. **50**, 424(2014)
3. D.O. Glushkov, G.V. Kuznetsov, P.A. Strizhak, Therm. Sci. **19**, 1541 (2015)
4. D.O. Glushkov, G.V. Kuznetsov, P.A. Strizhak, Math. Probl. Eng. **2014**, 920480 (2014)
5. G. Kuznetsov, D. Feoktistov, E. Orlova, Thermophys. Aeromech. **23**, 17 (2016)
6. G. Kuznetsov, D. Feoktistov, E. Orlova, J. Eng. Phys. Thermophys. **89**, 317 (2016)
7. D. Zaitsev, D. Kirichenko, O. Kabov, Tech. Phys. Lett. **41**, 551 (2015)
8. D.V. Zaitsev, A.M. Lozano, H. Auracher, O.A. Kabov, Microgravity Sci. Technol. **19**, 71 (2007)
9. S.Y. Misyura, J. Heat Transfer **138**, 111501 (2016)
10. S.Y. Misyura, Exp. Therm. Fluid Sci. **75**, 43 (2016)
11. R.S. Volkov, G.V. Kuznetsov, P.A. Strizhak, Int. J. Heat Mass Tran. **85**, 1 (2015)
12. O.V. Vysokomornaya, G.V. Kuznetsov, P.A. Strizhak, Fire Safety J. **70**, 61 (2014)
13. A. Sivkov, E. Naiden, A. Ivashutenko, I. Shanenkov, J. Magn. Magn. Mater. **405**, 158 (2016)
14. A. Sivkov, Y. Shanenkova, A. Saigash, I. Shanenkov, Surf. Coat. Technol. **292**, 63 (2016)
15. E.G. Orlova, D.V. Feoktistov, K.O. Ponomarev, EPJ Web Conf. **93**, 012011 (2015)
16. E.G. Orlova, D.V. Feoktistov, K.A. Batishcheva, IOP Conference Ser.: Mater Sci. Eng. **93**, 012011 (2015)
17. D.V. Feoktistov, E.G. Orlova, A.G. Islamova, MATEC Web Conf. **37**, 01020 (2015)

MATEC Web of Conferences 92, 01070 (2017) DOI: 10.1051/matecconf/20179201070
Thermophysical Basis of Energy Technologies - 2016

Research of efficiency of the organic Rankine cycle on a mathematical model

N. Galashov[1,*], *S*. Tsibulskiy[1], *A*. Gabdullina[1], *D*. Melnikov[1], and *A*. Kiselev[1]

[1]National Research Tomsk Polytechnic University, 634050 Tomsk, Russia

Abstract. The object of the study are the organic Rankine cycle. The purpose of research is to evaluate the impact on the net efficiency of the initial and final properties of the cycle at work on a saturated and superheated steam. Investigations were carried out on the basis of a mathematical model, in which the thermodynamic properties of materials are determined on the basis of "REFPROP". On the basis of the available scientific publications on the use of working fluids in an organic Rankine cycle analysis was selected ozone-safe pentane. A mathematical model has been developed on condition that condenser is used as air cooler which allows the substance to condense at a temperature below 0 °C. Numerical study on the mathematical model shown that net efficiency at work on pentane linearly depends on the condensation temperature and parabolically depends on the initial temperature with the saturated steam. During work at the superheated steam efficiency strongly depends on both the initial temperature and of the initial pressure. With rising initial temperature is necessary to gradually increase the initial pressure under certain conditions.

1 Introduction

Organic Rankine Cycle (ORC) – is a cycle, the working fluid which is a low boiling substance (LBS). The use of the working substance LBS allows you to solve a number of problems that are available at the water as the working body: to reduce the temperature of heat removal by the use of air cooled condensers in the winter and by this increase the efficiency of the cycle; due to the high density to reduce the size, weight and cost of facilities; by condensing steam at high pressures to reduce or solve air inflow in the condenser and improve condenser heat exchange therein. Application of air cooled condensers in the ORC can also eliminate or significantly reduce the consumption of water for technological purposes, it is very urgent task both in industry and in the energy sector. Therefore, in recent years in the scientific literature appeared a large number of studies devoted to the use ORC [1-9]. Basically, ORC is used to generate electricity through the use of low-grade waste heat, but as shown in [10] it can be used in a combined cycle gas turbine (CCGT) and diagram of powerful generating thermal and nuclear power plants.

* Corresponding author: gal@tpu.ru

Economic, environmental and safety measure of ORC are strongly dependent on the type of LBS. Environmentally friendly LBS determined ozone depletion potential ODP and global warming potential GWP.

2 Subject of research

The object of study is power installation with air cooled condenser, working on the basis of the Organic Rankine cycle, diagram is shown in Figure 1.

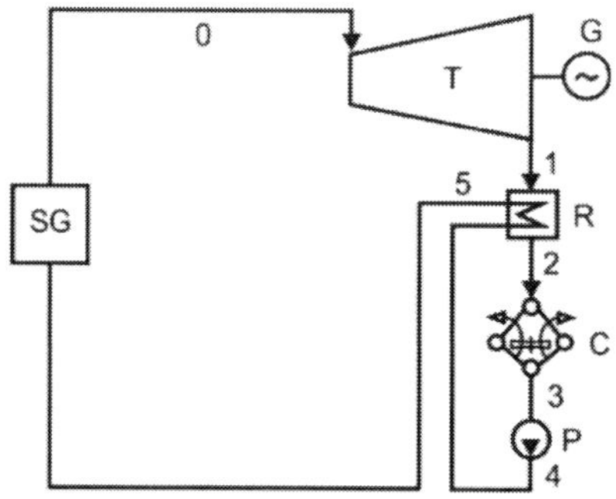

Fig. 1. Diagram installation: SG - a steam generator; T - turbine; G - generator; R - regenerator; C - condenser; P - pump; 0 - condition steam before the turbine; 1 - the output of the turbine; 2 - for the regenerator; 3 - liquid condition of the condenser; 4 - downstream of the pump; 5 - for the regenerator.

Consider the principle of installation work shown in Figure 1. In the steam generator is carried out heating, evaporation and possible overheating of the LBS. Then the LBS steam with properties at 0 enters turbine T where does work and set in motion generator. At the outlet of steam turbine at 1 depending on the properties the LBS is adjudged to wet or superheated steam. If steam is superheated state, that in the regenerator it is cooled to state 2, which is 1 – 5 °C above the saturation temperature at the pressure in the condenser. In the condenser steam condenses and condensate in the state 3 is pumped through the regenerator to the steam generator. In this case the condensate is heated in the pump due to compression work, and in a regenerator by cooling steam. Condensate properties of downstream from the pump characterize the state 4, and state of the regenerator 5.

Application air-cooled condenser allows the condensed LBS steam at a temperature below 0 °C, it conditions is impossible to have a water-cooled condenser.

3 Mathematical model calculation installation of the ORC

The mathematical model for calculating the properties and efficiency ORC with air cooled condenser is designed on the basis of Figure 1.

The initial data are given:

- the working fluid;
- temperature t_0 and pressure P_0 at the turbine inlet;
- the temperature in the condenser t_C;
- step-up ratio of the pressure in the pump k_P;
- step-up ratio of pressure increase in the regenerator k_R;
- steam superheating at the outlet from the regenerator Δt_R;
- efficiency of the regenerator η_R;
- efficiency of the pump η_P;

- net efficiency of the turbine during operation with superheated steam η_{oiSS}.

Calculation algorithm

If only given temperature t_0, the steam pressure is saturated and equals:

$$P_0 = f(t_0). \quad (1)$$

Enthalpy and entropy:

$$h_0 = f(t_0),\ s_0 = f(t_0). \quad (2), (3)$$

During operation on the ORC superheated steam initial parameters of the working substance:

$$h_0 = f(t_0, P_0),\ s_0 = f(t_0, P_0). \quad (2a), (3a)$$

Steam pressure in the condenser:

$$P_C = f(t_C). \quad (4)$$

Steam pressure at the outlet from the turbine in view of the pressure loss in the regenerator:

$$P_1 = k_R P_C. \quad (5)$$

Steam enthalpy at the turbine outlet at isentropic process:

$$h_{1t} = f(P_1, S_0). \quad (6)$$

Work 1 kg of steam in the turbine at isentropic process:

$$H_0 = h_0 - h_{1t}. \quad (7)$$

The enthalpies of liquid and steam in the saturated state at the pressure P_1:

$$h_1' = f(P_1),\ h_1'' = f(P_1) \quad (8), (9)$$

The degree of dryness of the steam at the end of the isentropic process in the turbine:

$$x_1 = (h_{1t} - h_1')/(h_1'' - h_1') \quad (10)$$

Coefficient of energy loss from moisture in the turbine:

$$\text{if } x_1 > 1, \text{ then } k_{ms} = 1, \text{ another } k_{ms} = [1 - 0.8(1-0.15)(1-x_1)/2(H_{ms}/H_0)] \quad (11)$$

Turbine internal efficiency with account for the loss of energy from the humidity:

$$\eta_{oi} = k_{ms}\eta_{oi.SS}. \quad (12)$$

The enthalpy of steam at the turbine outlet considering energy loss:

$$h_1 = h_0 - \eta_{oi}H_0. \quad (13)$$

Work done by 1kg steam in the turbine process with account for the loss of energy:

$$H_i = h_0 - h_1. \quad (14)$$

Steam temperature at the turbine outlet:

$$t_1 = f(P_1, h_1). \quad (15)$$

If $t_1 - t_C < 5$, then $\Delta t_R = 0$.
Temperature and enthalpy of steam at the outlet of the regenerator:

$$t_2 = t_C + \Delta t_R,\ h_2 = f(P_C, t_R). \quad (16), (17)$$

The change of the steam enthalpy in the regenerator:

$$\Delta h_R = h_1 - h_2. \quad (18)$$

The enthalpy and the entropy of the liquid at the outlet of the condenser:

$$h_3' = f(P_C),\ S_3' = f(P_C). \quad (19), (20)$$

The pressure of the pump:

$$P_4 = k_P P_0. \quad (21)$$

The enthalpy of the pump in the isentropic compression process:

$$h_4 = f(P_4, S_3'). \quad (22)$$

Increasing entropy in the pump:

$$\Delta h_P = (h_4 - h_3')/\eta_P. \quad (23)$$

The enthalpy of liquid in the regenerator:

$$h_5 = h_3' + \Delta h_P + \eta_R \Delta h_R. \quad (24)$$

Net cycle efficiency, %:

$$\eta_i = 100[(h_0 - h_1) - \Delta h_P]/(h_0 - h_5). \quad (25)$$

The model is implemented mathematical program in the package Excel spreadsheets. The calculations of properties of all substances produced based on the functions of the database "REFPROP" [11, 12].

4 Experimental results and analysis

Investigations were carried out with the following initial data:
- the working fluid – pentane;
- temperature t_0 and pressure P_0 at the turbine inlet;
- condenser temperature t_C = -20 – 50 °C;
- coefficient of pressure increase in the pump $k_P = 1.15$;
- coefficient of pressure increase in the regenerator $k_R = 1.1$;
- superheat at the outlet of the regenerator $\Delta t_R = 3$°C;
- the efficiency of the regenerator $\eta_R = 0.98$;
- pump efficiency $\eta_P = 0.7$;
- net efficiency of the turbine during operation with superheated steam η_{oiSS}=0.87;

Pentane is selected for investigation as one of the substances, the most recommended in scientific works for use in the ORC.

Numerical studies have been conducted with operation for saturated steam from condensing temperature change from -20 to 50 °C and the initial temperature of 100 to 180 °C.

Fig. 2 shows the resulting the efficiency investigations.

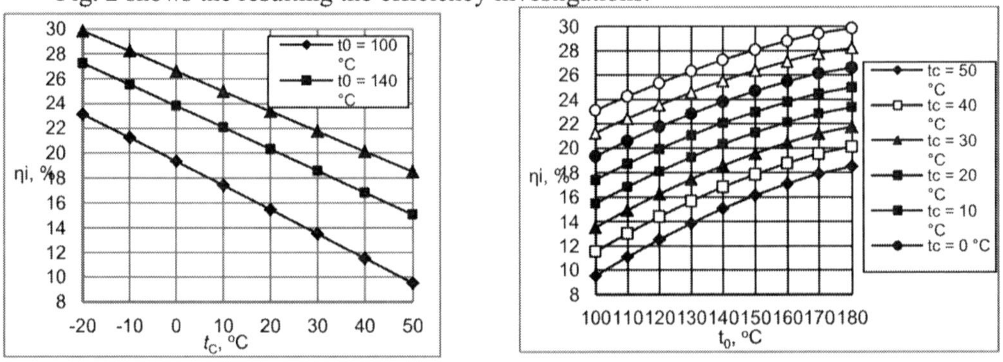

a b

Fig. 2. *a* – dependence of the net efficiency ORC from t_C and t_0, *b* – dependence of the net efficiency ORC from t_0 and t_C.

In fig. 2*a* can see that the efficiency has a linear function from t_C and nonlinear from t_0. With a decrease t_C on 10 °C efficiency increased by 0.8-1 %. In fig. 2*b* net efficiency takes the form of a parabolic dependence. By increasing t_0 the efficiency growth slows.

Fig. 3 shows results of investigations operation with superheated steam at a initial properties t_0, P_0 and the temperature in condenser 10 °C.

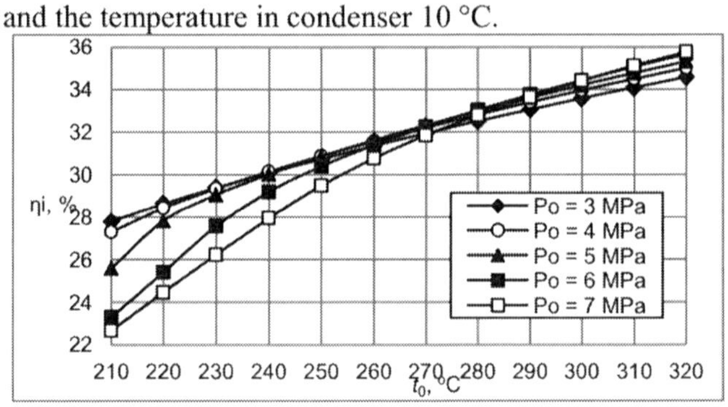

Fig. 3. Dependence of the net efficiency ORC from t_0 and P_0 with conditions t_C = 10 °C.

Fig. 3 shows that the net efficiency strongly depends on t_0 and P_0. At the same time, up to t_0<240 °C profitably operate at P_0 = 3 MPa, for t_0 =240 – 270 °C to operate at profitable P_0 = 4 MPa, and at t_0> 290 °C profitably operate at P_0 = 7 MPa.

Conclusion

At operate on saturated steam net efficiency of the cycle at work on pentane linearly depends on the condensation temperature and parabolically depends on the initial temperature.

Efficiency strongly depends on both the initial temperature and the initial pressure at work with superheated steam. With increasing initial temperature need to gradually increase the initial pressure under certain conditions.

The research was realized with financial support of Minobrnauki of Russia in framework of FTP "Research and development in prior direction of scientific-technological complex of Russia in 2014-2020 years", unique R&D identifier RFMEFI58114X0001.

References

1. M. Soffiato, C.A. Frangopoulos, G. Manente, S. Rech, A. Lazzaretto, Energy Convers. Manage **92**, 12 (2015)
2. B.C. Choi, Y.M. Kim, Energy **58**, 404 (2013)
3. U. Larsen, L. Pierobon, F. Haglind, C. Gabrieli, Energy **55**, 803 (2013)
4. G. Shu, L. Liu, H. Tian, H. Wei, G. Yu, Appl. Energ. **113**, 1188 (2014)
5. A. Toffolo, A. Lazzaretto, G. Manente, M. Paci, Appl. Energ. **121**, 219 (2014)
6. M. Yang, R. Yeh, Energy Convers Manage **88**, 12 (2014)
7. T. Guo, H.X. Wang, S.J. Zhang, Energy **36**, 2639 (2011)
8. A. Lazzaretto, G. Manente, Int. J. Thermodyn. **17**, 3 (2014)
9. L. Branchini, A. De Pascale, A. Peretto, Appl. Therm. Eng. **61**, 129 (2013)
10. N. Galashov, S. Tsybulsky, Power Techn. Eng. **48**, 5 (2015)
11. E. Lemmon, M. Huber, M. McLinden, Reference fluid thermodynamic and transport properties-REFPROP, standard reference database 23, version 8.0, National Institute of Standard and Technology (2007)
12. N. Galashov, S. Tsibulskiy, T. Serova, EPJ Web Conf. **110**, 01068 (2016)

Simulation of viscous melt drop crystallization by the SPH method

Alexander V. Gerasimov[1] , *Roman* O. Cherepanov[1], *Raisa* A. Krektuleva[2*] and *Vladimir* N. Barashkov[3]

[1]National Research Tomsk State University, 634050 Tomsk, Russia
[2]National Research Tomsk Polytechnic University, 634050 Tomsk, Russia
[3]Tomsk State University of Architecture and Building, 634050 Tomsk, Russia

Abstract. Heat and mass transfer during absorption on a film of lithium bromide water solution flowing by a cooled wall in the steam atmosphere is numerically investigated in this paper. The self-similar solutions are using as the initial conditions for solving the problem beyond the entrance region. The key criteria characterizing heat and mass transfer in the film absorption with uniform velocity profile and with a constant thickness have been determined.

1 Problem statement

Numerical simulation of gas-thermal spray coatings is a difficult task, the urgency of which is due to the widespread application of technologies for the deposition of hardening coatings. The interaction of sprayed particles with the surface of a processed material is three-dimensional, has a wide range of impact velocities (from 100 to 2000 m/s) and a high strain degree. The melted particles interacting with the surface can contain solid inclusions, which makes it difficult to simulate the behavior of this media.
This paper represents the simulation results of melted viscous metal drops impacting with a cold surface, as well as the crystallization process using the SPH method.

2 Computational method

The paper considers a variant of the SPH method, namely a symmetrized method of smoothed particles. The SPH approximation of the unsteady heat transfer equations is used to calculate thermal conductivity.

The symmetrized SPH method is conservative concerning the momentum, almost conservative concerning the energy, has a first-order time accuracy, but it has a low approximation accuracy of spatial derivatives (of order zero in the general case). The method allows free boundary problems to be solved.

The method is based on nuclear approximation:

* Corresponding author: rakrekt@tpu.ru

$$f(\vec{x}) \cong \int f(\vec{y}) W(\vec{y}-\vec{x},h) d\vec{y} \quad (1)$$

where $\vec{x},\vec{y}$ is the radius vectors (spatial coordinates), W is the smoothing function (usually a B-spline) [1], h is the smoothing parameter that determines the area of support for the smoothing function. The approximation (1) of partial spatial derivatives is given by:

$$f(\vec{x})_{,\alpha} = \frac{\partial f(\vec{x})}{\partial x_\alpha} \cong \int f(\vec{y}) W(\vec{y}-\vec{x},h)_{,\alpha} d\vec{y} \quad (2)$$

The discrete form of the approximation (2) has the form:

$$f^n_{,\alpha} = \sum_k f^k \frac{m^k}{\rho^k} W^{nk}_{,\alpha} \quad (3)$$

The equations of the viscous compressible fluid motion have the form:

$$\rho \dot{v}_i = -\sigma_{ij,j}, \quad (4)$$

$$\sigma_{ij} = -P\delta_{ij} + \mu\dot{\varepsilon}_{ij}, \quad (5)$$

$$P = P(\rho, E), \quad (6)$$

$$\dot{\varepsilon}_{ij} = \left(v_{i,j} + v_{j,i}\right)/2; \quad (7)$$

where ρ is the density of the medium, v_i is the components of the velocity vector, σ_{ij} is the stress tensor components, P is the pressure, μ is the viscosity of the medium, E is the internal energy.

The energy balance equation is given in the form:

$$\dot{E} = \sigma_{ij}\dot{\varepsilon}_{ij} + \left(\lambda T_{,j}\right)_{,j} \quad (8)$$

The temperature is considered to be an internal energy function:

$$\dot{T} = \dot{E}/(C_P \rho) \quad (9)$$

The approximation of equation (4) with the use of (3) does not meet the law of conservation of momentum and energy (nonconservative concerning the momentum and energy). To restore conservatism, a transformed form of the equation (4) is used:

$$\rho \dot{v}_i = -\left(\sigma_{ij,j} + \left[1_{,j}\right] \cdot \sigma_{ij}\right) \quad (10)$$

In (10) the gradient of one (mathematically equal to zero) is used to adjust the SPH approximation of the stress divergence and make it symmetrical [2], thus ensuring the equality of forces acting between a pair of SPH nodes. After applying the SPH-approximation (3) to (10) we have:

$$\dot{v}^n_i = \frac{1}{m^n} \sum_k \frac{m^n m^k}{\rho^n \rho^k} \left(\sigma^k_{ij} + \sigma^n_{ij}\right) W^{nk}_{,j} \quad (11)$$

To calculate the heat transfer, the heat flow divergence is calculated in a similar manner:

$$\dot{T}^n = \frac{1}{C_P^n}\left[\frac{\sigma_{ij}^n \varepsilon_{ij}^n}{\rho^n} + \frac{1}{m^n}\sum_k \frac{m^n m^k}{\rho^n \rho^k}\left(q_i^k + q_i^n\right)W_{,i}^{nk}\right]$$
$$q_i^n = \lambda^n \sum_k \frac{m^k}{\rho^k} T^k W_{,i}^{nk} \tag{12}$$

In both cases, the symmetrization procedure (10) allows the energy conservation to be (almost) achieved, that is the calculated acceleration and strain rate values meet the law of conservation of energy:

$$\sum_k m^k v_i^k a_i^k = \sum_k \sigma_{ij}^k \dot{\varepsilon}_{ij}^k V^k \tag{13}$$

However, the total conservatism of the scheme is broken by integrating the equations of motion in time, since the error is accumulated per the time step when calculating the kinetic energy component:

$$\Delta E_K = \sum_k \Delta t \cdot m^k v_i^k a_i^k + \sum_k \frac{\Delta t^2}{2} m^k a_i^k a_i^k \tag{14}$$

The second term in (14) introduces an error in the calculation of the total energy balance. The magnitude of this error was controlled in the computations; the total losses during the transition of kinetic energy into potential one did not exceed 4% in the computations at initial moments of time after the impact at the initial velocity of particles up to 2000 m/s.

The Stefan problem was solved by the shock-capturing method. For this purpose, the thermal energy of each computational point was calculated at each time step. When achieving the melting temperature, the temperature growth was stopped, and the internal energy was accumulated to reach the value of the latent melting heat, after which the material was considered to be completely melted and was calculated using a model of an ideal viscous compressible fluid. If the internal energy dropped below the melting energy, the material was considered to be solid and was calculated by the model of an ideal elastic-plastic body.

The scheme described was used to simulate the interaction of melted metal drops with a cold surface.

The impact of a melted metal drop with an iron substrate was considered to be a test problem, and the metal parameters were taken for the 36NHTU steel from the reference book [3]: the density was 7.87 g/cm^3, the melt viscosity was 5.4 mPa/s, T of melting was 1280 K, Q of melting was 84 J/g, the heat capacity was 0.56 J/g, $\lambda = 28$ W/(m K); the elastic metal properties were constant and did not depend on the temperature for a first approximation: E = 200 GPa, G = 70 GPa, $Y_{0,\text{dynamic}} = 1.2$ GPa.

The computations were performed for three impact velocities: 100, 150 and 200 m/s, the initial temperature of the drop was 1380K, and the diameter of the drop was 8 microns. The process was computed until the maximum particle velocity dropped up to 50 m/s. The cross sections of the drops are shown in Figure 1 for the final stages of impact.

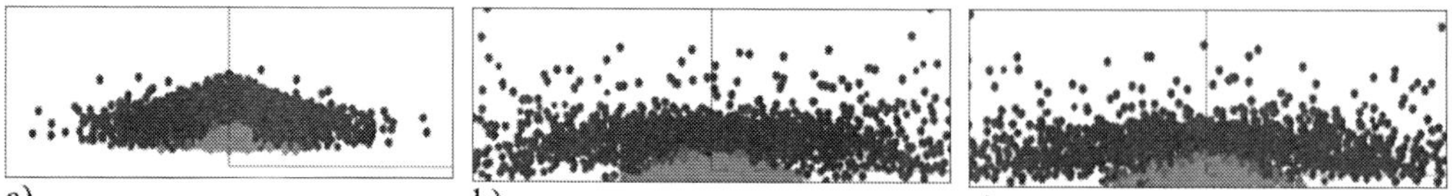

a) b) c)

Fig. 1. Cross sections of drops at the final stages of the process, (a) 100 m/s, (b) 150 m/s, (c) 200 m/s, the red color denotes the melt, the green color denotes the crystallized material.

3 Conclusion

The computations obtained by the model described allowed us to find that the impact velocity of about 150 m/s was critical. At lower velocities the major part of the drop maintains uniformity and, after cooling, a larger portion of the drop mass forms a hardening coating on the surface. At higher velocities the drop is significantly failed and forms a cloud of disintegrated material, and less than 20% of the original melt mass is crystallized on the surface. The disadvantage of this model is that the surface tension forces are not considered, and therefore these computations are very approximate and can characterize the process only qualitatively.

The study was funded by the Russian Science Foundation (project No.16-19-10264).

References

1. L.B. Lucy. Astron. J. **82**, 1013 (1977)
2. Shaofan Li, Wing Kam Liu, *Meshfree Particle Methods*, (Springer Science & Business Media, Germany , 2007)
3. I.K. Kikoin, Table of Physical Quantities. Reference Book (Atomizdat, Moscow, 1976)

Estimation of energy efficiency of residential buildings

Sergey Glushkov[1,*], *Nikolay* Kachalov[1], *Elena* Senkiv[1], *Victoriya* Glushkova[2]

[1]National Research Tomsk Polytechnic University, 634050 Tomsk, Russia
[2]Tomsk State Pedagogical University, 634061 Tomsk, Russia

Abstract. Increasing energy performance of the residential buildings by means of reducing heat consumption on the heating and ventilation is the last segment in the system of energy resources saving. The first segments in the energy saving process are heat producing and transportation over the main lines and outside distribution networks. In the period from 2006 to 2013. by means of the heat-supply schemes optimization and modernization of the heating systems. using expensive (200–300 $US per 1 m) though hugely effective preliminary coated pipes. the economy reached 2.7 mln tons of fuel equivalent. Considering the multi-stage and multifactorial nature (electricity. heat and water supply) of the residential sector energy saving. the reasonable estimate of the efficiency of the saving of residential buildings energy should be performed in tons of fuel equivalent per unit of time.

1 Introduction

Optimization of the use of energy-efficient solutions for construction and operation of residential buildings is not just technically. but also economically a difficult problem. The concept of heating optimization includes various options for energy saving. including reduction of heat losses in the main and distribution heat networks. increase of the use of CHP heat productivity. thus increasing their overall efficiency and reduce specific fuel consumption [1].

2 Analysis of the energy consumption of new residential areas

The data representing a change of requirements to reduction of energy consumption in the USA over the period of 2010-2030 [2]. which are shown in table 1. are of a great interest for the development of the process of increasing energy efficiency of the residential buildings in Siberian region. As table 1 shows. energy consumption (excluding electricity) in the US residential buildings constructed in 2007. is about 20-30% less compared with those in the Siberian region (90 + 70 = 160 kW*h / m^2*year) (in the USA - 104.5-128.3 kW*h / m^2*year). It should be noted that as early as in 1999-2001 US standard allowed

* Corresponding author: glushkov@tpu.ru

energy consumption (excluding electricity) in the amount of 147.3-163 kW*h / m^2*year. By 2015 this difference increased to about 35-50% (56.9-72.7 kW*h / m^2*year in the USA and 110 kW*h / m^2*year in Siberia) for practically the same climatic conditions.

Table 1. Requirements to Lowering power consumption (kW*h / m^2*year) in the USA.

Year	Standard				Indicator on energy-efficiency design ASHRAE5
	ASHRAE 90.1[1]	ASHRAE 90.1[2]	ASHRAE 189.1		
			1[3]	2[4]	
2007	128.3	–	104.5	154.8	117.9
2010	96.7	154.8	89	129.4	84.4
2013	85.3	132.7	78	114	75.8
2015	72.7	113.8	56.9	100.6	56.9
2020	56.9	75.9	40	57.5	0
2025	44.2	56.9	19.9	27.8	0
2030	31.6	50.6	0	0	0

[1]Without taking into account the consumption of electricity by the equipment users (eg. office equipment. food processors. etc.).
[2]Taking into account the consumption of electricity by the equipment users.
[3] Aimed to reduce energy consumption by 30% compared with standard 90.1-2004 (without taking into account electrical energy consumption by users of equipment).
[4] Aims to reduce energy consumption by 30% compared with standard 90.1-2004 (taking into account users of the electrical energy consumption of the equipment).
[5] Aimed to reduce energy consumption by 30% compared to standard 90.1 for 2007-2009; by 50% - for the period of 2009-2011; transition to buildings with zero energy consumption - 2013-2015

Transition of new residential buildings to heat consumption in the amount 19.9-44.2 kW*h / m^2*year) (the USA) by 2025 requires development and production in the sphere of obtaining and use of the maximum sizes of the RES (renewable energy sources) from the exhaust warm air and water. This is particularly important for the Siberian region.

Improving energy efficiency of residential buildings by reducing heat costs for heating and ventilation is the last element in the system of energy saving. The first elements in this process are energy efficiency of producing heat and its transportation by main lines and outside distribution networks. Optimizing heating schemes and modernization of heating networks using expensive (200-300 US dollars per 1 m) but very effective pre-insulated pipes since 2006 till 2013 resulted in savings. The total loss of heating energy in the utilities sector in Siberia remains high.

In terms of energy consumption of housing and communal sector (~ 40%) the energy balance of the residential district of Tomsk approximately corresponds to the USA. where residential and commercial buildings consume more than 40% of fossil energy resources. more than 70% of electricity and 50% of natural gas [2]. Approximately 1/3 of the total CO_2 emissions in the atmosphere of the United States is generated by power supply of residential and commercial buildings.

It should be noted that in the USA energy performance of new buildings deteriorates by 30% during the first three-four years of use [2]. This problem is typical for Siberia. Therefore. it is necessary to pay more attention to the process of commissioning and changeover of heating systems during operation

3 The energy efficiency rating of the essential services system inside the building

Energy efficiency of separate buildings does not guarantee high energy efficiency of entire residential areas. as modern essential services of residential building requires the use of various energy carriers (hot and cold water. electricity. gas). The delivery of energy with the necessary parameters to the consumers requires significant additional energy expenses. demanding additional charges in foreign currency. In such formulation the problem of energy efficiency of residential areas. even in the areas of European countries has not been previously considered.

The objectives of the proposed program are:

• Evaluation of energy consumption. spent on the production of various types of energy used to provide essential services system of both separate buildings and a group of buildings;

• accounting losses in producing and transportation of various types of energy to consumers.

The problems are technically solved by means of assessing the energy efficiency of the essential services system inside the building. including the characteristics of different types of energy sources. systems of its delivery to customers and in-house consumption of various types of energy. They consist in the fact that consumption of all types of energy (electricity. heating. gas. water). used for the building. obtained from external distribution networks. leads to the initial fuel consumption or equivalent energy spent on the production of the used energy. which is compared with the current standards.

Energy efficiency of buildings should be considered as the final element of the overall energy-saving system. which includes main lines and distribution networks for the delivery of various types of energy to the consumer. as well as generators of all kinds of energy. All energy generators (boilers. CHP plants. power plants. etc.). as well as a network for the delivery of all types of energy to consumers with certain operating efficiency. on which significantly and sometimes decisively the overall energy efficiency of the entire essential services system for residential buildings is dependent. Here we should also add the flow of electricity for pumping in heating systems and air in the quartz-family-recuperators heat exchangers "air – air". It is necessary to point out that the electric power is generated with conversion factor of about 0.3 from fuel compared to 0.8-0.9 for the thermal energy. In addition. it is necessary to take into account energy consumption for cleaning. pressure build-up and delivery of tap water [3].

Currently. the main indicator of the efficiency of the heat consumption for heating the building is the annual consumption (kW*h / m^2). referred to 1 m^2 of the total area of the apartment. This indicator is useful for all types of energy when transferring their characteristics to the appropriate dimension. Taking into account the dependence of the total energy of the system of essential services of the building on the sources of energy supply. then kilogram of fuel equivalent corresponding to the old technical units of 7.000 kcal. which is equivalent to 8.14 kW*h. can be used as the index. Therefore. consumption of energy source. called the fuel (in kW*h / m^2*year or in kg of oil equivalent/ m^2*year) can serve as general indicator of energy efficiency of life support systems in buildings.

Primary energy evaluation is performed separately for each of the types of energy (electricity. heat. water. gas). In case of domestic use of natural gas. which is ready fuel. its preparation and delivery can be disregarded. as they are carried out outside the general system of essential services of the considered objects.

Methods of evaluating of the overall energy efficiency of the building is based on the full consideration of all types of energy (electricity. heat. fuel gas. water) consumed from the outside distributive networks for the normative life support conditions. Then it is also necessary to take into account the loss of energy and transportation costs of each energy source from the source of its receipt to the consumer. The final step is the account of the energy conversion factor of fuel energy into the energy of some necessary for the sustenance of the building energy carrier. Often this is called efficiency factor.

In general. the total amount of energy consumed by the building during the year (referred to 1 m^2 of total area) from external energy distribution networks is determined by the expression

$$Q = Q_1 + Q_2 + Q_3 + Q_4,$$

where Q_1 is the annual electricity consumption of household appliances (refrigerator. TV. lighting. electronics. etc...). apartment and house flow boosters (apartment and house fans are heat exchangers. recuperators "air – air". pumps for provision of coolant circulating in an independent system. etc.); Q_2 - the annual consumption of electrical energy referred to 1 m^2 of total area. for heating and hot water. kW*h/m^2; Q_3 - the annual consumption of electrical energy referred to 1 m^2 of floor area. for the cold water. kW*h / m^2; Q_4 - annual consumption of gas energy. referred to 1 m^2 of total area. kW*h/m^2.

It should be noted that the return of any kind of energy at the expense of resources of the building (residential heat. heat recovery of exhaust air. solar cells to generate electricity. preliminary heating of the hot water by the warm waste. use of rainwater. etc.) reduces energy consumption from the external distribution networks. However. the given expression does not take into account the difference in the flow of natural energy resources. spent for the receipt and transmission of electric power and heat to the consumer. and this difference is quite significant.

Transportation loss of electricity for tap water during the operation of the electric pumps are about 1-2 kW*h / m3 depending on the distance of transportation [4]. As an example of calculation of the power consumption let us take a two-room apartment with total area S = 60 m^2. in which three people live. which roughly corresponds to the current average area of 20 m^2 per person.

To calculate the power consumption of running water let us take daily consumption of 140 liters of cold water and hot water to 70 liters. which generally corresponds to 210 liters per day per person and approximately equal to the average consumption. Power consumption of 1 m3 of water delivered to the consumer. as a rule. is 1-2 kW*h / m3. Heating 210 liters of water per day for the three people to about 60 ° C requires 14.62 kW*h for an apartment. or 0.2442 kW*h / m^2. which is 89 kW*h / m^2 for a year. Annual specific energy consumption for the supply of tap water with minimal energy consumption of 1 kW*h / m3 is at least 4 kW*h / m^2.

It is obvious that energy consumption for essential services in an old apartment is only a part of the cost of natural resources. the share of which for the new. so-called energy-efficient apartments. the sources of energy and transportation systems increases.

4 Conclusion

The real energy savings in the current transition to the energy efficiency of buildings is of 15-20%. which is clearly insufficient and points to the need of natural energy savings in all directions. Further reduction of heat consumption in residential buildings is possible through the use of effluent waste water heat with the temperature of 30-35 ° C.

The use of the roofs of residential buildings to produce electricity from the sun and wind is not yet economically viable.

The use of heat pumps can be economically justified in some cases. with the absence of the main lines and the necessary distribution networks in the vicinity.

References

1. S.A. Chistovich, *Avtomaticheskoye regulirovaniye raskhoda tepla v sistemakh teplosnabzheniya i otopleniya* (Stroyizdat Publ., Moscow, 1975) [in Russian]
2. L.D. Boguslavskiy, *Snizheniye raskhoda energii pri rabote sistem otopleniya. ventilyatsii i konditsionirovaniya vozdukha* (Stroyizdat Publ., Moscow, 1982) [in Russian]
3. J. Laustsen, *Energy Efficiency Requirements in Building Codes. Energy Efficiency Policies for New Buildings* (JEA Information Paper, Paris 2008)
4. S.N. Osipov, Effektivnyye metody proektirovaniya. stroitel'stva I ekspluatatsii sistem teplosnabzheniya: Sbornik, 33 (1990) [in Russian]

MATEC Web of Conferences 92, 01073 (2017) DOI: 10.1051/matecconf/20179201073
Thermophysical Basis of Energy Technologies - 2016

Modeling of flow in microchannel with bubbles layer on surface

Dmitriy Gluzdov[1,*]

[1]Kutateladze Institute of Thermophysics, 630090 Novosibirsk, Russia

Abstract. Results of 2D numerical solution of liquid flow in microchannel with bubbles layers on surface are presented. Bubbles layers are modeled by setting of bubble size and Navier slip condition. Calculations have been done using OpenFoam PISO method. The results of modeling compared with analytical solution.

1 Introduction

Microtechnologies play important role in the modern world. Power of the integrate circuits grows every year, and its cooling becomes more difficult. Traditional cooling systems will be not able to effectively cool modern microchips in the near future. The demand of the liquid cooling systems grows, which can be successfully integrated into microchips using microchannels. Microchannels are an inherent part of liquid cooling systems as well as one of main obstacles on the way to increase their effectiveness. The problem of decreasing hydraulic resistance in the microchannel flows is actual issue nowadays.

The one of possible solutions reported in [1,2,3]. Reducing of hydraulic resistance will occur if the boundary surface has a grooves surface with gas, which allows liquid to slip. However, creating such structure is costly, and it leads to problem with maintaining gas in grooves. Moreover, the question of stability and rupture of liquid films on such substrates arises [4,5,6].

Another possible solution to drug reduction is use of hydrophobic coatings for microchannel surfaces [7], where gas bubbles are collected. This gas bubbles have a height of several microns. The hydrophobic surface can significantly decrease adhesion; on the other hand, gas bubbles can reduce the volume for liquid flow, which can reduce a flow rate. So, the investigation to clarify the efficiency of using hydrophobic coatings is required.

The liquid flow in microchannel with Navier slip condition [8] and convex portions on hydrophobic parts modeling the configuration presented in [7] are simulated in 2D Navier–Stokes formulation. Calculations have been done using OpenFoam PISO method. The results of modeling compared with analytical solution.

* Corresponding author: gluzdov1@yandex.ru

2 Theory and the numerical solution

Two-dimensional problem is considered. The scheme for calculation of microchannel flow is presented in Fig. 1. Hatching shows the hydrophobic parts.

Calculations made in OpenFoam. OpenFoam is an open integrated platform for numerical solutions. Method PISO (Pressure Implicit with Splitting of Operator) is chosen for calculations of full Navier–Stokes equations with Navier slip condition and convex portions on hydrophobic parts. The analytical solution of Poiseuille flow with the same boundary condition is used as a test of this method. The time of establishing steady state flow in this test is used. Error of this method does not exceed of 1%.

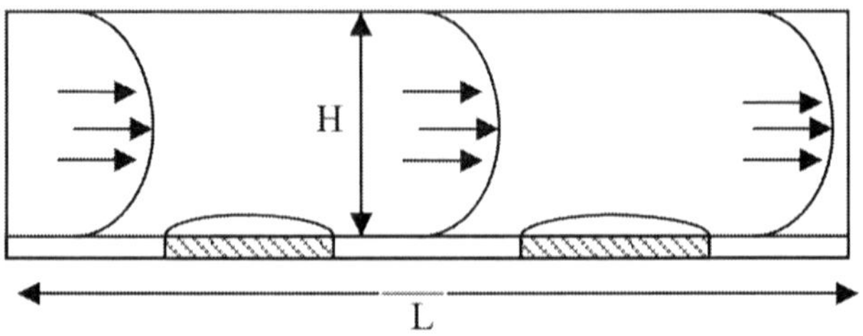

Fig. 1. Two-dimensional scheme for calculation of the flow in microchannel.

3 Results

Calculations are made for microchannel height H=0.5 mm and length L=100 mm. Liquid is water with temperature of 293K. The bubbles are approximated by arcs with height less than 0.05 mm. It was shown in [7] that the height of bubbles should be less than 30% of height channel for drug reduction.

Calculations are made for different slip lengths. The calculated volumetric flow rates and slip lengths are presented in Table 1. The volumetric flow rate can be increased by 27% for slip length equals to channel height compared with flow rate realized for slip length of 1% height of microchannel.

Table 1. The flow rates and slip lengths.

Slip length, mm	**Volumetric flow rate, $10^3\times mm^3/s$**
0.005	188.8
0.25	214.3
0.5	240.5

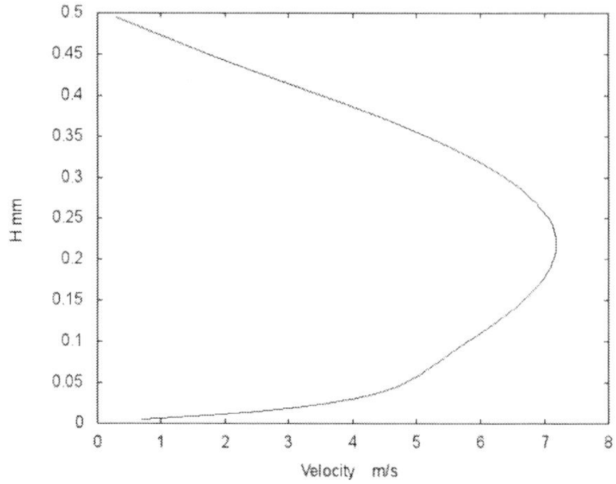

Fig. 2. The velocity profile on the no slipping surface.

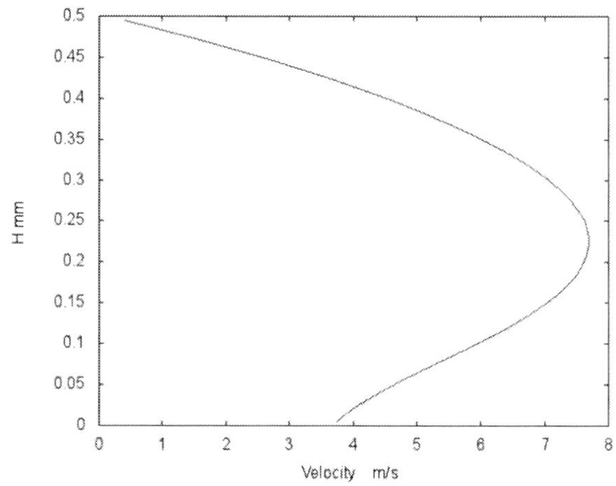

Fig. 3. The velocity profile on the hydrophobic surface.

Velocity profiles for the slip length of 0.5 mm are presented in Fig. 2 and Fig. 3. The maximum of a transverse component of velocity is less than a lengthwise component of velocity by three orders of magnitude. It means that the transverse component of velocity practically does not affect on the flow in microchannel with bubbles layer.

4 Conclusions

Results of 2D numerical solution of liquid flow in microchannel with bubbles layers on surface are presented. Bubbles layers are modeled by setting of bubble size and Navier slip condition. Calculations have been done using OpenFoam PISO method. The results of modeling compared with analytical solution.

The vertical velocity arising due to bubbles are not significant for the flow in the microchannel.

The volumetric flow rate can be increased by 27% if slip length equals to channel height compared with the flow rate realized for slip length of 1% of microchannel height.

This work was supported by a grant from the Russian Science Foundation (project Nr. 14-19-01755).

References

1. J. Ou, B. Perot, and J.P. Rothstein, Phys. Fluids. **16**, 4635 (2004)
2. J.P. Rothstein, Annu. Rev. Fluid Mech. **42**, 89 (2010)
3. E. Karatay, A. S. Haase, C. W. Visser, C. Sun, D. Lohse, P. A. Tsai, and R. G. H. Lammertnik, Proc. National Acad. Sci. USA **110**, 8422 (2013)
4. V.S. Ajaev, E.Y. Gatapova, O.A. Kabov, Phys. Fluids **25**, 122102 (2013)
5. V.S. Ajaev, E.Y. Gatapova, O.A. Kabov, Microgravity Sci. Techn. **24**, 33 (2012)
6. V.S. Ajaev, E.Y. Gatapova, O.A. Kabov, Adv. Colloid interface Sci. **228**, 92 (2016)
7. E.Y. Gatapova, V.S. Ajaev, O.A. Kabov, JETP Lett. **101**, 160 (2015)
8. A.V. Belyaev, O.I. Vinogradova, J. Fluid Mech. **652**, 489 (2010)

Utilization direction of industrial raw products built-up in power station ash dumps

Snejana A. Lihach[1,*], *Aygul* S. Ilyasova[1], *Roman* N. Kulesh[1], and *Valentina* I. Nikolaeva[1]

[1]National Research Tomsk Polytechnic University, 634050 Tomsk, Russia

Abstract. Nowadays hundreds million tons of ash and slag waste (ASW) is produced in Russia yearly. Large territories are needed in order to store such a big waste volume. Besides, it is necessary to conduct special engineering and ecological work at the design and usage stages of this structure. The goal of the research is to outline acceptable ASW utilization methods accumulated in coal burning power station ash dumps and to determine the order of activities to solve the problem. The research methods: experimental where Kansk-Achinsk and Kuznetsk coals are the object of the research. Besides, review of relevant to the problem literature and normative documentation was done to determine activities order, possible ways and limitations of the problem solving. We elucidated that ASW transportation to depleted coal quarries to restore them is essential to arrange in order to solve the problem of ASW utilization. As to new produced ASW, they should be divided into groups according to application field (mostly in construction). The groups correspond to boiler unit load operation. After coal combustion ash is stored in special places (reservoirs, silos). Therefore modern boiler unit might be seen as a production complex of steam and ash and slag material of an adequate quality.

1 Introduction

Nowadays hundreds million tons of ash and slag waste (ASW) is produced in Russia yearly. Large territories are needed in order to store such a big waste volume. Besides, it is necessary to conduct special engineering and ecological work at the design and usage stages of this structure. New territories withheld from commerce are allocated as ash dumps for ASW are filled.

For instance, about 6 million tons of ASW was stored up in the around 70 hectare territory of Mosenergo JSC thermal power plant (TPP) 22 over the years [1]. Ash dump use demands many problems to be solved: dust release from dry beaches causes air and soil pollution, high-mineralized water filtration through bot-tom soils leads to its mixing with surface and ground water [2].

Ash dumps of many stations, taking in account their age, work over designed lifetime what requires constant build-up of their reservoir capacity. Therefore stations bare the on-going cost of such ash dumps building and modernization. Some stations have two or even

* Corresponding author: ronikul@tpu.ru

three ash dumps (for example, it is planned to build ash dump No. 3 in the territory of Kuznetskaya PPT). In situations like this stations are obligated to pay land tax. Today the situation is getting worse due to power industry planned change-over to coal.

2 Survey

By address the problem of utilization first we need to classify all known types of utilization. After literature analysis we set aside the following directions:

- ASW extraction from ash dump followed by accompanied transportation and worked-out coal pit filling, or use in landscape work;
- ASW use in construction (ASW use in the industry is well characterized and found a great use);
- ASW use in farming as a soil desoxidant and other;
- ASW use for some elements recovery;
- Modern approaches to ASW utilization which mostly is in developing stages.

Large scale and variety of ASW types are evident. Actually, if we will raise ash from the rank of waste to the rank of industrial goods, raw material, placed product then ash capture will stop being forced and unprofitable and will turn to normal profitable industrial process [1].

There is a solution to the problem: producing operations which products are materials made of ASW or their components; or products of their deeper reprocessing. However, most ASW use variants are impose stringent requirements to initial stock parameters.

A study of ASW, accumulated at ash dump of a single power station, is needed in order to determine acceptable ways for its utilization. First of all, we are interested in an elementary composition of the waste and its evenness through an ash dump area; then we need to determine physical characteristics (such as fusibility properties, density, granulometric composition, specific surface area and other). Obtained data is a base for studying waste passport and determination ever or not ASW could be used in different utilization cycles. Choice of the final utilization method is made after the study for an environment hazard and radiation hazard class has been completed

Initially ASW transition into thermal power station product is attributed to the material passport preparation and approval in certain ecological organizations. The passport contains information about elementary composition, physic-chemical and other properties. When the passport is put together it is necessary to take into consideration product characteristics dependence on many factors such as: combustion fuel type, way of combustion, furnace conditions, boiler load, fly-ash collector type and other.

As can be seen from the above large-scale survey is needed to be conducted. The survey should result in product characteristics and mentioned above properties chart in their various combinations.

Product properties identification should also comply to target groups. For instance, it is very important to determine fly-ash properties against fresh and mature concrete, concrete physical form [3, 4], concrete production and handling as well as prove-in performance and environment compatibility [5] when the fly-ash is used as a concrete additive. Besides, important to know the difference between substance matter properties (elementary, physical, and mineralogical), material structure properties (ignition losses, particle distribution, and density), and specific properties (effect on concrete properties: filling effect, bearing effect etc.) and user benefits (time and cost saving, safety improvement etc.).

DOI: 10.1051/matecconf/20179201074

3 Laboratory research

We examined the ash dump and studied coal and ASW samples of power stations which burn Kansk-Achinsk and Kuznetsk coal in an attempt to evaluate ASW usability as an industrial raw material. Bench-marks results are presented in table 1.

The analysis proved there are significant descriptors differences of tested ASW samples from ash dumps. As it was noticed earlier, most industries impose stringent requirements to initial stock parameters and therefore the elucidated ASW cannot be utilized in conventional industries (due to a great unevenness in CaO, CaO_{free}, results determination, ignition losses and other main descriptors which are important for construction industry).

However, "fresh" ash and slag show relative perdurability of tested descriptors which are vary in dependence of boiler loading and some other conditions. As can be seen from the above ash and slag research in dependence of boiler loading might be the basis for properties passport preparation in dependence of its work condition, and modern boiler unit we should recognize as a complex of required quality steam and ash and slag materials (ASM) production. Constant composition and properties, proved by the passport of station production, are marked apart ASM from ASW.

Table 1. Kansk-Achinsk and Kuznetsk coal basin coals comparison.

No	Discription		Kuznetsk coal basin	Kansk-Achinsk and Kuznetsk coal basin, Irsha-Borodinskoe field
1	Grade of fuel		1CC	B2
2	As-received basis	W^p, %	9.0	33.0
3		A^p, %	18.2	6.0
4		$S^p_к + S^p_{op}$, %	0.3	0.2
5		C^p, %	61.5	43.7
6		H^p, %	3.7	3.0
7		N^p, %	1.5	0.6
8		O^p, %	5.8	13.5
9	Lower calorific value, ccal/kg		5 700	3 740
10	Ash on the dry basis, %		20.0	9.0
11	Maximum humidity, %		--	36.0
12	Maximum ash content, %		25.0	15.0
13	Maximum sulphur content, %		--	0.5
14	Hygroscopic moisture, %		1,6	12.0
15	normalized moisture, %*10^3кг/ккал		1,58	8.82
16	Normalized ash content, %*10^3кг/ккал		3,20	1.61
17	Dry ash free volatile-matter content , %		30,0	48.0
18	Ash fusion point, °C	Initial deformation	1 100-1 500	1 180
19		Weaking initiation	1 240-1 500	1 210
20		Fluid initiation	1 280-1 500	1 230
21	Ash chemical composition on sulphates absent basis	SiO_2, %	60.0	47.0
22		Al_2O_3, %	20.8	13.0
23		TiO_2, %	0.83	--
24		Fe_2O_3, %	10.4	8.0
25		CaO, %	2.9	26.0
26		MgO, %	1.4	5.0
27		K_2O, %	2.61	0.5
28		Na_2O, %	0.6	0.5

That is to say we think that stations ASW utilization should be managed in two directions: first of all accumulated in ash dumps ASW utilization (this ASW show great heterogeneity of chemical and other descriptors); secondly – "fresh" ASW utilization (with their classification and division based on consumer group and composition).

For the purpose of the challenge implementation it is reasonable to utilize ASW according to the fist point of the classification presented above (ASW transportation into worked-out coal quarry). Besides it is important to take into consideration acute situation with ash dump reservoir filing at some stations. As consequence of that we see an urgent necessity to take measures to utilize big volumes of ASW, which is hardly possible unless the way described above will be brought to life.

The ASW utilization method described above consists of several important stages: cycle of ASW ex-traction from ash dump followed by loading into auto transport, cycle of reloading into railway transport, and also ASW loading point at coal quarry.

Evaluation of this method should be done with attention to one critical part: distance between ash dump to detraining point of railway transport (usually this point placed near a railway station, i.e. place where coal id delivered to), railway distance between station to coal quarry (economical indicators will rise sharply if empty after coal transportation rail cars will be used), and seasonal prevalence (i.e. capability to of ASW shipping in a winter).

In the setting of available information for every station being examined we will be able to calculate cost and liquidity of accumulated ASW transportation. This is a good way to keep region environmental balance because in this case new ash dumps construction is not needed. Therefore there will be no environmental problems and related expenses for a station.

According to the scenario described above we can see that there are two possible ways: either comprehensive utilization of ASW which are located in ash dump, or curtain ASW volume control with cycle productivity variation in terms of surveying ASW volume.

For all new ASM we think it makes sense to divide them into groups in accordance with their application (mostly in construction). In terms of boiler unit load operation it is essential to store them in special places (reservoirs, silos). Several silos will be needed for ash. They should be compatible to product parameters and selling destinations. Slag proposed to be stored in special places in ash dumps; slag shipment to a consumer will be carried out with aid of equipment participated in accumulated ASW extraction.

We offer comprehensive utilization complex of station ASW. This way might become an alternative to new ash dumps building which needs sizable financial investments (according to some evaluation ash dump building costs over one billions of rubles).

It is necessary to mention that most Thermal Power Stations (TPS) were designed without regard to possible industrial use of ASW. That is why they don't have equipment for ASW transportation and storage, and ash pneumatic conveyor system line up is expensive. Besides, operation of such systems also brings some problems: for instance transportation line sealing happens quite often if pneumatic conveyor system is used for high-free calcium oxide (CaO_{free}), Kansk-Achinsk coal due to CaO_{free} reaction with atmospheric moisture. Ash transportation demands special equipment (cement truck and other). What is more spillage and dusting are impossible to avoid during loading and transportation processes.

Ash packaging in plastic bags is possible to avoid such problems. Ash transportation from boiler shop to packaging shop might be performed with use of some leak-proof tanks or vessels. This scheme has some advantages:

- ability to save some product properties (for instance sorption) by using vacuum packaging;
- scheme simplicity: it is fast and easy to erect, there is no need in additional machinery and pipelines;
- there is no need for special transportation equipment (ash packed in plastic bags might be shipped by regular transport and load-unload by hand).

4 Conclusions

In conclusion it is worth to say that TPS ASW utilization cycles only started to find their application in Russia. Germany on the other hand already utilizes almost 100% of its ASW [6, 7].

Engineering of the utilization process is in need of government legislature support. Existing industry regulatory documents do not provide us with the whole ASW utilization cycle information, and some-times even contradict with each other [7, 8]. Stations which produce ASW also need to make some steps forward because ASW utilization efficiency depends on their policies on the matter. We think that those steps are product certification and marketing research for possible ASM customer acquisition.

Summarize we would like to say that environmental advancement of power industry suggests complex problem solving which includes systematic scientific and technical, technological, standard and methodological and management approach.

References

1. M.M. Kozlov, V.D. Ilyin, Y.K. Celikovskyi, Elektricheskie stantsi **11**, 34 (2005) [in Russian]
2. J. Hot, M. Sow, C.Tribout, M. Cyr, Constr. Build. Mater. **10**, 218 (2016)
3. N. Nordin, M.M.A.B Abdullah, M.F.M. Tahir, A.V. Sandu, K. Hussin, Int. J. Conservation Sci. **7**, 161 (2016)
4. M. Pasetto, N. Baldo, Constr. Build. Mater. **108**, 112 (2016)
5. L.C. Ram, R.E. Mastp, N.K. Srivastava, J. George, V.A. Selvi. T.B. Das, S.K. Pal, S. Maity, D. Mohanty, Environ. Monit. Assess. **187**, 14 (2015)
6. Y.K. Celikovskyi, Energetik, 10 (2006) [in Russian]
7. H.-P. Backes, D. Brandenburger, M. Meibner, VGB PowerTech **92**, 45 (2012)
8. V.L. Shulman, Power Plants **5**, 29 (2006)

MATEC Web of Conferences **92**, 01075 (2017) DOI: 10.1051/matecconf/20179201075
Thermophysical Basis of Energy Technologies - 2016

Modeling the heat and mass transfer in the pores of the thermal protection carbon-carbon frame during the gas-phase deposition of silicon carbide

Sergey V. Reznik[1], *Konstantin* V. Mikhailovskiy[1], and *Pavel* V. Prosuntsov[1,*]

Bauman Moscow State Technical University, 105005, Moscow, Russia

Abstract. This paper is dedicated to the numerical simulation of the gas-phase deposition into the porous carbon-carbon frame for the gradient thermal protection of reentry vehicles. The paper presents the specifics of creating the representative volume elements of the porous carbon-carbon frames based on the microstructure data from the computed tomography scanner, electronic scanning microscope and porosimetry results. The finite-element models of the representative volume elements are created in order to obtain the net independent solution, including the special finite elements at the phase interfaces. The finite-element and finite-volume method was used for calculation, implemented in MSC.Digimat and ANSYS software packages. The effect of the reagent parameters on the gas-phase deposition process, uniformity and rate of silicon carbide deposition was simulated parametrically, and the results are presented.

1 Introduction

The rise in the international space programs and the development of new generation reentry space vehicles are calling for the design of heat-resistant thermal protection. The selection of the thermal protection type is affected by the reusability factor, the character and the level of thermal loads. For instance, disposable structures (like reentry vehicles or solid propellant motors) employ ablative thermal protection coating (TPC) from polymer composite materials (CM) [1, 2]. The protective mechanisms in TPC trigger the endothermic reactions of melting, destruction, evaporation, filtration of the destruction products through the porous frame and their injection into the boundary layer, the film of the melted substances flowing over the surface. Moreover, the heat transfer through the coating is reduced by the formation of the porous frame and the increase in the radiative ability of the surface. The drawbacks of the ablative TPC include the physical-chemical transformation in CM resulting in irreversible changes in the aerodynamic and thermal protection properties affecting their reusability.

* Corresponding author: pavel.prosuntsov@mail.ru

Due to technical, technological and economic reasons, reusable large spacecraft do not employ ablative TPC, since they alter their shape, dimensions in the operation process, and lose their heat-insulating ability as a result of physical chemical transformations. The perspective airspace vehicles require TPC which can withstand fluctuating power loads, high-velocity reactive gas and erosion particles flows. Carbon-ceramic composite materials (CCCM) are very attractive in this respect, having low density, adequate strength and stiffness properties, small thermal coefficient of linear expansion, capable of working at temperatures above 2000 K, at high heating rates [3-5]. The mass of the thermal protection systems can be reduced by using porous CCCM.

The projects in the gradient TPC and materials at NASA Glenn Research Laboratory and NASA [6-8] demonstrated the possibility of manufacturing these materials to enable the required thermal mode of the spacecraft. The TPC in question were based on CCCM or ceramic matrix composite material with the silicon carbide fiber frame, modified with the high-melt zirconium-based carbide systems.

The complexity and time-intensity of the TPC parts fabrication require the mathematical simulation to be used in order to reduce the time and costs when selecting technological parameters. The main operation in the TPC parts fabrication process is the gas-phase deposition of silicon carbide into the porous carbon-carbon frame lasting hundreds of hours and determining the final physical mechanical, thermal physical and optical properties consistent with the operation requirements. Physical mathematical models of this gas-phase deposition process are further complicated by its non-stationary, multidimensional nature, also the chemical kinetics of the hard phase formation from the gaseous monomethylsilane must be taken into account. The analysis of the current research indicated that the majority of the papers focus on the gas deposition process in a simplified setting [9-12]. As a rule, the problems were solved in a one-dimensional or two-dimensional setting, the porous medium was viewed as a set of cylindrical pores, and the radiative heat transfer in the pores was not taken into account.

2 Problem description

This research aims to develop the algorithm predicting on the micro level the process of SiC gas-phase deposition into the porous carbon-carbon (C-C) frame to create the aerospace thermal protection.

The predicting algorithm includes several stages: microstructural investigations with an electronic scanning microscope and CT scanner, as well as a porosimeter; a 3D geometrical model of the pore space based on microstructural CT; a geometric model of the representative volume element for the C-C frame; a geometric model of the C-C frame pores; finite-element models of the representative volume elements to obtain the net-independent solution; specifying the thermal physical and optical properties of the C-C frame from the porous C-C composite material before and after the gas-phase deposition process, of the furnace heating unit, the tooling and the reaction medium; specifying the technological parameters and the gas-phase deposition process modes; specifying the boundary conditions on the monomethylsilane decomposition kinetics.

The reference data on the porous CCCM structure were obtained from the microstructural analysis of the witness specimen on an optical microscope measuring the dimensions of the non-woven strands. The sample was also subjected to the analysis on the Brucker Skyscan 1172 micro-CT scanner, which uses a multitude of X-Ray images at various angles to create a 3D model of the sample. The 3D model analysis enables detailed investigation of the porous CCCM frame microstructure resulting in satisfactory geometric models of the representative volume elements of the C-C frame.

The physical model of the SiC deposition in the CCCM pores takes into account the following specifics: the gas-phase reactor working compartment and the porous CCCM position; vacuuming the gas-phase reactor and heating it up to the required temperature; the reagent (monomethylsilane) injection into the working chamber; transferring the reagent and the partial interaction products in the gas phase to the heated surfaces of the porous CCCM; the gas-phase components diffusion through the CCCM pores; chemical reactions (decomposition) of the reagent onto the heated surfaces of the CCCM pores; SiC matrix formation; the reaction products desorption; transferring the desorbed products from the surface into the reaction chamber and the pumping duct. In this paper, the problem under consideration is solved using the ANSYS CFX software package and the user-specific programming language.

Since the chemical reaction rate is primarily determined by the level and distribution of temperature in the reactor chamber, and the thermal field inside the chamber is not known in advance, the problem in this paper is formulated with the account for the conductive, convective and radiative heat transfer. As a result, the mathematical model of the gas-phase deposition for SiC into the porous CCCM part includes the equation of the multi-component flotation (with macro and micro level description of heterogeneous chemical description) and the equation of the conjugate heat transfer reaction in complex geometry systems and in the pore volume. The mathematical model is based on the Navier-Stokes equations for the multi-component reacting gas mixture, as well as on the chemical kinetics equations (in particular, takes into account Arrhenius equation for the reacting surfaces) [13, 14].

The thermal physical characteristics of the reacting medium and the gas-phase reactor consistent with the operation temperature are presented in [15, 16]. The following assumptions are introduced for the mathematical simulation of the deposition process: the reaction medium is optically transparent; the radiative heat transfer surfaces are grey (emissivity unaffected by the wavelength); the surfaces are diffusely radiating and diffusely absorbing; the changes in the thermal physical properties in the representative volume elements of the porous CCCM are not taken into account when the SiC matrix is formed.

3 The results of the heat and mass transfer mathematical simulation in the pore space

The first stage of research was concerned with the conjugate heat transfer during SiC gas-phase deposition in the gas-phase reactor with a CCCM blank, the second stage dealt with the mass transfer in the reaction chamber of the gas-phase reactor, the third stage dealt with the microlevel heat and mass transfer in the pore space. In order to simulate the heat and mass transfer during the gas-phase deposition of SiC in the CCCM pores, geometric models were created, which were divided into hexagonal and prismatic finite elements and volumes, so that a net-independent solution was obtained with acceptable degree of discretization. The total number of elements was 1,200,000 (Fig. 1). The elements in the model were bound by the contact condition of the adjacent surface elements.

The boundary conditions for the monomethylsilane concentration fields, pressure, renewal rate and thermal state in the near-surface layer of the porous CCCM blank were obtained at the second stage of simulation. It is known that the structure, density and the residual porosity of the CCCM are directly affected by the technological parameters of the process (temperature, curing time), while the homogeneity of the ceramic matrix is affected by the physical chemical and thermal physical properties of the precursor. Consequently, this paper included parametric calculation in order to estimate how the uncertainty in the precursor properties influences the technological parameters of CCCM production. The variable properties are the dynamic viscosity and the thermal conductivity coefficient. These parameters are considered critical for determining the features of heat and mass transfer in the porous medium.

Fig. 1. Finite-element model of the CCCM pore representative volume element.

Below are some of the results of the mass and heat calculation in the CCCM pores during the SiC gas-phase deposition at 880K temperature in the reaction chamber and $0.43{\cdot}10^{-5}$ Pa·s dynamic viscosity. Figs. 2 and 3 show the calculation results for the velocity field and SiC growth rate in the pore space.

As a result of the mass and heat transfer calculation in the CCCM pore space during the SiC gas-phase condensation onto the carbon frame with the varied precursor properties, it was demonstrated that :

– when the dynamic viscosity of monomethylsilane changes from $0.43{\cdot}10^{-5}$ Pa·s to $6.88{\cdot}10^{-5}$ Pa·s, the maximum deposition rate changes by less than 0.5%, but the mass and heat transfer is less intensive and the precursor velocity in the pore space decreases by more than a factor of ten;

– when monomethylsilane heat conductivity coefficient changes from 0.0113 W/(m·K) to 0.1816 W/(m·K), the maximum deposition rate changes by less than 0.5%, which is determined by the precursor flow only slightly influencing the intensity of the heat transfer with the pore surfaces

4 Conclusion

1. As a result of the research, the algorithm for the mathematical simulation of SiC gas-phase deposition into the porous CCM was developed, which enabled:

– parametric modelling of SiC gas-phase deposition into the representative volume elements of the porous CCCM;

– determining that the change in monomethylsilane dynamic viscosity from $0.43{\cdot}10^{-5}$ Pa·s to $6.88{\cdot}10^{-5}$ Pa·s alters the maximum deposition rate by less than 0.5%, but the mass and heat transfer becomes less intensive and the precursor velocity in the pore decreases by more than a factor of ten;

– establishing that the change in monomethylsilane heat conductivity coefficient from 0.0113 W/(m·K) to 0.1816 W/(m·K) alters the maximum deposition rate by less than 0.5%, which is determined by the precursor flow only slightly influencing the intensity of the heat transfer with the pore surfaces.

2. The results of the parametric calculations help to select and validate the technological parameters for SiC gas-phase deposition to densify the porous blanks for aerospace thermal protection parts.

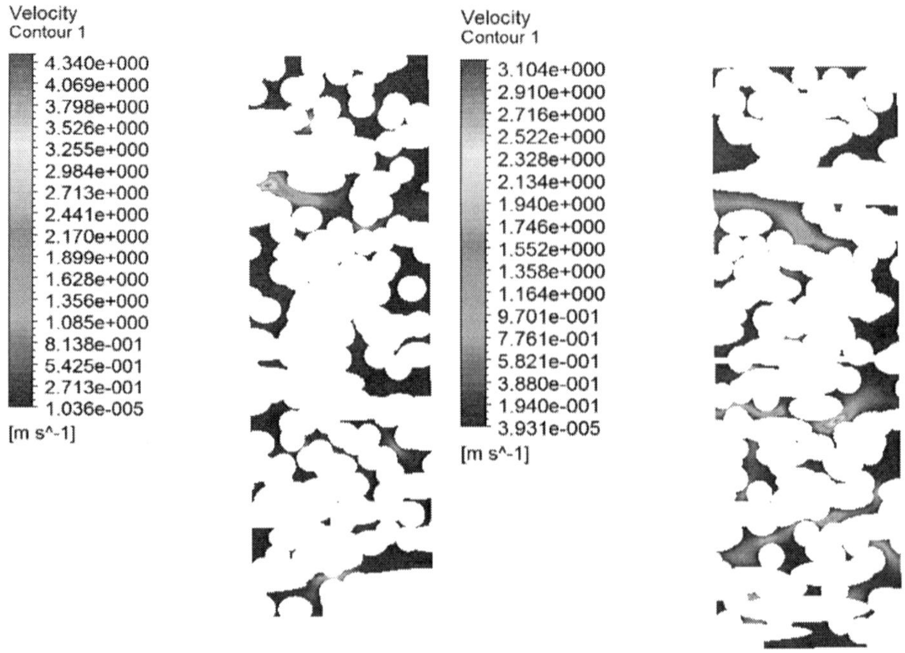

a b

Fig. 2. Velocity fields in the pore space of the CCCM representative volume element during the SiC gas-phase deposition, m/s: a – 10 microns from the end surface; b – 45 microns from the end surface.

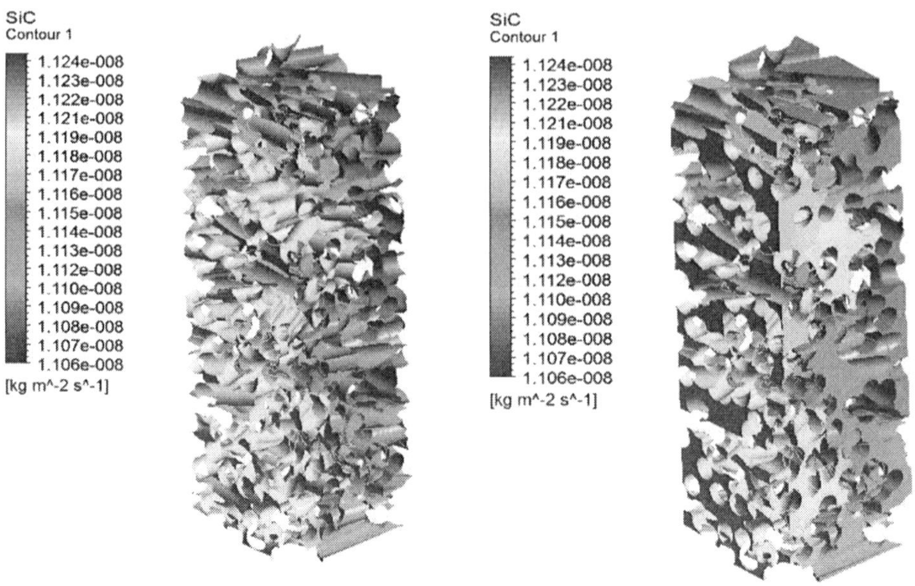

a b

Fig. 3. SiC deposition rate on the reacting surfaces in the CCCM representative volume element, kg/($m^2 \cdot s$): a – in the pore zone; b – in the near-surface zone.

.

Some results of this paper were obtained as part of the grant agreement № 14.577.21.0099 by the Russian Federation Ministry of Education and Science (unique identifier for Applied Scientific Research — RFMEFI57714X0008) in the framework of the priority area "Transport and space systems" of the federal target programme "Research and development in the priority areas of the Russian science and technology development for 2014-2020".

References

1. T.A. Heppenheimer, *Facing the Heat Barrier: A History of Hypersonics* (Washington D.C., 2007)
2. Y.K. Chen, F.S. Milos, Journal of Spacecraft and Rockets **36**, 475 (1999)
3. B. Laub, M.J. Wright, E. Venkatapathy, 6th International Planetary Probe Workshop, Atlanta, Georgia Short Course on Extreme Environments Technologies (2008)
4. N. Bansal, *Handbook of Ceramic Composites* (Springer Science, New York, 2005)
5. S.V. Reznik, K. V. Mikhaylovskiy, S. O. Yurchenko, *Heat-resistant composite materials and their use in reusable space vehicles* (Bauman MSTU Publishing House, Moscow, 2011)
6. S. Kumara, S.P. Mahulikarb, Journal of Thermal Science and Engineering Applications **10**, 021003 (2015)
7. M. Ferraiuolo, O. Manca, International Journal of Thermal Sciences **53**, 56 (2012)
8. E. Kolawa, T.S. Balint, G. Birurand, *Extreme Environment Technologies for Future Space Science Missions. Technical Report JPL D–32832, National Aeronautical and Space Administration* (Washington D.C., 2007)
9. T.R. Seth, F.J. Klavs, J. Appl. Phys. **83**, 524 (1998)
10. T.S. Cale, M.O. Bloomfield, D.F. Richards, K.E. Jansen, M.K. Gobbert, Comp. Mater. Scie. **23**, 3 (2002)
11. M.K. Gobbert, S.C. Timothy, Surf. Coat. Tech. **201**, 8830 (2007)
12. N. Cheimarios, G. Kokkoris, A.G. Boudouvis, Appl. Numer. Math. **67**, 78 (2013)
13. S.V. Reznik, K.V. Mikhailovsky, All Materials. Encyclopedic Reference Book **1** (2013)
14. K.V. Mikhaylovskiy, Blanking Productions in Mechanical Engineering, **3** (2011)
15. R.F. Gibson, *Principles of Composite Material Mechanics* (CRC Press, New York, 2011)
16. V.I. Kostikov, A.N. Varenkov, *Overhigh-temperature composite materials* (Intermet ENGINEERING, Moscow, 2003)

MATEC Web of Conferences 92, 01076 (2017) DOI: 10.1051/matecconf/20179201076
Thermophysical Basis of Energy Technologies - 2016

Experimental check of possibility of research of repeated inclusion of the open gas generator in the water environment

Sergey Goldaev[1], *Vitaly* Barsukov[2], *Sergey* Basalaev[2*], *Natalya* Minkova[2]

[1]National Research Tomsk Polytechnic University, 634050 Tomsk, Russia
[2]National Research Tomsk State University, 634050 Tomsk, Russia

Abstract. Experimental study of repeated burning interruption with the subsequent ignition in the water environment for bibasic firm fuel as a part of a model open solid propellant gas generator is presented. Video filming of burning zone mobile localizer providing process of repeated inclusion of a gas generator are given. Some parameters of the processes proceeding in these conditions are defined

1 Introduction

One of ways of giving of lifting force to the sunk or emergency objects, as we know, is use of the different tanks filled by compressed air. Attempts to use for this purpose more effective gas-generating systems on solid fuel have the essential shortcoming connected with difficulty to carry out the dosed pressurization in any portions of products of burning [1-4]. So, in one of the last large inventions in this area it is offered to use set from six separate gas generators allowing to carry out pressurization respectively in six receptions [5].

Solution of the problem of pressurization of tanks of buoyancy is represented more preferable if to use source of working body with possibility of repeated inclusion. Such sources were developed generally in relation to missile systems [6-9].

Analysis of scientific and technical literature shows that the majority of currently known solid sources of gases such as rocket motors or to multiple use inclusion of gas generators has common disadvantages, namely the complexity of the design and the limited number of inclusions. In one cases the number of starts unambiguously is defined by quantity of heat-insulated sections of fuel, in others, at implementation of extinguishing of charge, – amount of igniters. Besides, all known solid-fuel sources of gases, especially and with repeated inclusion, in case of their use in underwater conditions, have restrictions on the working depth of immersion about 5…6 km.

It is possible to get rid of these shortcomings at the expense of enough simple technical means in specific conditions of burning of unitary solid fuel directly in the water environment. So, in work [8] the way repeated interruption of burning and repeated ignition of sample of solid fuel in the water environment and for its implementation is offered the

* Corresponding author: tarm@niipmm.tsu.ru

key diagram of open solid-fuel gas generator with the mobile localizer of zone of gasification (MLZG) is provided. On the basis of this method in [8] implemented parametric analysis of controlled lifting the wreck deep pontoon, blown by means of the gas generator with multiple inclusion. At the same time, the offered way of underwater burning of unitary solid fuel demands further improvement.

The purpose of this work is experimental study of the processes involved in the ignition and the interruption of the combustion and the subsequent ignition in the water environment checkers dibasic solid fuel being a part of model open solid propellant gas generator multiple inclusion.

2 Device description

As the object of the study was used a model version of the open solid fuel gas generator incorporating multiple, depicted in Fig. 1 in the operating position, at the start of the next ignition fuel checkers. Gas generator consists of a shaped support *1* with a mounted thereon a cylindrical checker dibasic solid fuel *2*, thereon which is set up warming movable localizer of zone of gasification. MLZG contains a container *3* with an internal electric heater, made in the form of ceramic rod *4*. On the ceramic rod is wound a spiral filament *5* with lead wires *6*. Electric heater at the side surface and the upper end is insulated asbestos *7*. Rest of the space of the container *3* are filled with epoxy resin *8*. At the bottom of the container *3* to weight located wire winding *9* and the external asbestos insulation *10*. To the bottom of the container *3* is attached cylinder *11*, which has a side windows *13*. Lower part of the cylinder 11 are formed as guide 14.

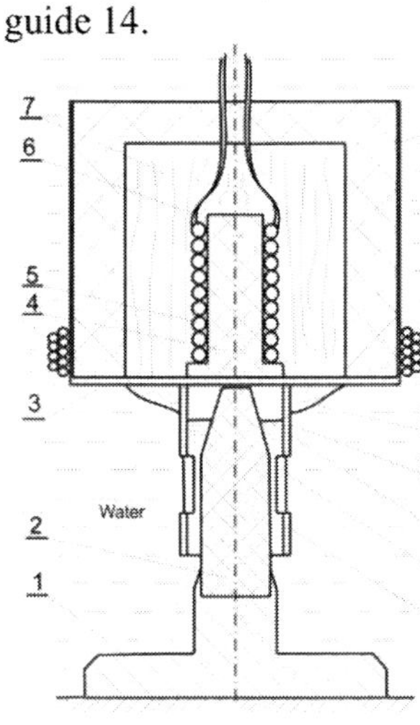

Fig. 1. Scheme model of an open solid-fuel gas generator in the beginning of the second ignition fuel checkers placed in the water environment.

3 Testing of the model device

When tested in a model gas generator was placed in a water-filled container with transparent walls. At placement of installation under water special attention to that the beaker *11* has been completely filled with water, including the working volume of beaker (height of working part of beaker made 17.6 mm). This ensured the condition of the first ignition of fuel does directly in the water environment.

The samples used in the experiment was a cylindrical grain of dibasic solid fuel with a diameter of 20 mm and height 60 mm. Electric power delivered to the filament *5* was 350 W.

Process was recorded on video digital camera "NIKON" with a frequency of 60 frames per second. The separate video footage illustrating provisions MLZG in the course of functioning of model solid-fuel gas generator is given in fig. 2. Under the video frames, given an appropriate time for the various stages of operation: the ignition mode, the combustion mode, the withdrawal mode MLZG.

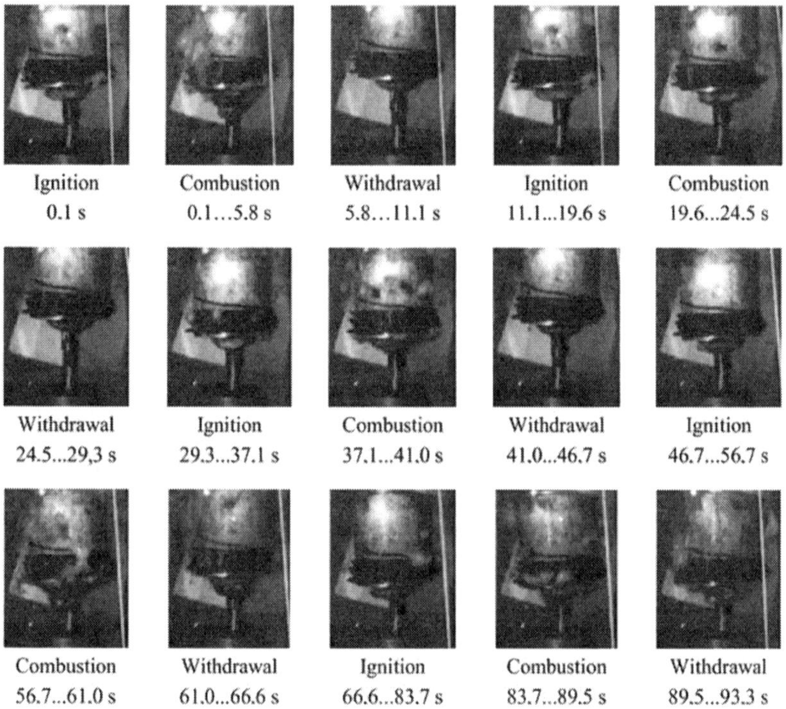

Fig. 2. Characteristic video frames illustrating the position MLSG in the process of functioning of model solid propellant gas generator with the ignition, combustion and extinction of fuel.

According to the video with an absolute error no more than 0.5 mm was determined according to time of the provisions MLSG in the process of multiple interrupts combustion and subsequent ignition of the fuel. Immediate results shown in Fig. 3 (experimental values indicated by dots), which some times lack of movement MLSG more ignition timing. This was due to a mechanical braking MLSG. Therefore, the ignition timing were corrected based on video frames, retaining the appearance of gas bubbles. The revised data is shown in Fig. 4. While in both figures the Y-axis was postponed distance from the upper plane of the base of the shaped support *1* to the lower edge of the guide *14*.

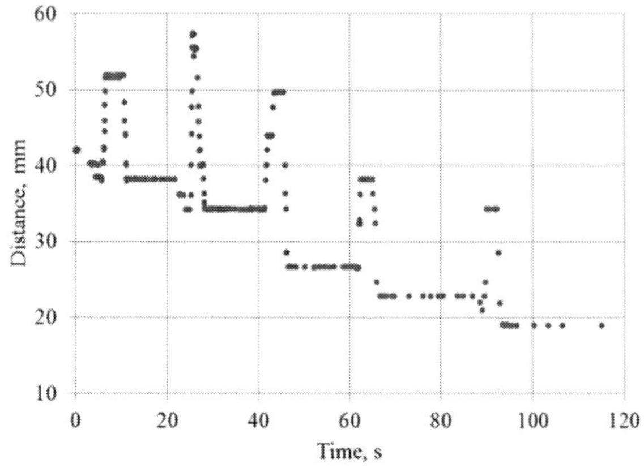

Fig. 3. Experimental points the position MLSG at different points in time in underwater ignition, combustion and extinction of fuel.

For illustration of idea of process it is more convenient to use the specified schedule in fig. 4. Inclusion of video camera has been made for saving of time of shooting at the moment when there were first bubbles of products of burning (point A in fig. 4). Then, in connection with the beginning of combustion, MLZG under the influence of curb weight moves with fuel combustion speed down. Further, (point B) MLZG withdrawal up, some time is maintained at any distance, and then again falls by fuel sample (point C). During warming up there is secondary ignition of fuel (point D) and therefore MLZG starts over again to move down (point E). Another lowering MLSG on the fuel checker in F point for the subsequent ignition. In the future, there is an alternation the ignition mode, the combustion mode, the withdrawal mode MLZG.

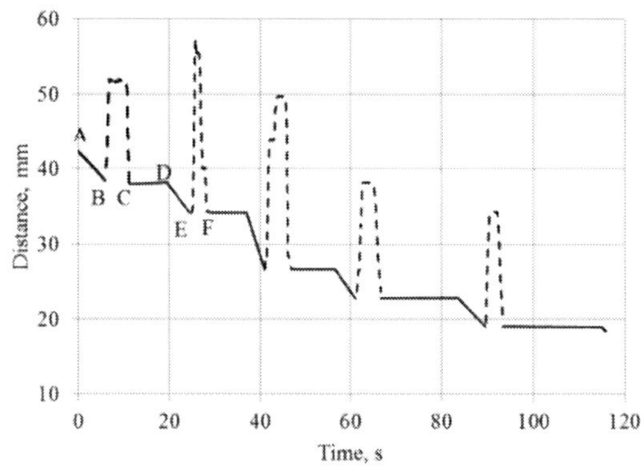

Fig. 4. Schedule of time dependence of situation MLSG at underwater ignition, combustion and extinction of fuel.

By results of video filming some parameters of functioning of gas generator which are provided in the enclosed table have been calculated.

Tests of gas generator on the different modes has shown rather wide spacing in time of the first ignition of fuel checker (≈ 10…20 s). Comparison of information received from fig. 2 and 3 allows to define ignition time at different stages of inclusion of gas generator. Time re-ignition determined with a relative error less than 3 %, were of the same order with the time of first ignition.

The first ignition is carried out at full filling of working volume of beaker with water. Therefore the part of heat energy is spent for its evaporation, at least, in the layers adjacent to MLZG bottom. Remains liquid is then forced out generated by combustion products.

DOI: 10.1051/matecconf/20179201076

Follows from the made experiments that repeated ignition is carried out ranging from 8.5 to 22.9 s.

Table 1. Some operation parameters of the model solid propellant gas generator in underwater ignition, combustion and extinction of fuel.

No.	Ignition time $t_з$, s	Combustion time t, s	MLZG motion speed U, mm/s	MLZG withdrawal speed, U_0, mm/s
1	10...20	5.7	0.68	8.2
2	8.5	5.0	0.79	14.3
3	7.8	3.9	1.91	6.6
4	10.0	4.3	0.90	14.6
5	17.1	5.9	0.64	13.4
6	22.9	-	-	-

The high speed of the movement MLZG at the third stage of combustion attracts attention. We may assume that this rate corresponds to the cone combustion mode.

MLZG withdrawal speed in averaged 11.4 mm/c, so the extinction time is in the range of 2.7 s. There is question of the motion speed of the front of extinction. Obviously, when driving MLSG up front fading fuel is moved at a speed close to the speed of rise of the water level adjacent to the sample. Some lag is in limits of time of change of personnel making less than 0.07 s.

Conclusion

Thus, studies conducted have revealed the peculiarities of the open model of the gas generator based on dibasic solid fuel in water with repeated starting and stopping. Shows an approximate estimate of the rate of motion MLZG change during the transition from the end surface gasification, the fuel combustion rate equal to the cone-shaped, depending on the conditions of the first stage of combustion. It is shown that the retraction of the movable localizer area gassing up at a speed of from 6.6 to 14.6 mm/s the extinction of the fuel is carried out for a time from 2.7 to 1.2 s, while the retard of extinction relative movement MLSG does not exceed 0.07 s [10]. Time and again the following ignitions is from 8.5 to 22.9 s.

The reported study was funded by RFBR according to the research project No. 16-03-00630a.

References

1. V. M. Bukalov, A.A. Narusbayev, *Design of nuclear submarines* (Shipbuilding, Leningrad, 1968)
2. I. Delgrende, Bull. A. Ssol Techn. mer etacronaut. **72**, 369 (1972)
3. A.S. 664873 USSR, MKI2 V63S 07/08. Rigid pontoon for rise sunk objects / E. Bril, V. N. Govorushkin, O. I. Kalinin, etc. Zayav. 05.01.78; Opubl. 30.05.79. Bulletin No. 20. 2 p.
4. E.V. Dobrozrakova, A.N. Zharinov, V. M. Kulyukin, Questions of defense equipment. It is gray **1**, 50 (1992)
5. The patent No. 2134212 Russian Federation System of emergency emersion of Malachite-1 submersibles, the device for blowing off of tanks of the main ballast at emergency emersion and way of emergency emersion/VA. Sirotkin, A.A. Tyurikov, N. I. Homenko, Yu.A. Zakharov. Zayavl. 25.08.98. Opubl. 10.08.99.

DOI: 10.1051/matecconf/20179201076

6. A.A. Shishkov, *Gas generators of missile systems* (Mechanical engineering, Moscow, 1981)
7. V.D. Barsukov, S.V. Goldaev, *Scuba unitary ignition and combustion of solid fuels. Theory, experiment, technical proposals* (Publishing house of Tomsk, Tomsk, 2003)
8. V.D. Barsukov, S.V. Goldaev, N.P. Min'kova, S.A. Basalaev, *Underwater burning solid fuels dibasic* (Publishing house of Tomsk, Tomsk, 2016)
9. V.D. Barsukov, S.V. Goldaev, N.P. Min'kova, S.A. Basalaev, Chem. phys. mesoscopy **14**, 327 (2012)
10. V.D. Barsukov, S.V. Goldaev, N.P. Min'kova, S.A. Basalaev, Chem. phys. mesoscopy **17**, 166 (2015)

AUTHOR INDEX

AUTHOR INDEX

EDP Sciences – Web of Conferences
17, Avenue du Hoggar
Parc d'Activité de Courtabouf
BP 112
F-91944 Les Ulis Cedex A
France

ISBN 978-1-5108-3509-2

CURRAN ASSOCIATES INC.
proceedings.com

Research, Invention, and Innovation Congress (RI2C 2023)

Held Online

E3S Web of Conferences Volume 428 (2023)

Bangkok, Thailand and Online
24-25 August 2023

edp sciences